*Gene Expression in Brain*

# *Gene Expression in Brain*

*edited by*

*CLAIRE ZOMZELY-NEURATH*

*and*

*WILLIAM A. WALKER*

*A WILEY-INTERSCIENCE PUBLICATION*

**JOHN WILEY & SONS**

*New York • Chichester • Brisbane • Toronto • Singapore*

***Library of Congress Cataloging in Publication Data:***

Main entry under title:

Gene expression in brain.

"A Wiley-Interscience publication."
Includes bibliographies and index.
1. Brain. 2. Brain chemistry. 3. Gene expression.
I. Zomzely-Neurath, Claire. II. Walker, William A.
(William Alfred) [DNLM: 1. Brain. 2. Gene Expression Regulation. QH 450 G3265]
QP376.G45 1984 591.1′88 84-11918
ISBN 0-471-86209-6

Printed in the United States of America

10 9 8 7 6 5 4 3 2 1

# Contributors

DAVID K. BATTER, Department of Cell Biology and Anatomy, Cornell University Medical College, New York, New York

IAN R. BROWN, Scarborough College, University of Toronto, West Hill, Ontario, Canada

ANTHONY T. CAMPAGNONI, Department of Chemistry, University of Maryland, College Park, Maryland

CHING-LING C. CHEN, Department of Biochemistry and the Center for Reproductive Sciences, Columbia University Medical School, New York, New York

LOUIS J. DEGENNARO, Department of Neurochemistry, Max Planck Institute for Psychiatry, Martinsried, German Federal Republic

MARIAN J. Q. EVINGER, Department of Biochemistry and the Center for Reproductive Sciences, Columbia University Medical School, New York, New York

MARILYN FISHER, Department of Anatomy, University of Utah School of Medicine, Salt Lake City, Utah

IRITH GINZBURG, Department of Neurobiology, The Weizmann Institute of Science, Rehovot, Israel

ANTHONY E. GIOIO, Department of Cell Biology and Anatomy, Cornell University Medical College, New York, New York

ILLANA GOZES, Department of Hormone Research, The Weizmann Institute of Science, Rehovot, Israel

PAUL GREENGARD, Department of Pharmacology, Yale University School of Medicine, New Haven, Connecticut

W. SUE T. GRIFFIN, Department of Cell Biology, University of Texas Health Science Center, Dallas, Texas

SELMA D. KANAZIR, Department of Pharmacology, Yale University School of Medicine, New Haven, Connecticut

BARRY B. KAPLAN, Department of Cell Biology and Anatomy, Cornell University Medical College, New York, New York

LESLIE P. KOZAK, The Jackson Laboratory, Bar Harbor, Maine

ROBERT M. LEWIS, Department of Pharmacology, Yale University School of Medicine, New Haven, Connecticut

URIEL Z. LITTAUER, Department of Neurobiology, The Weizmann Institute of Science, Rehovot, Israel

EARL MAYERI, Department of Physiology, University of California, San Francisco, California, and Department of Basic Sciences, California College of Podiatric Medicine, San Francisco, California

MARCELLE R. MORRISON, Department of Neurology, University of Texas Health Science Center, Dallas, Texas

JAMES L. ROBERTS, Department of Biochemistry and the Center for Reproductive Sciences, Columbia University Medical School, New York, New York

BARRY S. ROTHMAN, Department of Physiology, University of California, San Francisco, California

BETH S. SCHACHTER, Department of Obstetrics, Gynecology and Reproductive Science, Mount Sinai School of Medicine, New York, New York

RICHARD H. SCHELLER, Department of Biological Sciences, Stanford University, Stanford, California

WILLIAM C. WALLACE, Department of Pharmacology, Yale University School of Medicine, New Haven, Connecticut

# Preface

Gene expression per se has been studied extensively during the past two decades. The advances made in recombinant DNA technology have led to discoveries in molecular genetics that have revealed much about the basic mechanisms on which life depends. However, if one wants to approach complex functions, and the most complex and potentially most interesting is that of brain, one has no choice but to investigate this organ. Until recently molecular approaches to the analysis of the nervous system have only been utilized by a few investigators. This is not surprising since the brain is a complex system and has been, to a large extent, inaccessible to the application of the techniques used so effectively by molecular biologists in their investigations of simpler bacterial and eukaryotic organisms. However, the recent developments in recombinant DNA research and related methods now make it possible to study complex biological problems. Indeed, the area of molecular neurobiology is now being pursued actively by molecular biologists as well as neuroscientists using the powerful tools developed by the former. The coupling of the exciting developments in molecular biology and neuroscience should have a powerful impact not only on basic neuroscience but also on clinical medicine.

The approaches used by investigators in the area of molecular neurobiology are presented in this volume. A range of options exists for the methods used to obtain DNA for amplification by cloning. Some of these include the use of synthetic oligonucleotide probes as, for example, in the chapters on pro-opiomelanocortin, myelin basic protein, and vasoactive intestinal polypeptide (also present in brain), synthesis of complementary DNA (cDNA) from messenger RNAs (mRNA) present in relatively high abundance (tubulin and actin) to mRNA representing less than 1% of the total mRNA population (phenylethanolamine *N*-methyl transferase [PNMT]). The scope of the chapters in this volume ranges from studies on the bag cell neurons of *Aplysia* to the molecular biology of human brain showing that a variety of approaches can yield important information on genetic expression in the nervous system. The authors have all combined cellular and molecular biology in the research presented in these chapters.

The exciting area of molecular approaches to the study of gene expression in brain has enormous research potential and should make this volume of interest not only to neuroscientists but also to molecular biologists contemplating the study of molecular genetics in complex systems such as the brain. The purpose of this volume is to show what has been accomplished, the various approaches that can be used, and to promote research efforts in this area.

CLAIRE ZOMZELY-NEURATH
WILLIAM A. WALKER

*Honolulu, Hawaii*
*Syosset, New York*
*August 1984*

# Contents

# *Gene Expression in Brain*

# 1

# *The Construction and Identification of Recombinant DNA Probes for the Study of Gene Expression in Nervous Tissue*

***Barry B. Kaplan, Anthony E. Gioio, and David K. Batter***

The complexity of structure and function in the mammalian nervous system has, through the years, necessitated a multidisciplinary experimental approach to its study. Recently nucleic acid hybridization analysis of brain RNA populations has revealed that the number of different genes expressed in brain far exceeds that transcribed in other somatic tissues and organs. The remarkable diversity of gene expression in brain bears on many issues in neurobiology, including the degree to which this large amount of genetic information participates in the development and maintenance of brain structure and function. The findings of RNA-DNA hybridization studies with regard to the diversity of gene expression in different brain regions, neural cell types, and during development and aging have been reviewed (Chaudhari and Hahn, 1983; Dokas, 1983; Kaplan and Finch, 1982). While this approach can provide new insights into gene expression in brain, it is not in itself entirely satisfactory. The complexity of brain RNA populations, coupled with limits in the resolution of the hybridization "assay," preclude one from addressing many important questions concerning the regulation of gene expression in nervous tissue.

The advent of recombinant DNA technology has opened several new avenues of approach to the study of neural gene activity. In this chapter we describe strategies and procedures involved in the construction of cloned complementary DNA (cDNA) probes for specific proteins of neurobiological interest. To illustrate the types of results achieved, data will be given for the cloning of DNA complementary to mRNA coding for phenylethanolamine *N*-methyltransferase (PNMT), the final enzyme in the catecholamine biosynthetic pathway. The utility of cloned cDNA probes will also be discussed.

## 1. ISOLATION OF SPECIFIC mRNA

To date, detailed study of gene structure and regulation has focused on those systems which produce particular mRNAs in high abundance (i.e., a large number of copies per cell). Consequently, much of our present knowledge of eukaryotic gene expression is derived from analysis of mRNAs coding for highly specialized cell products. Outstanding examples are actin (Engel et al., 1982; Schwartz et al., 1980), albumin (Sala-Trepet et al., 1979; Gordon et al., 1978), amylase (MacDonald et al., 1980), fibrinogen (Crabtree and Kant, 1981), ovalbumin (Breathnach et al., 1977; Dugaiczyk et al., 1978; McReynolds et al., 1978; Roop et al., 1978), ovomucoid (Buell et al., 1979; Nordstrom et al., 1979), conalbumin (Lee et al., 1980), casein and $\alpha$-lactalbumin (Richards et al., 1981), globin (Efstratiadis et al., 1977; Kinniburgh et al., 1978; Tilghman et al., 1978), immunoglobulins (Adams et al., 1980; Gough et al., 1980; Schibler et al., 1978), vimentin (Zehner and Paterson, 1983), and vitellogenin (Ohno et al., 1980; Smith et al., 1979; Wahli et al., 1978). However, due to the heterogeneity of cell types comprising nervous tissue and the striking complexity of the mRNA population, mRNAs for brain-specific proteins will frequently be present in lower abundance. Therefore, the levels of these mRNAs will often require enrichment to facilitate the construction and identification of recombinant plasmids. One experimental approach to this problem is outlined below.

### 1.1. Polyribosome Preparation and Immunopurification

Recent refinements in recombinant DNA technology obviate the need to obtain specific mRNAs purified to homogeneity. Rather, one can use cloning procedures to select the desired recombinant plasmid from a cDNA library constructed from a poly(A+) mRNA population simply enriched for the appropriate mRNA.

Figure 1 outlines an immunological approach to the enrichment of specific mRNA. The method exploits the ability of antibodies raised against native protein to react with the corresponding nascent polypeptide chain on the polyribosome. The soluble antibody–polyribosome complex is subsequently isolated by binding to protein A–Sepharose (Schutz et al., 1977; Shapiro and Young, 1981). Nonreacted polyribosomes are removed by extensive washing, and immunoadsorbed mRNA eluted by disruption of polyribosomes with EDTA. Poly(A+) mRNA is then isolated from total polysomal RNA by affinity chromatography on oligo (dT)-cellulose.

The assumptions inherent in the strategy outlined in Figure 1 are threefold. First, that polyribosomes isolated by standard procedures contain intact nascent peptide chains. Second, that polyribosome integrity can be maintained throughout the extensive exposure to antibody and chromatographic fractionation required by the enrichment procedure. Finally, that nascent peptide chains contain a sufficient number of antigenic determinants to efficiently react with antibody. The validity of these assumptions was investigated as described below.

Data in Figure 2 demonstrate that polyribosomes isolated by sedimentation through 2.0 *M* sucrose contain nascent chains. To prevent ribosome "run-off," polyribosomes were prepared in the presence of cycloheximide, an inhibitor of chain elongation. In this experiment, approximately 90% of the radiolabeled amino acid was associated with polyribosomes sedimenting at $>$ 200 S.

Exposure of polyribosomes to antibody with subsequent chromatographic fractionation had no deleterious effect on polyribosome integrity as shown in Figure 3. Here, protein A–Sepharose chromatography was also used to remove trace ribonuclease (RNase) activity from an IgG fraction prepared from

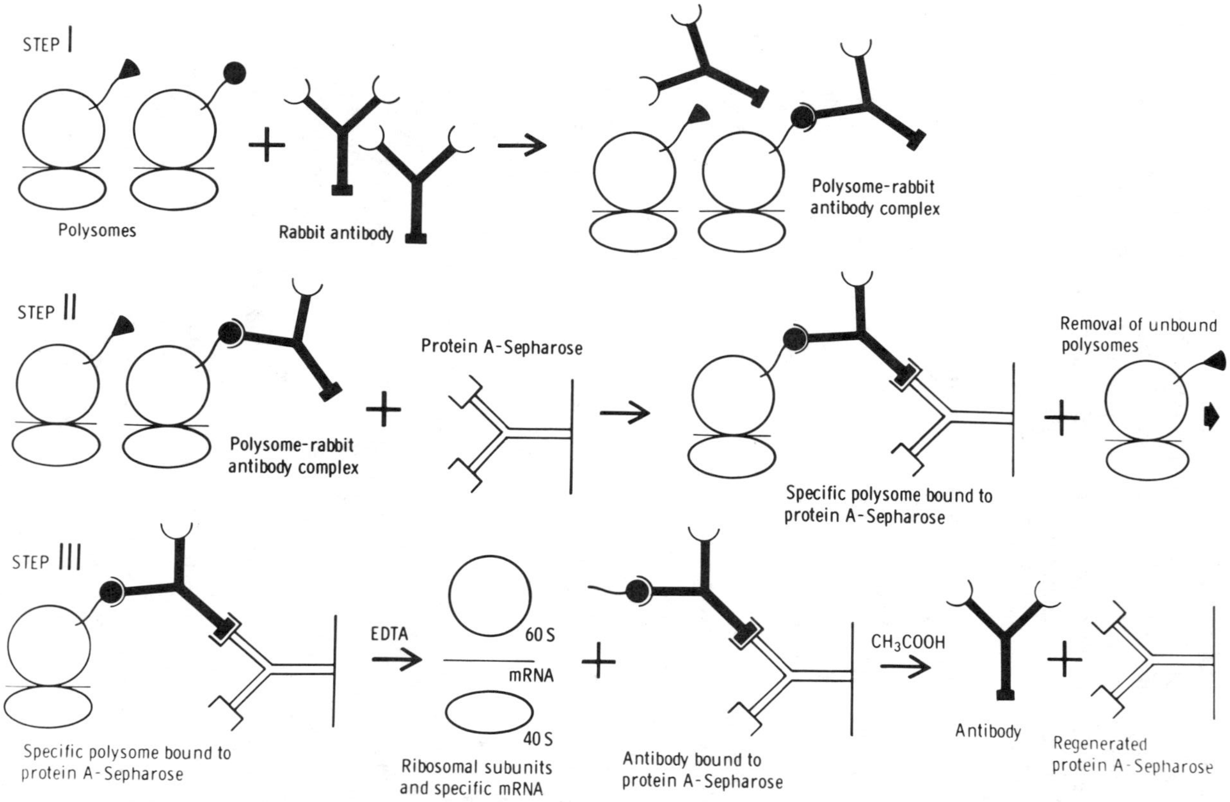


**Figure 1.** Principles underlying the immunopurification of specific mRNAs. Modified from Schütz et al., 1977.

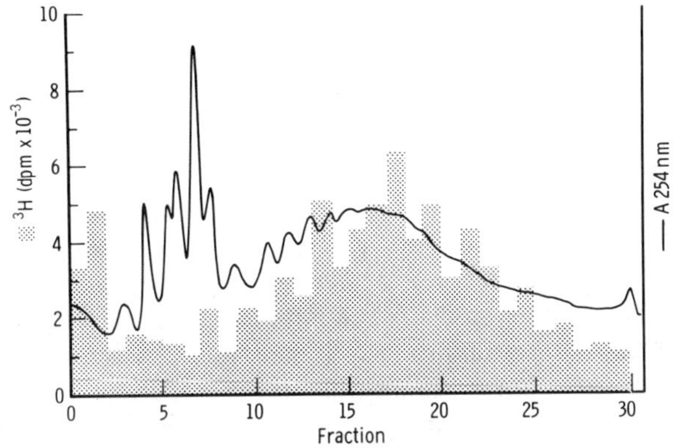


**Figure 2.** Sucrose density gradient profile of total pulse-labeled pheochromocytoma (PC-12) cell polyribosomes. PC-12 cells were grown in the presence of a [$^3$H]-labeled amino acid mixture (5 min) and polyribosomes isolated from a detergent-treated postmitochondrial supernatant by centrifugation through 2.0 *M* sucrose (Baetge et al., 1981). All solutions used in polysome isolation contained a 1000-fold excess of unlabeled amino acid and cycloheximide (2 $\mu$g/mL). Polyribosomes were resuspended and centrifuged through linear 15–45% sucrose gradients in a SW-41 rotor at 36,000 rpm for 75 min at 2°C. Migration is from left to right.

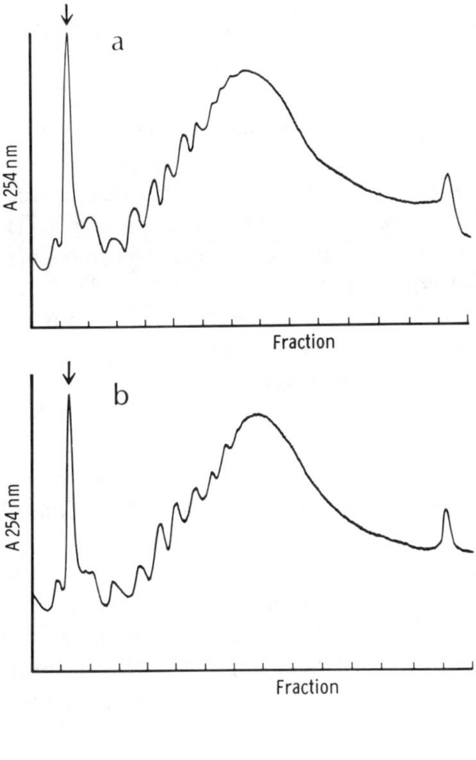


**Figure 3.** Sucrose density gradient profile of polyribosomes before (a) and after (b) immunoadsorption. Polyribosomes prepared from bovine adrenal medulla were reacted with protein A–Sepharose purified rabbit antibodies raised against PNMT (10 $\mu$g IgG/15 $A_{260}$ units of polyribosomes) for 2 hr at 4°C and subsequently passed over a protein A–Sepharose column (Shapiro and Young, 1981). Polyribosomes were displayed on linear sucrose gradients as described in Figure 2. Arrow indicates the position of the 80S monomer. Migration is from left to right.

monospecific rabbit antisera. The quality of the polyribosomes recovered in the column eluent was surprising, providing one an opportunity to reuse the sample for the isolation of a second mRNA (see below).

### 1.2. mRNA Translation

The efficacy of the immunopurification procedure is most frequently evaluated by *in vitro* translation using a micrococcal nuclease-treated reticulocyte or wheat germ cell-free system. The incorporation of radiolabeled amino acid into total mRNA translation products is monitored by acid precipitation and liquid scintillation counting. Amino acid incorporation into specific protein is estimated by immunoprecipitation using the appropriate antibody. The interpretation of data obtained from the cell-free translation assay assumes that all mRNAs are equally efficient in directing protein synthesis in a heterologous system and that the immunoprecipitation of specific translation products is quantitative.

The level of enrichment achieved by immunoadsorption of bovine adrenomedullary polyribosomes using rabbit monospecific antiserum prepared against cow adrenal PNMT (Park et al., 1982) has been evaluated. Data in Table 1 indicate that PNMT antibody immunoprecipitates 0.6% of the total radioactivity incorporated into acid-insoluble products by a cell-free system programmed with bovine adrenomedullary total mRNA. In comparison PNMT comprised approximately 30% of the translation products synthesized from PNMT-enriched mRNA, a 50-fold enrichment. In this experiment, polyribosomes that did not bind to protein A–Sepharose when challenged with PNMT antibody were recovered, reacted with antibodies raised against tyrosine hydroxylase (TH), and repassed over protein A–Sepharose (see Section 1.1). Data obtained from the *in vitro* translation of control and TH-enriched mRNA are also given in Table 1. The reaction of polyribosomes with TH antibody resulted in a 32-fold elevation of TH-mRNA levels. Based upon the control data in Table 1 and the levels of enrichment achieved in the immunoadsorbed mRNA, we estimate that 38 and 28% of the available PNMT- and TH-mRNA was recovered, respectively.

### 1.3. RNA-DNA Hybridization

The efficiency of the mRNA enrichment procedure can also be evaluated independently by RNA-DNA hybridization analysis. In this study complementary DNA of high specific activity ($2 \times 10^8$ dpm/$\mu$g DNA) is synthesized from total mRNA and mRNA enriched for a specific message *in vitro* using a viral reverse transcriptase (see Section 2.1). The cDNA probes are subsequently hybridized back to excesses of their template RNA. Hybridization is monitored by the formation of radiolabeled structures resistant to

**TABLE 1. CELL-FREE TRANSLATION OF BOVINE ADRENAL POLY(A+) mRNA[a]**

| RNA Sample | Total $^{3}$H Incorporation[b] (dpm/rx) | TH Incorporation[c] (dpm/rx % total dpm) | | PMNT Incorporation[c] (dpm/rx % total dpm) | | Enrichment (-fold) |
|---|---|---|---|---|---|---|
| Control mRNA[d] | 242,400 | 750 | 0.25 | 1,450 | 0.60 | — |
| PMNT-enriched mRNA[d] | 170,000 | — | — | 51,500 | 30.3 | 50.5 |
| TH-enriched mRNA | 296,800 | 23,600 | 8.0 | — | — | 32.0 |

[a]mRNA was translated in a cell-free system derived from rabbit recticulocytes as previously described (Baetge et. al., 1981) using leucine as radiolabeled amino acid.
[b]dpm corrected for $^{3}$H incorporated in control reactions containing no exogenous RNA.
[c]dpm corrected for $^{3}$H activity immunoprecipitated by IGg obtained from preimmune serum.
[d]data from Baetge et al. (1983).

hydrolysis by S1 nuclease, a single-strand-specific nuclease prepared from *Aspergillus oryzae*. Under conditions of RNA excess, the hybridization reaction follows pseudo-first-order kinetics. Thus, the larger the RNA sequence complexity (i.e., the greater the number of different mRNA species), the slower the reaction and the longer it takes to reach completion (for review see Kaplan, 1982).

Results of a typical RNA-cDNA hybridization experiment are given in Figure 4. The reaction of total cDNA to its template total adrenomedullary mRNA was relatively slow, spanning approximately five log orders of Cot (Fig. 4b). In contrast, cDNA synthesized from PNMT-enriched mRNA hybridized to its template RNA 100- to 1000-fold faster. According to ideal pseudo-first-order kinetics, the reaction of a single RNA species to its cDNA will reach completion within 1.5 log units of Cot [concentration of nucleic-acid (moles nucleotide per liter) × the time of incubation (sec)]. Data in Figure 4a show an inflection point in the hybridization kinetics of PNMT-cDNA 1.5 log units after initiation of the reaction. This kinetic component

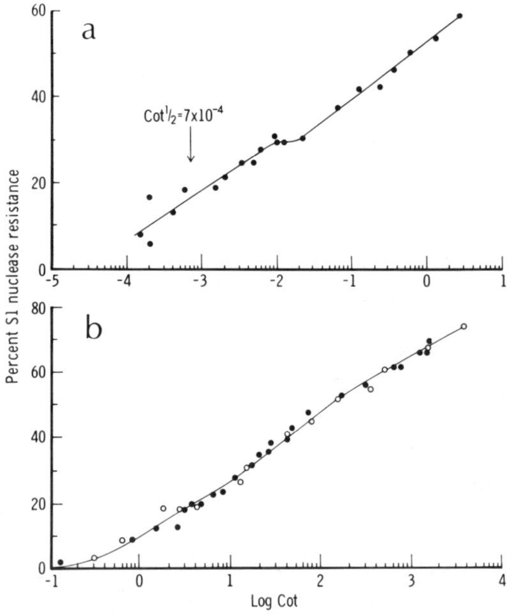


**Figure 4.** Kinetics of hybridization of cDNA to its template mRNA. [$^{32}$P]-labeled cDNA was hybridized back to its corresponding template polysomal poly(A+) mRNA prepared from bovine adrenal medulla. RNA concentrations ranged from 0.7 to 10.0 $\mu$g/mL. Reactions were carried out in 0.18 *M* $Na^+$ at 60°C and hybridization monitored by a S1 nuclease/DEAE filter disc assay (Maxwell et al., 1978). (a) cDNA to PNMT-enriched mRNA; (b) cDNA to total mRNA.

comprises approximately 25–30% of the cDNA mass ($Cot_{1/2} \sim 7 \times 10^{-4} M$ sec), a finding which agrees well with the *in vitro* translation data (see Section 1.2).

### 1.4. Discussion

The experimental approach outlined in Figure 1 permits the isolation of specific mRNAs that have no unusual physicochemical properties to exploit (e.g., size, base composition, or abundance). The efficacy of the approach lies in the use of protein A–Sepharose chromatography which obviates the need to immunoprecipitate the soluble antibody–polyribosome complex with a second antibody, thereby avoiding increased non-specific binding and physical trapping of polyribosomes in the precipitated complex. Additionally, the binding of the antibody–polyribosome complex to protein A–Sepharose facilitates the removal of contaminating polyribosomes by extensive washing of the adsorbed material. The IgG fraction of monospecific antisera, affinity-purified polyclonal antibodies, and monoclonal antibodies have all been employed successfully in the immunoadsorption of specific polyribosomes.

A significant advantage of this approach is that several different mRNAs can be prepared from a single polyribosome preparation (Section 1.3). Moreover, antibody used in the procedure can be recovered from the column (see Fig. 1) with little loss in reactivity. In addition to the preparation of mRNA enriched for PNMT and TH, this method has been used in the purification of mRNA for cystathione synthetase (Kraus and Rosenberg, 1982) and the heavy chain of the human histocompatibility antigen DR (Korman et al., 1982). These proteins are thought to comprise approximately 0.01–0.05% of the total protein population. It seems likely, therefore, that the method described will prove applicable to the purification of other moderately abundant mRNAs from brain.

## 2. CLONING OF SPECIFIC cDNA

### 2.1. Synthesis of Double-Stranded cDNA

A schematic diagram outlining the procedures employed in the construction of recombinant plasmids is given in Figure 5. Experimental protocols for each of the procedures illustrated have been well described (Maniatis et al., 1982; Schrader and O'Malley, 1982). Initially, single-stranded cDNA is synthesized from poly(A+) mRNA using a viral reverse transcriptase, deoxyribonucleotides as substrates, and oligo$(dT)_{10-18}$ as a primer [see Fig. 5 (1)].

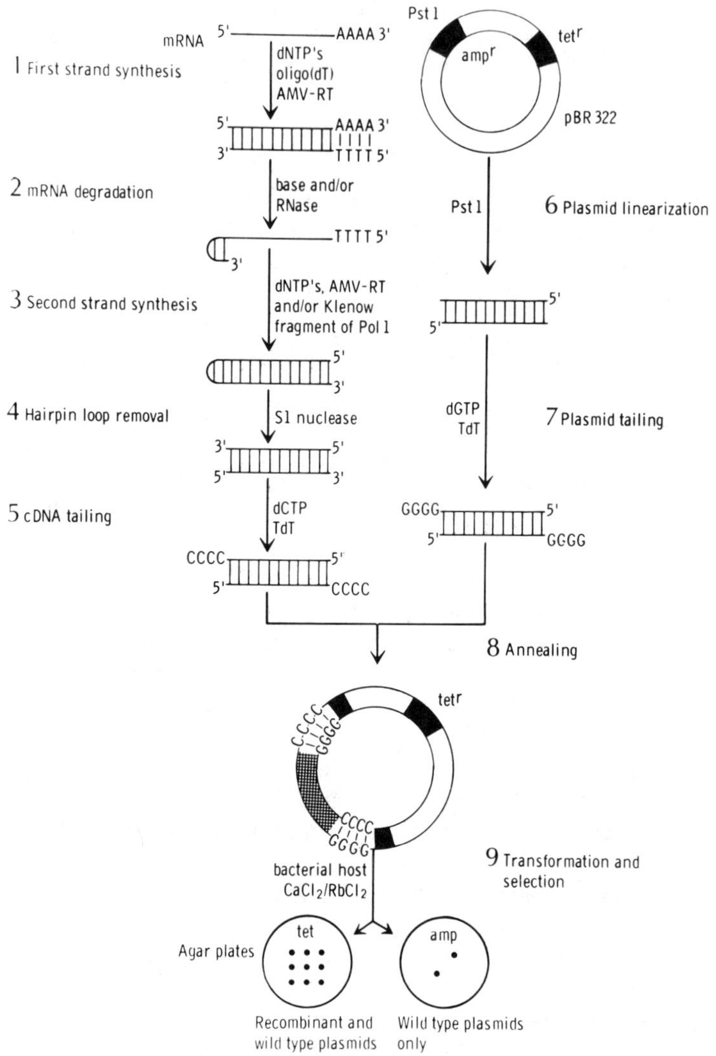


**Figure 5.** Strategy employed in the construction and selection of recombinant plasmids. Abbreviations: dNTP, deoxyribonucleotide triphosphate; AMV-RT, avian myeloblastosis virus reverse transcriptase; Pol 1, *E. coli* DNA polymerase 1; dCTP, deoxycytidine triphosphate; TdT, terminal deoxynucleotidyl transferase; amp$^r$, ampicillin resistance; tet$^r$, tetracycline resistance.

Inclusion of trace amounts of radiolabeled nucleotides facilitates determination of the mass and length of the cDNA synthesized. The quality of the reverse transcriptase, the integrity and secondary structure of the mRNA, and the reaction conditions all influence the length and yield of the primary cDNA transcript (for review see Efstratiadis and Villa-Komaroff, 1979). Using relatively high enzyme concentrations (Friedman and Rosbash, 1977) and appropriate RNase inhibitors, we routinely generate cDNA from heterogeneous mRNA populations 800–2000 nucleotides in length with a conversion efficiency of 70–90% (mass synthesized cDNA/mass mRNA template × 100).

Removal of the RNA template is achieved by alkaline hydrolysis or RNase treatment [Fig. 5 (2)]. The self-priming ability of the hairpin loop formed at the 3′ end of the cDNA transcript facilitates its conversion to a double-stranded (ds) molecule. Second-strand synthesis is achieved using either reverse transcriptase or the Klenow fragment of DNA polymerase I and deoxyribonucleotides [Fig. 5 (3)]. In some instances, both enzymes have been used sequentially in the same reaction, presumably increasing the size and yield of double-stranded cDNA (Wickens et al., 1978). The regions remaining single-stranded, including the hairpin loop, are specifically degraded with S1 nuclease yielding blunt-ended ds-cDNA molecules [see Fig. 5 (4)]. Small DNA fragments and nucleotides generated by S1 nuclease treatment are removed from the ds-cDNA preparation by sucrose density gradient centrifugation. Beginning with 2 μg poly(A+) mRNA, we generally recover approximately 0.1 μg ds-cDNA.

## 2.2. Construction of Recombinant Plasmids

Double-stranded cDNAs are generally amplified and selected for in plasmid vectors [see Fig. 5 (5–8)]. Of the various plasmids available, pBR322 has been most widely used (Sutcliffe, 1978). pBR322 (4.4-kb pairs) contains two selective markers, ampicillin and tetracycline resistance, and has been sequenced in its entirety. The plasmid also contains several useful unique sites that can be cleaved with type II restriction endonucleases, for example, *Eco R1* and *Pst 1*. To construct a recombinant molecule, the plasmid is first linearized with *Pst 1,* inactivating the gene for ampicillin resistance [Fig. 5 (6)]. The ds-cDNA is inserted into this site using the oligo(dG): oligo(dC)-tailing method described by Villa-Komaroff et al. (1978). The plasmid is tailed at the 3′-termini with deoxyguanosine residues using terminal deoxynucleotidyl transferase (TdT) in the presence of $Mg^{2+}$ (Deng and Wu, 1981). Commercial preparations of linearized, homopolymer-tailed pBR322 are now available. The ds-cDNA is similarly tailed with deoxycytidine residues using TdT in the presence of $Co^{2+}$ (Nelson and Brutlag,

1979). The optimal length of the homopolymer tails is apparently 10–15 nucleotides (Peacock et al., 1981). Circular recombinant molecules are formed by annealing the tailed plasmid and cDNAs in equimolar amounts at 55–60°C in the presence of 0.15 *M* $Na^+$. By design, the *Pst 1* sites are reconstructed at each plasmid–cDNA junction allowing subsequent removal of the cDNA (see Section 3.2). Ligation of the phosphodiester backbone is not required as this occurs in the bacterium after transformation.

## 2.3. Bacterial Transformation

Most methods for bacterial transformation derive from the observation that treatment of bacteria with calcium enhances their uptake of bacteriophage DNA (Mandel and Higa, 1970) or plasmid DNA (Cohen et al., 1972). Many variations in this basic technique have been described, all directed toward optimizing transformation efficiencies for different bacterial strains. Several studies analyzing the variables involved in the transformation of *E. coli* using chimeric plasmids have been conducted (see, e.g., Norgard et al., 1978; Peacock et al., 1981; Hanahan, 1983). Early transformation efficiencies achieved using calcium ions alone were on the order of $10^4$–$10^5$ colony-forming units (cfu) per microgram of plasmid DNA. Since transformation efficiencies obtained with recombinant plasmids are 1–2 orders of magnitude less than that achieved with "wild-type" plasmid, it was initially difficult to obtain reasonable numbers of clones in circumstances where the amount of mRNA was limited. Recently, procedures employing $Mg^{2+}$ or $Rb^+$ in addition to calcium have significantly improved transformation efficiency [Fig. 5 (9)]. Using the method described by Kushner (1978), we have been able to obtain $2.5 \times 10^7$ cfu/$\mu$g pBR322 in *E. coli* RR1, a readily available host strain. Efficiencies achieved with recombinant plasmids containing insertions 300–800 bp in length were $0.5–1.0 \times 10^6$ cfu/$\mu$g DNA.

## 2.4. Screening

Once within the host, the recombinant plasmids replicate and express the genes for drug resistance that permit survival in the presence of an antibiotic. Insertions in the *Pst 1* site of pBR322 result in transformed cells resistant to tetracycline ($tet^r$) but sensitive to ampicillin ($amp^s$). This forms the basis for the selection of bacterial colonies containing recombinant plasmids. In our hands approximately 95% of the transformed bacteria (i.e., $tet^r$) are amp sensitive (data not shown).

Prior to the identification of clones of interest (see Section 3.1) transformed bacteria are often first screened by colony hybridization (Grunstein and Hogness, 1975). In this type of experiment bacterial colonies are grown on solid supports (Hanahan and Meselson, 1980; Taub and Thompson, 1982) and filter-bound DNA hybridized with [$^{32}$P]-labeled cDNA prepared from the RNA initially used for cloning (i.e., template RNA for first-strand cDNA synthesis). Colonies containing recombinant plasmids are detected by autoradiography, the intensity of the signal being related to the abundance of the mRNA. The sensitivity of this technique is sufficient to detect clones that are complementary to as little as 0.1% of the cDNA probe (Gergen et al., 1979). Given the mRNA of interest is present in moderate abundance, colony hybridization can greatly reduce the number of clones that will require positive identification by nucleic acid sequencing or hybridization selection (see, e.g., Crabtree and Kant, 1981).

In instances where mRNA is brain-specific, colonies can be effectively screened by differential colony hybridization. For example, a positive selection is first performed by colony hybridization as described above. Positive colonies are subsequently hybridized to cDNA obtained from mRNA of a control, non-neural tissue. Bacterial colonies that yield positive signals with control cDNA are subsequently culled from the population. Differential screening has been used to great advantage in the identification of recombinant plasmids containing cDNA inserts for tyrosine hydroxylase mRNA (Lamouroux et al., 1982).

### 2.5. Discussion

A widely recognized drawback of the cloning procedure described is the inability to obtain cDNA fragments that contain the entire nucleotide sequence of the corresponding mRNA. This derives, in part, from the difficulty encountered in converting cDNA to full-length, double-stranded molecules and from the loss of sequences corresponding to the 5′ end of the mRNA, resulting from the S1 nuclease treatment required to eliminate the hairpin loop (see Fig. 5). The presence of the entire sequence would facilitate mapping mRNA coding and non-coding sequences onto the corresponding genomic DNA.

Recent modification of cDNA cloning procedures permits the recovery of recombinant plasmids with full- or nearly full-length cDNA inserts (Land et al., 1981; Okayama and Berg, 1982). One procedure uses a plasmid vector which itself serves as the primer for the first- and second-strand cDNA synthesis, thereby eliminating the need for S1 nuclease treatment of the cDNA (Okayama and Berg, 1982). Moreover, the procedure results in the preferential cloning of recombinants containing full-length cDNAs. This procedure has been applied successfully in the cloning of full-length cDNAs for $\alpha$- and $\beta$-globin (Okayama and Berg, 1982) and the $\gamma$-subunit of *Torpedo californica* acetylcholine receptor (Ballivet et al., 1982).

It bears emphasis that the cloning procedure described here represents but one approach to the cloning of moderately abundant mRNAs. One alternative approach involves the use of synthetic oligonucleotide probes. In this procedure a mixture of oligonucleotides complementary to all possible coding sequences of a small segment of a specific protein (three to five amino acids) can be used to prime the synthesis of the corresponding cDNA. Either the synthetic oligonucleotide itself (Wallace et al., 1981) or oligonucleotide-primed cDNA is used as a hybridization probe to select specific recombinant plasmids from a cDNA library (Agarwal et al., 1981; Chan et al., 1979; Goeddel et al., 1980).

Obviously, the applicability of this approach is limited to proteins of known amino acid sequence. Moreover, most amino acids are coded for by two or more codons, inhibiting one's ability to deduce with certainty the correct complementary sequence for a suitable DNA probe (but see Agarwal et al., 1981). Nonetheless, this procedure has been successfully applied in the selection of bacterial clones containing cDNA sequences for the enkephalin precursor (Comb et al., 1982; Gubler et al., 1982; Noda et al., 1982), nerve growth factor (Scott et al., 1983), and myelin basic protein (Roach et. al., 1983; Zeller et al., 1983).

## 3. IDENTIFICATION AND CHARACTERIZATION OF CLONES

### 3.1. Hybridization Selection

Given that sufficient amino acid sequence data is available for the protein of interest, the identity of selected clones can be unequivocably established by nucleic acid sequencing. Alternatively, clones can be screened and identified by hybridization selection (Bittner et al., 1974; Prives et al., 1974).

In this procedure, purified plasmid DNA is linearized with endonuclease, denatured, and immobilized on solid supports such as nitrocellulose filters (Harpold et al., 1978; McGrogan et al., 1979; Ricciardi et al., 1979) or diazobenzyloxymethylcellulose (DBM) paper (Goldberg et al., 1979). The filter-bound DNA is then hybridized to RNA under conditions which minimize RNA degradation. After extensive washing to remove nonspecifically bound RNA, the hybridized RNA is eluted by heat or formamide. The selected mRNA is recovered by ethanol precipitation and translated in a cell-free protein-synthesizing system. The resulting [$^{35}$S]-methionine-labeled peptides are subsequently identified by immunoprecipitation with specific antibody and SDS–polyacrylamide gel electrophoresis.

The results of a typical hybridization selection experiment are illustrated in Figure 6. Filter-bound plasmid DNA containing a cDNA insert for PNMT

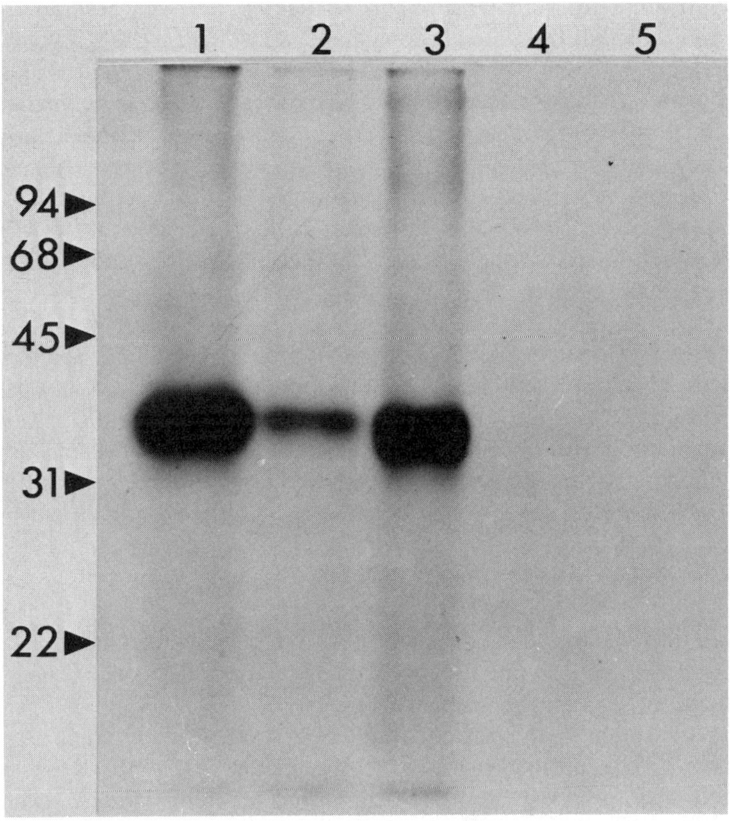


**Figure 6**. Translation of hybridization-selected PNMT-mRNA and immunoprecipitation of PNMT translation products. CsC1-purified plasmid DNA was linearized by treatment with *Eco R1*, denatured and bound to nitrocellulose filters (50–80 μg plasmid DNA/2.5 cm filter disc). Filter-bound DNA was hybridized to bovine adrenomedullary polysomal poly(A+) mRNA (80 μg/100 μL reaction) in 50% formamide at 42°C for 4–6 hr as described by Ploegh et al. (1980). Hybrid-selected and control mRNA were translated in a micrococcal nuclease-treated reticulocyte cell-free system in the presence of [$^{35}$S]-methionine. PNMT was immunoprecipitated from total translation products using specific antibody and formalin-fixed *Staphylococcus aureus* cells. Immunoprecipitated [$^{35}$S]-labeled translation products were fractionated on 12% SDS–polyacrylamide gels and visualized by fluorography. Lanes: 1, total poly(A+) mRNA; 2, mRNA hybridization-selected by pPNMT 1.0; 3, mRNA hybridization-selected by pPNMT 1.4; 4, mRNA selected by actin-pBR322; 5, no exogenous mRNA added. Insert size of recombinant plasmids PNMT 1.0 and PNMT 1.4 was 300 and 640 base pairs, respectively. Molecular weights, shown × $10^{-3}$, were calculated by reference to the mobilities of standard proteins: phosphorylase B (94,000), bovine serum albumin (68,000), ovalbumin (45,000), carbonic anhydrase (31,000), soybean trypsin inhibitor (22,000).

was hybridized to adrenomedullary poly(A+) mRNA. Hybridization-selected mRNA was translated in a reticulocyte cell-free system in the presence of [$^{35}$S]-methionine, and translation products immunoprecipitated by antibodies to PNMT displayed on a SDS–polyacrylamide gel. *In vitro* translation reactions programmed with total adrenomedullary poly(A+) mRNA served as a positive control. As shown in Figure 6, translation of mRNA hybrid selected by recombinant DNA yielded a 32,000-Da protein product, the molecular weight of PNMT (Park et al., 1982; Sabban et al., 1982). In contrast, no immunoprecipitable protein was obtained in control hybridization experiments using pBR322 DNA containing a 1000-bp cDNA insert to $\beta$-actin.

In instances where the mRNA of interest is present in moderate abundance, the sensitivity of the hybridization selection procedure is sufficient to permit the binding of DNA from several different clones to a single filter, thereby enabling one to screen multiple recombinants simultaneously (Parnes et al., 1981). Individual clones from groups yielding positive hybridization signals can then be screened separately.

### 3.2. Characterization of Cloned cDNA

The ability to remove inserted ds-cDNA from recombinant plasmids is a prerequisite for the generation of cDNA probes for molecular hybridization studies and nucleic acid sequencing. That the dC-dG tailing procedure (see Fig. 5) reconstructs *Pst 1* sites flanking the inserted cDNAs is demonstrated in Figure 7. Here, purified plasmid DNA containing a PNMT-cDNA insert was treated with *Pst 1* and the DNA fragments generated were fractionated by agarose gel electrophoresis. The DNA restriction fragments were visualized by ethidium bromide staining and ultraviolet fluorescence. As shown, *Pst 1* treatment linearized the plasmid (4.4 kb), generating a single low-molecular-weight DNA fragment approximately 300 base pairs in length. The ds-cDNA insert can be recovered from the gel by electroelution or standard extraction procedures (see, e.g., Volgelstein and Gillespie, 1979).

The abundance and size of the mRNA corresponding to a cloned cDNA fragment can be ascertained by northern blot analysis (Thomas, 1980). In these experiments poly(A+) mRNA is fractionated on agarose gels under denaturing conditions and transferred to a solid filter support, for example, nitrocellulose, DBM- or DPT (diazophenylthioether)-paper. Bound RNA is subsequently hybridized to a [$^{32}$P]-DNA probe labeled to high specific activity *in vitro* by nick-translation (Balman and Birnie, 1979; Maniatis et al., 1975). The hybridization of the DNA tracer to its complementary RNA is detected by autoradiography. The intensity of the signal is related to the abundance of the mRNA and mRNA size determined relative to the mobility of molecular weight standards.

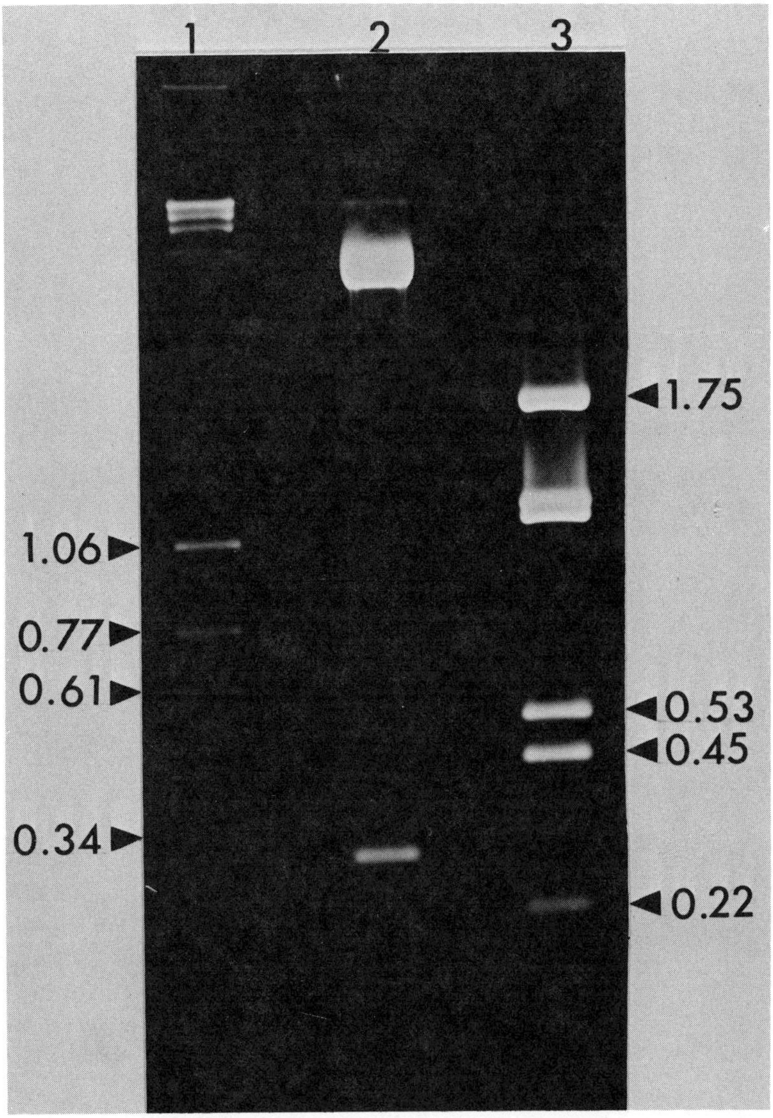


**Figure 7.** Size determination of PNMT-cDNA insert from recombinant plasmid. Purified plasmid was treated with *Pst 1* endonuclease (2.5 U enzyme/$\mu$g plasmid DNA) and restriction fragments fractionated on 1.5% agarose gels. Lanes: 1, lambda DNA-*Hind III*/$\Phi$X-174 RF DNA-*Hinc II* molecular weight standards; 2, *Pst 1*-treated recombinant plasmid (5.0) $\mu$g); 3, SV40 DNA-*Hind III* restriction fragments.

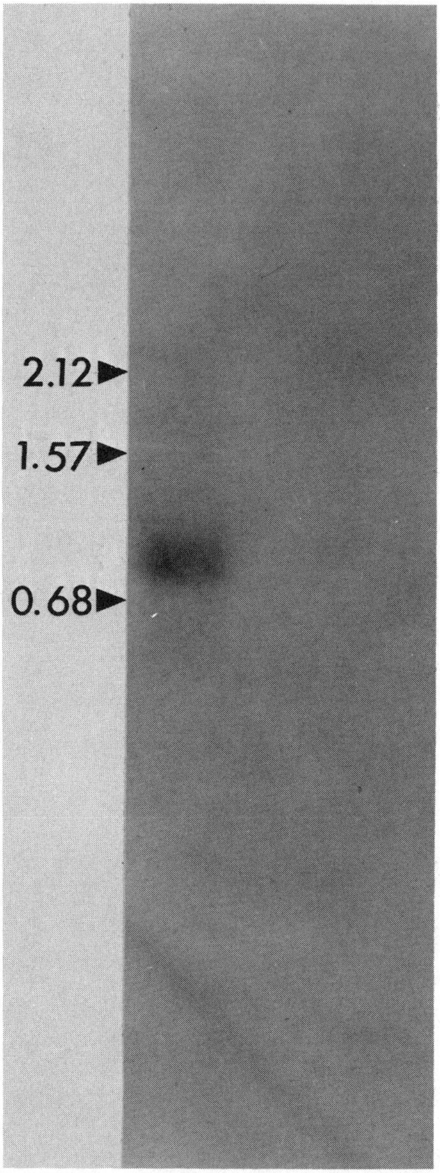


**Figure 8.** Northern blot analysis of adrenomedullary mRNA using plasmid containing PNMT-cDNA. Total bovine adrenomedullary poly(A+) mRNA was denatured and fractionated on a 1.2% agarose-formaldehyde gel and transferred to diazophenylthioether (DPT)-paper by electroblotting. [$^{32}$P]-labeled PNMT-pBR322 was hybridized to paper-bound RNA for 72 hr at 42°C in the presence of 50% formamide. After extensive washing using stringent criteria, the blot was visualized by autoradiography. Molecular weights (kb) were calculated by reference to *Rsa 1* restriction fragments of pBR322.

The result of a typical northern blot analysis is given in Figure 8. Here, bovine adrenomedullary polysomal poly(A+) mRNA was fractionated by agarose gel electrophoresis, transferred to DPT-paper and hybridized to [$^{32}$P]-labeled PNMT-pBR322 ($2.0 \times 10^8$ cpm/$\mu$g DNA). The DNA probe hybridized to a mRNA approximately 1050 nucleotides in length. Considering a 3′ end poly(A) tract of 100–150 nucleotides, the size of this mRNA is sufficient to code for a $M_r$ 32,000 peptide, the molecular weight of PNMT. With minor modification northern blot analysis can also be used to identify a specific mRNA population, help establish the tissue specificity of a particular mRNA, and monitor alterations in the levels of specific mRNAs in a semiquantitative fashion (Chen et al., 1983; White and Bancroft, 1982 Kafatos et al., 1979).

## 4. CONCLUDING REMARKS

In the past, recombinant DNA technology has provided experimental tools that have proven instrumental in enhancing our understanding of eukaryotic gene structure and expression (reviewed by Breathnach and Chambon, 1981). By its very nature, the cloning procedure can be used to purify to homogeneity nucleic acid sequences coding for brain-specific proteins present in relatively low abundance. Ironically, information concerning the structure of these proteins may first derive from the sequencing of recombinant DNA molecules. Useful antibodies may also be raised against synthetic peptide fragments derived from the nucleic acid sequencing data (see, e.g., Green et al., 1982; Sutcliffe et al., 1983).

In the near future cloned cDNA probes will undoubtedly facilitate study of the transcriptional regulation of brain-specific gene products. For example, judicious use of cDNA probes for key enzymes in the biogenic amine biosynthetic pathways will ultimately help determine regulatory events involved in the selection of neurotransmitter phenotype. The construction of cloned cDNA probes also permits the identification and isolation of chromosomal DNA fragments containing the specific gene-coding sequences from a genomic library, thereby enabling one to initiate detailed structural analysis. Future improvements in *in situ* hybridization techniques will most certainly allow the localization of brain-specific genes within the mammalian karyotype.

Clearly, recombinant DNA technology, in itself, cannot be used to address all the fundamental problems in the neurosciences. Nonetheless, the power of the technique will enable one to finally approach areas of long-standing neurobiological interest at the molecular genetic level.

## ACKNOWLEDGMENTS

The authors thank Dr. E. Edward Baetge for his participation in the experiments dealing with mRNA purification and Dr. Tong H. Joh for providing PNMT antibody. Studies described in the chapter were supported by a National Institutes of Health Grant, HD 11392.

## REFERENCES

Adams, J. M., N. M. Gough, E. A. Webb, B. M. Tyler, J. Jackson, and S. Cory, *Biochemistry* **19,** 2711 (1980).

Agarwal, K. L., T. Brunstedt, and B. E. Noyes, *J. Biol. Chem.* **256,** 1023 (1981).

Baetge, E. E., B. B. Kaplan, D. J. Reis, and T. H. Joh, *Proc. Natl. Acad. Sci. USA* **78,** 1269 (1981).

Baetge, E. E., H. M. Moon, B. B. Kaplan, D. H. Park, D. J. Reis, and T. H. Joh, *Neurochem. Intern.* **5,** 611 (1983).

Ballivet, M., J. Patrick, J. Lee, and S. Heineman, *Proc. Natl. Acad. Sci. USA* **79,** 4466 (1982).

Balman, A. and G. D. Birnie, *Biochim. Biophys. Acta* **561,** 155 (1979).

Bittner, W., Z. Veres-Molnar, and M. Green, *Proc. Natl. Acad. Sci. USA* **71,** 2951 (1974).

Breathnach, R. and P. Chambon, *Ann. Rev. Biochem.* **50,** 349 (1981).

Breathnach, R., J. L. Mandel, and P. Chambon, *Nature (London)* **270,** 314 (1977).

Buell, G. N., M. D. Wickens, J. Carbon, and R. T. Schimke, *J. Biol. Chem.* **254,** 9277 (1979).

Chan, S. J., B. E. Noyes, K. L. Agarwal, and D. F. Steiner, *Proc. Natl. Acad. Sci. USA* **76,** 5036 (1979).

Chaudhari, N. and W. E. Hahn, *Science* **220,** 924 (1983).

Chen, C. L. C., F. T. Dionne, and J. L. Roberts, *Proc. Natl. Acad. Sci. USA* **80,** 2211 (1983).

Cohen, S. N., A. C. Y. Chang, and L. Hsu, *Proc. Natl. Acad. Sci. USA* **69,** 2110 (1972).

Comb, M., P. H. Seeburg, J. Adelman, L. Eiden, and E. Herbert, *Nature (London)* **295,** 663 (1982).

Crabtree, G. R. and J. A. Kant, *J. Biol. Chem.* **256,** 9718 (1981).

Deng, G. and R. Wu, *Nucleic Acids Res.* **9,** 4173 (1981).

Dokas, L. A., *Brain Res. Rev.* **5,** 178 (1983).

Dugaiczyk, A., S. L. C. Woo, E. C. Lai, M. L. Mace, L. McReynolds, and B. W. O'Malley, *Nature (London)* **274,** 328 (1978).

Efstratiadis, A. and L. Villa-Komaroff, in J. Setlow and A. Hollaender, Eds., *Genetic Engineering,* Vol. 1, Plenum, New York, p. 15 (1979).

Efstratiadis, A., F. C. Kafatos, and T. Maniatis, *Cell* **10,** 571 (1977).

Engel, J., P. Gunning, and L. Kedes, *Mol. Cell. Biol.* **2,** 674 (1982).

Friedman, E. Y. and M. Rosbash, *Nucleic Acids Res.* **4,** 3455 (1977).

Gergen, J. P., R. H. Stern, and P. C. Wensink, *Nucleic Acids Res.* **7,** 2115 (1979).

Goeddel, D. V., H. M. Shepard, E. Yelverton, D. Leung, and R. Crea, *Nucleic Acids Res.* **8,** 4057 (1980).

Golberg, M. L., R. P. Lifton, G. R. Stark, and J. G. Williams, *Methods Enzymol.* **68,** 206 (1979).

Gough, N. M., E. A. Webb, S. Cory, and J. M. Adams, *Biochemistry* **19,** 2702 (1980).

Gordon, J. I., A. T. H. Burns, J. L. Christmann, and R. G. Deeley, *J. Biol. Chem.* **253,** 8629 (1978).

Green, N., H. Alexander, A. Olson, F. Alexander, T. S. Shinnick, J. G. Sutcliffe, and R. A. Lerner, *Cell* **28,** 477 (1982).

Grunstein, M. and D. S. Hogness, *Proc. Natl. Acad. Sci. USA* **74,** 560 (1975).

Gubler, U., P. Seeburg, B. J. Hoffman, L. P. Gage, and S. Udenfriend, *Nature (London)* **295,** 206 (1982).

Hanahan, D., *J. Mol. Biol.* **166,** 557 (1983).

Hanahan, D. and M. Meselson, *Gene* **10,** 63 (1980).

Harpold, M. M., P. R. Dobner, R. M. Evans, and F. C. Bancroft, *Nucleic Acids Res.* **5,** 2039 (1978).

Kafatos, F. C., C. W. Jones, and A. Efstratiadis, *Nucleic Acids Res.* **7,** 1541 (1979).

Kaplan, B. B., in A. Lajtha, Ed., *The Handbook of Neurochemistry,* Vol. 2, Plenum, New York, p. 1 (1982).

Kaplan, B. B. and C. E. Finch, in I. R. Brown, Ed., *Molecular Approaches to Neurobiology,* Academic, New York, p. 71 (1982).

Kinniburg, A. J., J. E. Mertz, and J. Ross, *Cell* **14,** **681** (1978).

Korman, A. J., P. J. Knudsen, J. F. Kaufman, and J. L. Strominger, *Proc. Natl. Acad. Sci. USA* **79,** 1844 (1982).

Kraus, J. P. and L. E. Rosenberg, *Proc. Natl. Acad. Sci. USA* **79,** 4015 (1982).

Kushner, S. R., in H. W. Boyer and S. Nicosia, Eds., *Genetic Engineering,* Elsevier-North Holland, New York, p. 17 (1978).

Lamouroux, A., N. F. Biguet, D. Samolyk, A. Privat, J. C. Salomon, J. F. Pujol, and J. Mallet, *Proc. Natl. Acad. Sci. USA* **79,** 2881 (1982).

Land, H., M. Grez, H. Hauser, W. Lindenmaier, and G. Schütz, *Nucleic Acids Res.* **9,** 2251 (1981).

Lee, D. C., G. S. McKnight, and R. D. Palmiter, *J. Biol. Chem.* **255,** 1442 (1980).

MacDonald, R. J., M. M. Crerar, W. F. Swain, R. L. Pictet, G. Thomas, and W. J. Rutter, *Nature (London)* **287,** 117 (1080).

Mandel, M. and A. Higa, *J. Mol. Biol.* **53,** 159 (1970).

Maniatis, T., A. Jeffrey, and D. G. Kleid, *Proc. Natl. Acad. Sci. USA* **72,** 1184 (1975).

Maniatis, T., E. F. Fritsch, and J. Sambrook, Eds., *Molecular Cloning: A Laboratory Manual,* Cold Spring Harbor Laboratory, Cold Spring Harbor, NY (1982).

Maxwell, I. H., J. Van Ness, and W. E. Hahn, *Nucleic Acids Res.* **5,** 2033 (1978).

McGrogan, M., D. J. Spector, C. J. Goldenberg, D. Halbert, and H. J. Raskas, *Nucleic Acids Res.* **6,** 593 (1979).

McReynolds, L. A., B. W. O'Malley, A. D. Mischet, J. E. Fothergill, D. Givol, S. Fields, M. Robertson, and G. G. Brownlee, *Nature (London)* **273,** 723 (1978).

Nelson, T. and D. Brutlag, *Methods Enzymol.* **68,** 473 (1979).

Noda, M., H. Furutani, H. Takahashi, M. Toyosato, T. Hirose, S. Inayama, S. Nakanishi, and S. Numa, *Nature (London)* **295,** 202 (1982).

Nordstrom, J. L., D. R. Roop, M. J. Tsai, and B. W. O'Malley, *Nature (London)* **278,** 328 (1979).

Norgard, M. V., K. Keem, and J. J. Monahan, *Gene* **3,** 279 (1978).

Ohno, T., P. J. Cozens, A. C. B. Cato, and J-P. Jost, *Biochim. Biophys. Acta* **606,** 34 (1980).

Okayama, H. and P. Berg, *Mol. Cell Biol.* **2,** 161 (1982).

Park, D. H., E. E. Baetge, B. B. Kaplan, V. R. Albert, D. J. Reis, and T. H. Joh, *J. Neurochem.* **38,** 410 (1982).

Parnes, J. R., E. Velan, A. Felsenfeld, L. Ramanathan, U. Ferrini, E. Appella, and J. G. Seidman, *Proc. Natl. Acad. Sci. USA* **78,** 2253 (1981).

Peacock, S. L., C. M. McIver, and J. J. Monahan, *Biochem. Biophys. Acta* **655,** 243 (1981).

Ploegh, H. L., H. T. Orr, and J. L. Strominger, *Proc. Natl. Acad. Sci. USA* **77,** 6081 (1980).

Prives, C. L., H. Aviv, B. M. Paterson, and B. E. Roberts, *Proc. Natl. Acad. Sci. USA* **71,** 302 (1974).

Ricciardi, R. P., J. S. Miller, and B. E. Roberts, *Proc. Natl. Acad. Sci. USA* **76,** 4927 (1979).

Richards, D. A., D. E. Blackburn, and J. M. Rosen, *J. Biol. Chem.* **256,** 533 (1981).

Roach, A., K. Boylan, S. Horvath, S. B. Prusiner, and L. E. Hood, Cell 34, 799 (1983).

Roop, D. R., J. L. Nordstrom, S. Y. Tsai, J. J. Tsai, and B. W. O'Malley, *Cell* **15,** 671 (1978).

Sabban, E., M. Goldstein, M. C. Bohn, and I. B. Black, *Proc. Natl. Acad. Sci. USA* **79,** 4823 (1982).

Schibler, V., K. B. Marcu, and R. D. Perry, *Cell* **15,** 1495 (1978).

Schrader, W. T. and B. W. O'Malley, Eds., *Laboratory Methods Manual for Hormone Action and Molecular Endocrinology,* 6th ed., Houston Biological Assoc., Houston, TX (1982).

Schwartz, R. J., J. A. Haron, K. N. Rothblum, and A. Dugaiczyk, *Biochemistry* **19,** 5883 (1980).

Schütz, G., S. Kieval, B. Groner, A. E. Sippel, D. T. Kurtz, and P. Fiegelson, *Nucleic Acids Res.* **4,** 71 (1977).

Scott, J., M. Selby, M. Urdea, M. Quiroga, G. I. Bell, and W. J. Rutter, *Nature (London)* **302,** 538 (1983).

Shapiro, S. Z. and J. R. Young, *J. Biol. Chem.* **256,** 2495 (1981).

Smith, D. F., P. F. Searle, and J. G. Williams, *Nucleic Acids Res.* **6,** 487 (1979).

Sutcliffe, J. G., *Nucleic Acid Res.* **5,** 2721 (1978).

Sutcliffe, J. G., R. J. Milner, T. S. Shinnick, and F. E. Bloom, *Cell* **33,** 671 (1983).

Taub, F. and E. B. Thompson, *Anal. Biochem.* **126,** 222 (1982).

Thomas, P., *Proc. Natl. Acad. Sci. USA* **77,** 5201 (1980).

Tilghman, S., P. J. Curtis, D. C. Tiemeyer, P. Leder, and C. Weissman, *Proc. Natl. Acad. Sci. USA* **75,** 1309 (1978).

Villa-Komaroff, L., A. Efstratiadis, S. Broome, P. Lomedico, R. Tizard, S. P. Naber, W. L. Chick, and W. Gilbert, *Proc. Natl. Acad. Sci. USA* **75,** 3727 (1978).

Vogelstein, B., and D. Gillespie, *Proc. Natl. Acad. Sci. USA* **76,** 615 (1979).

Wahli, W., G. U. Ryffel, T. Wyler, R. B. Saggi, R. Weber, and I. B. Dawid, *Dev. Biol.* **67,** 371 (1978).

Wahli, W., G. U. Ryffel, T. Wyler, R. B. and I. B. Dawid, *Dev. Biol.* **67,** 371 (1978).

Wallace, R. B., M. J. Johnson, T. Hirose, T. Miyake, E. H. Kawashima, and K. Itakura, *Nucleic Acids Res.* **9,** 879 (1981).

White, B. A. and F. C. Bancroft, *J. Biol. Chem.* **257,** 8569 (1982).

Wickens, M. P., G. N. Buell, and R. T. Schimke, *J. Biol. Chem.* **253,** 2483 (1978).

Zehner, Z. E., and B. M. Paterson, *Proc. Natl. Acad. Sci. USA* **80,** 911 (1983).

Zeller, N. K., M. J. Hunkeler, A. T. Campagnoni, J. Sprague, and R. A. Lazzarini, *Proc. Natl. Acad. Sci. USA* **81,** 18 (1984).

# 2

# *Peptide Hormone Gene Expression in the Brain*

**_Marian J. Q. Evinger, Ching-Ling C. Chen, Beth S. Schachter, and James L. Roberts_**

Recent developments in molecular biology have provided insight into the study of peptide hormone gene expression in the brain. In this chapter we will discuss how the principles of molecular biology and recombinant DNA methodologies can be employed to characterize the production of specific peptide hormones in neural tissue. With this technology the mRNAs encoding specific polypeptides may now be identified and quantified at their sites of synthesis. More importantly, the neurobiologist can combine these methods with conventional techniques to study the synthesis, regulation, and intercellular actions of these peptides.

The study of gene expression in the brain poses a set of unique problems, foremost among them being the extensive heterogeneity of neural tissue. Using immunological and histochemical techniques, neurobiologists have categorized hundreds of different cell types in the brain on the basis of the various cellular components they express. Great heterogeneity also exists in cellular morphology as evidenced by the spectrum of cells ranging from the unipolar interneurons to the highly arborized Purkinje cells. It therefore becomes readily apparent that the neurobiologist must first selectively define a restricted system for the study of gene expression in the brain.

## 1. PEPTIDE HORMONES AS A MODEL SYSTEM

The neurosecretory peptide hormones are well suited for the study of gene expression because extensive information has already been gathered on the distribution of many of these polypeptides in the brain. Using well-characterized antisera and purified peptides, immunohistochemists have assembled cytological maps localizing both the perikarya and terminal fields for many of the cells producing these hormones. Additionally, in certain instances the sites of hormone synthesis have been established by studies monitoring the incorporation of amino acids into specific polypeptides. It is in this manner that the presence of greater than 40 peptide hormones in the brain has thus far been demonstrated.

The production and release of brain peptide hormones is controlled by specific behavioral and hormonal effectors. For example, opiate peptides are released in large quantities in response to pain stimuli (Millan, 1981). In many cases the corresponding chemical stimuli have been identified, thereby making it feasible to examine the modulation of peptide hormone gene expression at the biochemical level. In a process analogous to that for other peptide hormones neuropeptide hormones are synthesized as larger precursor proteins which are then subjected to complex co- and post-translational modifications including removal of the hydrophobic signal peptide, proteolytic cleavages, and less frequently glycosylation, acetylation, sulfation, phosphorylation, amidation, or pyrolization (Eipper and Mains, 1980). Usually these post-translational modifications are necessary for the induction of specific biological activities from inactive precursor molecules.

The pro-opiomelanocortin (POMC) peptide system will be discussed extensively in this review because its synthesis provides an ideal example of the peptide hormone translational and modification processes just described, and it has been the most thoroughly studied peptide hormone gene of the mammalian brain. POMC is the protein precursor to several different bioactive peptides (Eipper and Mains, 1980; Roberts et al., 1982). Like many other peptide hormones, POMC appears to be expressed in cells scattered throughout the central and peripheral nervous systems. Although produced in large amounts by the anterior and neurointermediate lobes (AL and NIL) of the pituitary gland, POMC is synthesized by only a few thousand cells in the mammalian brain (Watson et al., 1977; Zimmerman et al., 1978). Hence the limitations in studying POMC expression in the brain are indicative of the degree of difficulty encountered in examining the regulation of many other neurosecretory peptide hormones.

### 1.1. Problems of Specificity in the Study of Peptide Hormones in the Brain

The methodologies developed for examining proteins in other tissues may be directly applied to peptide hormones synthesized by the brain. However, because of the great complexity of neural tissue, the molecular biologist must refine these methodologies to achieve adequate levels of sensitivity and specificity. Nucleic acid hybridization studies have revealed that the brain contains an extremely complex mixture of mRNA(s). Using either single-copy DNA or reverse-transcribed mRNA as a probe (Bantle and Hahn, 1976; Chaudhari and Hahn, 1983; Chikaraishi, 1979), it has been observed by $R_0t$ analysis that the pool of brain mRNAs possesses a complexity approximately fivefold greater than that of kidney or liver, thereby suggesting that about five times more unique-sequence genomic DNA is expressed by the brain than by these other tissues. Furthermore, the polysomal mRNA of the brain contains a high proportion (as much as 50%) of non-polyadenylated mRNA. Hybridization and cell-free translation experiments have suggested that some of these non-polyadenylated mRNAs may code for a class of proteins separate from those translated from polyadenylated (poly(A+)) RNAs (Chaudhari and Hahn, 1983). Although their functional significance remains to be demonstrated, the contribution of non-polyadenylated mRNAs must be considered in calculating the overall complexity of brain mRNAs (Dokas, 1983).

## 2. ISOLATION OF SPECIFIC DNAs FOR THE STUDY OF GENE EXPRESSION

As discussed previously, the complexity of brain mRNA coupled with the extremely low abundance of polypeptides frequently restricts the detection of their corresponding mRNAs in neural tissue. In our demonstration of a recombinant DNA approach we shall focus on those experimental techniques and designs that should prove particularly useful in the study of the neurosecretory peptides produced by the brain.

### 2.1. Detection of Specific mRNAs

The specificity conferred by coupling *in vitro* translations with biological assays or immunoprecipitations has provided the greatest advantage in the isolation, characterization, and cloning of specific mRNAs. Correspondingly, initial efforts to examine the synthesis of specific brain polypeptides have often relied on detection of translated polypeptides as proof of the presence of a specific mRNA. The identification of an mRNA by translation permits a direct estimate of the minimum size of the RNA as well as its relative abundance. Oligo-deoxythymidylate (dT) cellulose affinity chromatography, polysome immunoprecipitation, and fractionation on sucrose gradients or by preparative gel electrophoresis are procedures that may significantly enhance the proportion of a specific mRNA (Aviv and Leder, 1972; Rosen and Monahan, 1976; Sehgal, 1981). Such enrichment often facilitates subsequent analysis of the peptide product. Translation systems using wheat germ extracts, reticulocyte lysates, or *Xenopus* oocytes (Comstock et al., 1980) have been employed for the synthesis of peptide products that can then be identified by precipitation with antisera, comigration in two or more dimensions, measurement of biological activity, or identification of peptide fragments generated by digestion with proteolytic enzymes. Incorporation of radioactive amino acids into polypeptides synthesized *in vitro* has facilitated the detection of several peptide hormone messages. The relative levels of mRNAs for the $\alpha$ and $\beta$ subunits for the glycoprotein family containing thyroid-stimulating hormone (Gurr et al., 1983), chorionic gonadotropin (Chatterjee et al., 1976; Fiddes and Goodman, 1981), luteinizing hormone (LH) and follicle-stimulating hormone (Godine et al., 1981), and for the neurophysins (Giudice and Chaiken, 1979; Schmale and Richter, 1981) were assessed by their immunoprecipitated *in vitro* translation products. Additionally, the presence of multiple cysteine residues (permitting [$^{35}$S]-labeling to high specific radioactivity) and localized synthesis in the supraoptic nuclei (permitting enriched regional dissections) are two factors that have facilitated the study of mRNAs coding for vasopressin, oxytocin, and their respective neurophysins (Russell et al., 1980). Translation products may also be recognized following fractionation in two dimensions, first by isoelectric focusing and then by SDS–polyacrylamide gel electrophoresis (O'Farrell, 1975) as, for example, demonstrated by the identification of POMC synthesized *in vitro* (Civelli et al., 1982; Roberts et al., 1978).

Examining the biological activity of its corresponding polypeptide provides yet another means by which a specific mRNA may be detected. The ability to induce LH release from primary pituitary cultures is an assay used frequently for substances possessing the properties of gonadotropin-releasing hormone (GnRH, also termed *luteinizing-hormone-releasing hormone*). Similarly, a bioassay for peptidases which proteolytically degrade peptide hormones measures the hydrolysis of oligopeptide substrates by extracts of neural tissue (Advis et al., 1982). Finally, the ability to establish the presence of a particular mRNA on the basis of the relative amounts of specific proteolytic fragments produced after *in vitro* translation is demonstrated by the synthesis of rat dopamine $\beta$-hydroxylase in the brain (O'Malley et al., 1983) and the $\alpha$ subunit of human chorionic gonadotropin in the placenta (Daniels-McQueen et al., 1978).

At this time, however, antisera, bioassays, or identifiable peptide fragments do not exist for the majority of brain peptide hormones, particularly for their precursor forms. The inherent problems and approaches applicable to the study of peptide hormones in the brain in this far more common situation will be discussed in Section 2.2.

## 2.2. Use of Non-neuronal Tissues as mRNA Sources

The synthesis of a specific brain peptide hormone in another tissue may provide the additional advantage of cloning its respective mRNA from a site where it is expressed in greater abundance. An outstanding example of this approach is the use of anglerfish pancreatic islets as a highly enriched source of somatostatin relative to the hypothalamus for the isolation of a somatostatin cDNA (Hobart et al., 1980; Taylor et al., 1981). The detection and localization in brain tissue of several peptide hormones, notably POMC (Civelli et al., 1982), Met- and Leu-enkephalin (Comb et al., 1982; Gubler et al., 1981), preproenkephalin A (Noda et al., 1982), and relaxin (Hudson et al., 1981) have been facilitated by the preparation of cDNAs from tissues in which the peptide hormones are expressed in greater abundance (even if different forms of the peptide or its mRNA are expressed by the other tissues).

The availability of established and easily maintained cell lines that overproduce a specific polypeptide product presents another approach. For example, $\beta$-endorphin of the POMC complex was cloned from the mRNA of the AtT20 mouse tumor cell line (Roberts et al., 1979b). Similarly, a human pheochromocytoma line provided an abundant source of mRNA for Met- and Leu-enkephalin (Comb et al., 1982).

A particularly fortuitous circumstance exists if one segment of a peptide hormone possesses sufficient homology to function as a probe for another peptide hormone. For example, even limited sequence homology such as that observed among the $\beta$ subunits of the gonadotropins may permit the selection of a related sequence. A successful application of this approach is demonstrated by Fiddes (personal communication) who has employed the 82% homology between nucleic acid sequences for the isolation of the $\beta$ subunit of luteinizing hormone using the sequence for the corresponding $\beta$ subunit of chorionic gonadotropin. In like manner, specific cDNAs may be used to select genetic variants of a particular product as demonstrated by the use of the growth hormone probe (Seeburg, 1982) to isolate a similar but nonetheless distinct variant.

### 2.3. Selection of Oligonucleotide Sequences

Presently, in the study of many brain peptide hormones only the amino acid sequence of the mature (i.e., biologically active) form of the polypeptide has been determined. Knowledge of this peptide sequence nonetheless permits the prediction of a deoxyribonucleotide sequence complementary to its corresponding mRNA. The technique offering the greatest versatility for detecting and characterizing peptide hormones in the brain has involved use of specific oligodeoxyribonucleotides, frequently termed *oligonucleotides*. As described subsequently, these oligonucleotides may be effectively utilized in three major ways: (1) as primers for reverse transcription of a specific mRNA into cDNA, (2) as probes for the detection of specific cDNAs that have been reverse transcribed from poly(A+) RNAs via an oligo(dT) primer, and (3) as probes for the sequence of a specific gene contained within a genomic DNA library.

Construction of an oligonucleotide depends on several factors. The primary consideration focuses on the incorporation of the greatest specificity possible in the cDNA sequence—that is, selecting a sequence displaying maximal homology to its corresponding mRNA. In the event that a previously isolated DNA segment is unavailable for restriction modification, synthetic oligonucleotides can be constructed in which case the selection of their sequences will be based on the amino acid sequence of the peptide. Virtually any oligonucleotide sequence desired can now be constructed as a result of the many refinements in the chemistry and methodology of DNA synthesis. This is graphically demonstrated by the assembly of complete cDNA sequences for insulin (Crea et al., 1978) and human fibroblast interferon (Edge et al., 1981). Several methods (recently reviewed by Atkinson, 1983) currently exist for synthetic DNA chemistries, including the phosphodiester, phosphotriester, and phosphite methods (Khorana et al., 1976; Markham et al., 1980; Matteucci and Caruthers, 1981; Narang et al., 1979). More importantly, the availability of synthetic oligonucleotide assembly procedures have now placed this technology within the grasp of a much wider range of investigators.

In the assembly of a synthetic oligodeoxyribonucleotide complement of a specific mRNA, the degeneracy of the genetic code results in ambiguity of codon choices at the third position for many amino acids and at the first and second positions for a few amino acids. For this reason, it is advantageous to select regions of the peptide where the codon degeneracy for the amino acids is minimal, preferably those regions containing Met and Trp for which single codon choices exist. The availability of two codon choices for nine other amino acids (Asp, Asn, Cys, His, Phe, Tyr, Glu, Gln, and Lys) should also be considered in the selection of a sequence. One successful application of this approach is demonstrated by the fact that only eight permutations were necessary for the construction of the decahexamer encoding a portion of the proenkephalin peptide $NH_2$-Trp-Trp-Met-Asp-Tyr-Gln-COOH (Gubler et al., 1981). Although the high degree of degeneracy in the third codon positions increases the number of sequence permutations for a given region, the selection of T and G for third codon positions is recommended because dT:dG and dG:rU base mismatches do not exert destabilizing effects on DNA:RNA hybrid formation (Gillam et al., 1975). However, with other factors being equivalent, the greater thermal stability of GC hybrids relative to AT pairs recommends the selection of regions containing a higher G + C content. Furthermore, the natural frequency of mRNA codon selection for other mammalian amino acids may influence the choice of a primer sequence. A catalog compiling codon usages for nearly 200 mRNAs (Grantham et al., 1981) can now assist in the predictions of codon choices for specific single-copy genes.

The selection of oligonucleotide sequences may be influenced by the post-translational modifications in the mature peptide. Because the amidations of carboxy terminal amino acids as seen, for example, in melittin (Suchanek and Kreil, 1977) and gastrin (Noyes et al., 1979) are in all likelihood generated by transamidase reactions, it is reasonable to predict codons for a Gly residue at the carboxy terminal of these peptides. Similarly, post-translational cyclization of glutaminyl residues to pyroGlu (Schally et al., 1971) in thyroid-releasing hormone and GnRH (LHRH) indicates codon selections for Gln at the amino termini of these peptides. Moreover, the general observation that all peptide hormone precursors examined to date contain pairs of basic amino acids (Lys-Lys, Lys-Arg, Arg-Lys, or Arg-Arg) flanking the peptide (Steiner et al., 1971) should prove beneficial for the prediction of oligonucleotide sequences complementary to the terminal regions of peptides.

When numerous possibilities exist for predicting the sequence of an oligonucleotide, a choice must be made between construction of the "most probable" sequence and one containing all the permutations of such a sequence. The primary consideration in selecting the "most probable" sequence is a decision whether the sequence will provide sufficient thermal stability for the hybridization of DNAs and RNAs which may exhibit only partial homology. Alternatively, the choice to construct all sequence permutations acknowledges the fact that the correct sequence represents only a fraction ($1/N$) of a pool of highly similar homologues. It should be noted that both approaches have been employed successfully. For the prediction of the most probable cDNA sequence for the amino terminal region of human fibroblast interferon mRNA, codon choices were reduced initially by the choice of triplets most likely to show maximal homology with analogous mouse interferon codons then further reduced by considering the known frequencies of codon usages in other human single-copy genes (Houghton et al., 1980). Although this rationale reduced the number of potential coding sequences to two (differing in the third residue of a serine), one sequence differing only in a single codon near the midpoint failed to prime interferon cDNA synthesis. On the other hand the 11-mer (sequence of 11 nucleotides) predicted for rat relaxin corresponded to the correct sequence of its cDNA (Hudson et al., 1981).

Mixtures of oligonucleotide permutations have also provided specific cDNA priming. For example, pools of thirty-two 16-mers were used for isolation of a cDNA for 3-phosphoglycerate kinase (Singer-Sam et al., 1983). Other examples include the use of pooled probes to obtain cDNAs for rat angiotensinogen (Ohkubo et al., 1983) and for the $\alpha$ subunit of bovine pituitary glycoprotein hormones (Nilson et al., 1983). However, the destabilizing effect exerted by a single base mismatch between a 17-base oligonucleotide and $\phi$X 174 DNA was equivalent to a perfectly matched duplex 3 base pairs shorter (Wallace et al., 1979). In summary, it appears that the prediction of the most probable sequence for a cDNA sequence offers the most expeditious and economical approach to oligonucleotide construction. It should be noted, however, that extremely low abundance mRNAs or DNAs may be detected only by hybridization to their exact complement, thereby restricting such an approach to pools of synthetic oligonucleotides.

A consideration of the thermodynamic parameters involved in the hybridization of oligonucleotides is critical to the successful application of synthetic primers or probes for the detection of mRNAs encoding brain peptide hormones. Only the major factors affecting hybridization will be summarized here since an extensive treatment of these parameters is provided by M. Smith (1983). The length, nucleic acid composition, and primary nucleotide sequence of the oligonucleotide are the major contributors to the formation of stable hybrids. Eight nucleotides have been determined to be the minimum length necessary for forming and maintaining hybrids with thermal stability then increasing proportionately up to nucleotide lengths of 200 base pairs (the point at which duplex stability becomes independent of length). Hybrid stability is also dependent upon the nucleotide base composition as demonstrated by the 12°C increase in $T_m$ observed when three dA:dT base pairs are replaced by three dG:dC base pairs in a nonanucleotide. The position of a base mismatch in an oligonucleotide significantly affects the thermal stability of the hybrid. For example, a central mismatch destabilizes a duplex by an amount equivalent to removing at least two dA:dT pairs.

Maintenance of oligonucleotide hybrid structures is primarily influenced by ionic strength and temperature. Since duplex stability is directly proportional to ionic strength [$T_m$ = 69.3°C + 0.41 (G + C)%], hybridizations with oligonucleotides are often performed at high ionic strength (e.g., 0.6–1.0 *M* $Na^+$). The presence of divalent cations, usually $Mg^{2+}$, will also compensate somewhat for decreased salt concentrations. The sensitivity of this methodology is demonstrated by the ability of Conner et al. (1983) to distinguish a single base mismatch in the hybridization of two 19-mers to $\beta$-globin alleles based on the selection of appropriate stringencies for hybridization and washing. Other workers dealing with multiple permutations of oligomers must carefully assess the thermal stabilities required for each of the components represented in pooled probes, particularly where the primers coding for a single peptide sequence may span 37–75% G + C composition. In these cases the condition stabilizing a high G + C complex containing one or two mismatches may be more stringent than those needed to maintain perfect hybrids of more A + T-rich duplexes.

### 2.4. Applications of Oligonucleotides for Cloning Specific Brain Peptide mRNAs

As evidenced by the many examples in the literature, oligonucleotides can function as specific primers for cDNA synthesis. Of particular note are those used for cloning very low abundance mRNAs. Gastrin mRNA from porcine antral mucosa represents less than 0.6% of the poly (A+)-enriched RNA (less than 0.005% of the total RNA; Noyes et al., 1979). Human fibroblast interferon mRNA is present as approximately 0.03% of poly (A+) RNA (Houghton et al., 1980). The isolation of the cDNA for corticotropin-releasing hormone (CRH) (Furutani et al., 1983) provides a striking example of the successful application of this approach. In this case oligonucleotides were used to prime from a message whose level was too low to be detected on a Northern blot, hence too low even for an estimation of its relative abundance.

Generally, oligonucleotides labeled to a high specific radioactivity have functioned successfully as primers when employed in 5- to 40-fold molar excess for reverse transcription from poly (A+)-enriched mRNAs. If adequate specificity of priming is achieved, a single (or few) cDNA species may be resolved by gel electrophoresis (see, e.g., Noyes et al., 1979). Extraction of this cDNA with the subsequent construction of a recombinant in a plasmid vector can then be accomplished by well-established cloning techniques. As might be expected for a low-abundance mRNA, relatively large amounts of mRNA must be isolated initially in order to prepare sufficient cDNA (i.e., enough cDNA to ultimately produce a reasonable number of recombinants). Also, unless the sequence of the primer represents the carboxy terminal of the peptide, the initial cDNA represents only a portion of the mRNA; it must in turn be used to isolate a full-length cDNA. Notwithstanding the initial requirements of RNA and tissue, oligonucleotide priming confers above all great specificity in the initial preparation of a recombinant.

In the event that a given oligonucleotide does not function adequately as a primer for cDNA synthesis, it may nonetheless prove to be a useful probe for hybridization. Hybridization of radioactively labeled oligonucleotides to cDNAs fixed on nitrocellulose filters (Grunstein and Hogness, 1975) provides a rapid and convenient method for screening a large population of recombinants prepared from oligo(dT)-primed poly(A+) RNAs. The two major variations of this procedure include screening the recombinants with oligonucleotides themselves and screening with primer-extended cDNAs. Screening recombinant colonies with oligonucleotides is frequently complicated by high levels of non-specific background. Although preadsorption with excess DNA or RNA reduces this background substantially, the conditions of stringency often determine whether a specific signal will be clearly detected above background.

Although still a relatively uncommon application, oligonucleotides may be used as limited extension primers. Since reverse transcription terminates upon incorporation of a dideoxynucleotide, a population of cDNAs may be extended then terminated at a discrete nucleotide size class. The increased stability of this longer primer and the incorporation of radiolabel produces a population of labeled cDNAs in amounts suffucient to complete the subsequent cloning and transformation procedures. Sood et al. (1981) specifically primed human histocompatibility antigen HLA-B with an 11-mer but controlled the amount of extension to a 30-nucleotide fragment by the incorporation of a single dideoxynucleotide triphosphate. Similarly, Das et al. (1983) used a synthetic oligomer to prime dideoxy-terminated cDNA synthesis from the HLA-DR and B-chain mRNAs. Not only were specific cDNAs identified for each chain but the 175-nucleotide cDNA prepared by primer extension of the 20-mer could be subsequently used for isolation of the $\alpha$-chain genomic fragment.

Alternatively, oligonucleotides may be used to generate probes to detect tissue-specific expression of individual mRNAs. By contrasting the hybridization of primer-extended cDNAs synthesized from mRNAs of tissues expressing a specific peptide with the cDNAs from tissues which do not express this same product, it is possible to establish a plus–minus screening paradigm. Furthermore, primer-extended cDNAs may be used under conditions of higher stringency, thereby resulting in greater hybrid specificity than that achieved solely with oligonucleotides. This approach has been employed successfuly with the alpha 2*u* globulin gene family (Unterman et al., 1981). One caveat lies in the choice of a control, for example, non-expressing tissue. Any production of even low levels of a peptide hormone mRNA in the control tissue precludes its utility as a true tissue negative. For example, even the extremely low levels of POMC mRNA produced by the liver, cortex, and cerebellum (Civelli et al., 1982) would invalidate the use of mRNAs from these tissues as negatives in screening a hypothalamic cDNA library for POMC mRNA.

As indicated previously, another valuable function of synthetic oligonucleotides is their application as probes for hybridization. Frequently two pools of oligonucleotides are synthesized containing sequences complementary to two separate portions of a specific mRNA. Subsequent screening efforts can be substantially reduced by pursuing only those recombinants demonstrating hybridization to both of the region-specific oligonucleotides. As demonstrated for human $\alpha$-interferon, the subset of cDNAs hybridizing to both pools of oligonucleotides which had been constructed as complements to two separate regions of the message could then be examined individually for the ability to hybrid-arrest translation of the interferon protein (Nagata et al., 1980).

A somewhat more limited application of oligonucleotides involves their use as hybridization probes for the identification of a specific clone from DNA libraries, including both genomic (Maniatis et al., 1978) and cDNA (Young and Davis, 1983) libraries constructed in the bacteriophage lambda. Although examples of the use of oligonucleotides to screen phage cDNA libraries are not yet available, genomic fragments for several trypanosome variable-surface glycoprotein genes could be detected by hybridization to a probe of 16 nucleotides (Merritt et al., 1983). The major difficulty encountered in screening a genomic library is the enormous number of recombinants that must be analyzed. Typically, 1.5–3 genome equivalents must be examined to ensure detection of a specific sequence (Maniatis et al., 1982). The length of the oligonucleotide constitutes the most important factor in the detection of a specific fragment. As emphasized previously, the hybridization and washing conditions must be optimized and the probe size, base composition, and pool size must likewise be carefully selected before probing genomic DNAs fixed to nitrocellulose filters (Benton and Davis, 1977). Once again the availability of probes constructed to two separate regions of the sequence significantly enhances the specificity of selection.

In summary, one of the best approaches to date for the cloning of peptide hormones of the brain utilizes specific oligonucleotides. The prediction of the correct nucleic acid sequence, the ability to construct an oligonucleotide probe of sufficient size and specificity, and the relative abundance of a specific message are the primary factors for determining whether oligonucleotides can be expediently used as primers or as probes. Furthermore, the ability to control hybridization conditions confers a great deal of specificity which can be employed when using oligonucleotides for screening a comparatively limited number of recombinants contained either within a cDNA library or within the full chromosomal complement of a genomic DNA library.

## 3. IDENTIFICATION AND QUANTITATION OF SPECIFIC mRNAs

Many neurosecretory peptide hormones have been localized in both multiple sites throughout the brain and in non-neural tissues of the body. Of the many examples reported, one can selectively note the identification of POMC-derived peptides in extrapituitary sites such as the hypothalamus (Krieger et al., 1980; Liotta et al., 1979, 1980; Watson et al., 1977; Zimmerman et al., 1978) and placenta (Liotta and Krieger, 1980), of somatostatin not only in the hypothalamus (Lauber et al., 1979) but also in the pancreas (Daniels-McQueen et al., 1978), and of enkephalins in the hypothalamus (Barchas et al., 1978; Schultsberg et al., 1978) as well as in the adrenal medulla (Kilpatrick et al., 1981).

In many cases the synthesis of peptide hormones in various tissues has been established biochemically by tracing the incorporation of radiolabeled amino acids into specific peptide hormones. One of the first studies to demonstrate localized synthesis of POMC in the brain was the work of Krieger and colleagues, who prepared dispersed monolayer cultures from hypothalami of cows (Liotta et al., 1979) and young rats (Liotta et al., 1980). Cultures maintained in the presence of radioactive amino acids were shown to incorporate radioactivity into high-molecular-weight proteins which contained both adrenocorticotropin (ACTH) and $\beta$-endorphin antigenic determinants within the same molecule. The difficulty of this approach is that such analyses of amino acid incorporation are often not sufficiently sensitive for polypeptides representing less than 0.01% of the total protein synthesis or for those proteins subject to rapid post-translational cleavage or degradation. Particularly for neural tissue, the composition of a "representative" population of cells may be biased by the methods used for preparation of monolayer cultures. Such procedures may consequently result in the selection of only those cells which adhere to culture dishes or in which cell viability or the patterns of macromolecular synthesis have been altered.

These problems may be partially surmounted by culturing tissue explants rather than dispersed cells. Characteristic patterns of morphological differentiation are exhibited by fragments of rodent fetal and neonatal brain tissue following several weeks of culture *in vitro* (Toran-Allerand, 1979). The hormonal responses of the cultured fragments are typical of analogous brain regions (Toran-Allerand, 1980). Moreover, these explants can rapidly incorporate [$^{35}$S]-methionine into polypeptides, thereby indicating their capacity for protein synthesis (Schachter and Toran-Allerand, 1982). Although the major limitation to this approach has been the inability to culture adult neural tissue for long periods, nonetheless, organotypic culture clearly provides a useful system for investigating the regulation of low-abundance proteins.

An alternative approach for studying *de novo* synthesis of specific proteins in intact alert animals involves the infusion of radiolabeled amino acids into specific regions of the brain via a surgically implanted cannula. Using this technique Krause et al. (1982) were able to examine the biosynthesis of GnRH (LHRH) in hypothalamic regions following the infusion of radioactively labeled amino acids into the third ventricle of the rat brain. A particular advantage of direct infusion studies resides in the ability to study behavioral and pharmacological effects on peptide hormone biosynthesis in the brain of unrestrained animals.

Incorporation studies are valuable for establishing the location of specific proteins in tissues. However, because peptide hormones are secretory proteins, one must consider that localization can result either from synthesis *in situ* or by uptake of the hormone from extracellular fluid. By identifying a particular mRNA, oligonucleotides and cloned DNA probes complementary to specific mRNA sequences confer the ability to distinguish between *de novo* synthesis and uptake of specific proteins. By using such DNAs as hybridization probes, very small quantities of specific mRNA sequences can be detected within a heterogeneous population of RNA species. For such studies unlimited quantities of the DNA sequences can be obtained from recombinant bacterial plasmids or phage and radiolabeled to very high specific activity. These cDNAs can thereby become valuable probes for the detection of RNAs isolated from the tissues of interest.

### 3.1. Hybridization Histochemistry

Hybridization histochemistry (*in situ* hybridization) was developed in response to the need to examine the synthesis of specific gene products on a cellular basis. This technique permitted for the first time the detection of specific mRNA transcripts synthesized by individual cells within a population of heterogeneous cells (Brahic and Haase, 1978; McWilliams and Boime, 1980; Moen and Palmiter, 1980; Singer and Ward, 1982). Accordingly, the methodology of hybridization histochemistry has been adapted by neurobiologists for the localization of peptide hormone mRNA transcripts within cells of the central nervous system and pituitary. In a manner analogous to the detection of a polypeptide by an antibody, hybridization histochemistry permits mRNA sequences to be detected by hybridization of a labeled cDNA probe with fixed, mounted tissue sections. The appearance of silver grains on the tissue following autoradiography indicates hybridization of the cDNA probe, thereby establishing the presence of a specific mRNA within that cell.

The identification of POMC-producing cells in the pituitary and brain has provided a striking example of the application of hybridization histochemistry. In particular, the rat pituitary has proven to be an ideal tissue for characterizing the experimental variables of this technique because it is comprised of three anatomically distinct regions, each of which contains a different distribution of peptide hormone-producing cell types. As discussed previously, *de novo* synthesis of the POMC peptide in the AL and NIL has been demonstrated by *in vitro* culture (Chen et al., 1983; Rosa et al., 1980; Schachter et al., 1982) and by immunocytochemical analyses which have detected POMC-derived peptides in most cells of the intermediate lobe, in a few percent of the cells of the anterior lobe, and not at all in the posterior lobe (Nakane et al., 1972).

Knowledge of the levels and sites of POMC synthesis in the pituitary has made it possible to define the experimental parameters required for hybridization histochemistry with POMC mRNA. Using a [$^{3}$H]-labeled POMC probe, labeled cells were clearly identified in both the anterior and intermediate lobes of the pituitary (Gee and Roberts, 1983). Consistent with peptide and immunocytochemical analyses, a high density of POMC-mRNA-containing cells was observed in the NIL in a relative abundance 10-fold greater than that observed in the anterior lobe. Additionally, cells in both the AL and NIL showed significantly more grains over cytoplasm than nuclei consistent with the concept that POMC mRNA is transported to the cytoplasm shortly after it is synthesized. To verify that those pituitary cells hybridizing to the POMC probe also contained POMC-derived peptides, alternate tissue sections were reacted with an antibody to ACTH and localization established using the horseradish peroxidase immunocytochemical method (Sternberger, 1970).

A more rapid detection of the histospecific distribution of relatively abundant mRNA sequences (e.g., POMC mRNA in the NIL of the pituitary) may be accomplished by hybridizing tissue sections with [$^{32}$P]-labeled DNA probes and then exposing the sections to X-ray film. The high specific activity of the [$^{32}$P]-labeled probe allows rapid detection of the [$^{32}$P] signal not only on X-ray film but also on emulsion-coated tissue sections. One does not expect to see the tight cluster of silver grains in the emulsion that are observed with [$^{3}$H] autoradiography because the $\beta$-particle emitted by [$^{32}$P] has a relatively long path length. However, because the cells are larger in many regions of the adult brain and significantly less densely packed than pituitary cells, it is possible to achieve good grain localization over individual cells when brain sections are probed with [$^{32}$P]-labeled cDNA.

A demonstration of hybridization histochemistry with [$^{32}$P]-labeled probes is clearly depicted in the autoradiograms shown in Figure 1. In this experiment cDNA probes complementary to prolactin (PRL), growth hormone, and POMC mRNAs were hybridized with adjacent sections prepared from the pituitary of an adult female rat. In agreement with previous quantitations of the levels of these particular mRNAs (Schacter and Durgerian and Schachter and Roberts, unpublished observations; see also Baker and Gross, 1978; Civelli et al., 1982; Cooke et al., 1980; Schachter et al., 1982), the PRL signal was the strongest and confined to the anterior lobe while the growth hormone signal was less intense but also confined to the anterior lobe. Conversely, the POMC signal was quite strong over the intermediate lobe but faintly detectable over the anterior pituitary.

It has furthermore been possible to study POMC-producing neurons in adult rat hypothalamus by *in situ* hybridization histochemistry (Gee et al., 1983; see also Fig. 2). Cells containing a large number of silver grains over their cytoplasm were seen primarily in the periarcuate region; few labeled cells were found outside this region (Fig. 2). Here again tissue sections adjacent to those tested for POMC mRNA were subjected to immunocytochemical analysis, thereby confirming that those cells scored as positives by *in situ* hybridization also contained POMC-derived peptides.

In this section we have described several techniques for identifying the site(s) of peptide hormone gene expression in the brain. Definitive proof that a specific protein occurs in a tissue can be accomplished by demonstrating the presence of the corresponding radiolabeled or immunoreactive polypeptide. The application of cDNA probes to identify the cells containing mRNAs that code for specific proteins has been discussed as a means for avoiding the limitations of sensitivity and specificity inherent in peptide detection. The ability to identify the sites of synthesis for specific mRNAs quite naturally leads to the desire to measure the levels of such transcripts. Initially, a relative evaluation of an RNA's abundance may be obtained by densitometric analysis of autoradiograms. However, such measurements are currently only semiquantitative. A far better assessment of mRNA levels is obtained by measuring the extent of hybridization of cDNA probes to mRNA populations which have been isolated from tissues of interest.

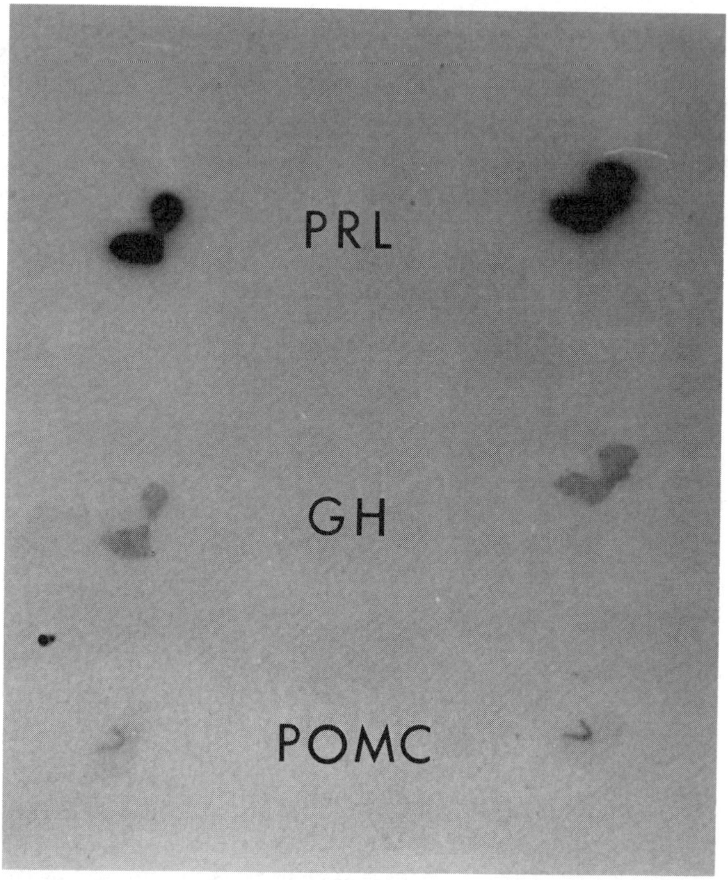


**Figure 1.** Hybridization histochemistry with [$^{32}$P]-labeled cDNAs. [$^{32}$P]-labeled probes containing prolactin (PRL), growth hormone (GH), and POMC cDNAs were hybridized *in situ* with pituitary sections from an adult female Sprague-Dawley rat. Pituitary tissue was fixed with 4% paraformaldehyde before sectioning with a cryostat microtome. Three adjacent 6-$\mu$m sections each containing a portion of the anterior, intermediate, and neural lobes were mounted on gelatin-coated microscope slides for hybridization histochemistry according to the procedure of Gee and Roberts (1983). Hybridization probes complementary to the respective mRNA sequences were prepared using restriction fragments isolated from purified plasmid cDNAs. Each insert was nick-translated using [$^{32}$P]-labeled deoxyribonucleotides to a specific activity of 2–4 $\times$ $10^8$ cpm/$\mu$g. Each tissue section was incubated with 25,000 cpm of [$^{32}$P]-labeled DNA, washed extensively, and dehydrated through a graded series of ethanols. This figure shows autoradiograms produced by exposure to X-ray film for 16 hr.

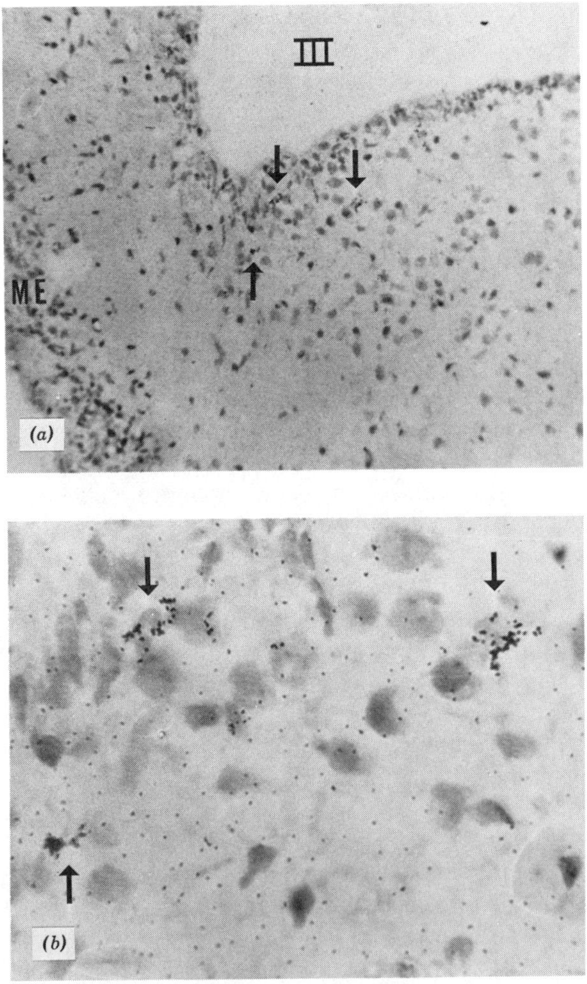


**Figure 2.** *In situ* hybridization histochemistry of POMC mRNA in the brain. Rat hypothalamus was perfused with 4% paraformaldehyde and then sectioned (8 $\mu$m) with a cyrostat microtome. Following treatment with 0.2 *N* HCl, sections were hybridized overnight with a [$^{32}$P]-labeled POMC cDNA probe (25,000 cpm; specific activity, $2 \times 10^8$ cpm/$\mu$g DNA). Unhybridized probe was removed by washing overnight with $0.5 \times$ SCC. The section was coated with Kodak NP2 emulsion and exposed for 17 days. The slide was developed and then counterstained with eosin and hemotoxolin. (a) The periarcuate region of the rat hypothalamus shows silver grains dispersed over several neurons in agreement with the known distribution of the POMC-containing cells in this region. (b) In this higher magnification of the regions shown in (a) the cellular localization of autoradiographic grains can clearly be seen in several individual neurons.

### 3.2. Quantitation of Specific mRNAs Isolated from Tissues

To quantitate the levels of a specific mRNA within any given tissue, known amounts of total RNA and trace amounts of radiolabeled DNA are hybridized. The concentrations of the specific mRNA species in each total RNA preparation are then reflected in either the rate at which the DNA probe forms hybrids or the absolute amount of DNA hybridized depending upon the particular experimental strategy employed (Britten and Kohne, 1968).

For example, POMC mRNA levels in the anterior and intermediate lobes of the rat pituitary were initially measured by the RNA-driven solution hybridization method. In these studies a trace amount of [$^{32}$P]-DNA complementary to POMC mRNA (Roberts et al., 1979b) was annealed in solution to pituitary RNA. Hybridization was terminated at successive intervals, and the amount of hybridized DNA in each sample determined following digestion by S1 nuclease. The amount of undigested, and hence hybridized, DNA was measured by scintillation spectrophotometry and expressed as the percentage of counts per minute in DNA bound to ion exchange filters. The relative abundance of the POMC mRNA in each total RNA preparation was directly proportional to the rate at which the tracer DNA formed hybrids. Accordingly, the results from these experiments demonstrated that the relative abundance of POMC mRNA in the intermediate lobe was 10-fold greater than that in the anterior pituitary (Schachter and Roberts, unpublished observation). Similar results were obtained when the amount of input RNA was varied instead of varying the time of incubation.

While solution hybridization can provide very accurate measurements of the concentration of a specific mRNA sequence within a heterogeneous population, quantitation of low-abundance mRNAs can be difficult. Furthermore, such situations require the preparation of single-stranded DNA probes (Civelli et al., 1982; O'Farrell et al., 1980). There exists, however, a technically simpler means of mRNA quantitation which not only overcomes the need for single-stranded DNA probes but also involves fewer sample manipulations than are required for analysis of hybrids formed in solution. In this case the hybridization is conducted after the RNA is coupled to a paper support. Coupling can be accomplished either by covalent bonding through linkage to diazotized filters (Seed, 1982) or by mechanical coupling by baking the RNA onto nitrocellulose sheets (Thomas, 1980). Filters may then be incubated with single-stranded or strand-separated double-stranded radiolabeled DNA. After the unhybridized DNA has been removed by extensive washing (Wahl et al., 1979), the amount of DNA hybridized to each RNA sample may be determined directly by scintillation counting of filters.

A variation of the filter hybridization technique, the dot blot hybridization procedure (White and Bancroft, 1982) provides an effective means for examining very small amounts of RNA and DNA, particularly those levels encountered in regulation studies of specific mRNAs in the pituitary and brain. This sensitive and convenient hybridization procedure (described in Fig. 6) has several advantages over solution hybridization, the major one being the ability to assess levels of a specific mRNA indirectly. The X-ray autoradiogram produced by a hybridized sample may be compared densitometrically to autoradiograms of filters containing standard preparations of labeled RNAs. Additionally, the dot blot procedure does not require the preparation of single-stranded cDNA as a hybridization probe. The inclusion of dextran sulfate (Wahl et al., 1979) in the hybridization buffer coupled with the use of an intensifying screen during autoradiography will enhance the signal intensity sufficiently to permit the detection of RNAs present at the picogram level. Furthermore, the system requires a very small amount of cells or tissue. POMC mRNA in the AL could be detected using as few as $2.5 \times 10^4$ cells, an amount equivalent to one-hundredth of a rat pituitary, thereby demonstrating the value of this assay when the amount of tissue is a limiting factor (Chen et al., 1983).

White and Bancroft have recently adapted the dot blot hybridization technique to measure the mRNA levels in denatured tissue cytosols. They were able to assess cytoplasmic PRL mRNA levels in tumor cells by spotting formaldehyde-denatured cytosols directly onto nitrocellulose for dot blot filter hybridization (White and Bancroft, 1982). It is expected that this technique may prove to be particularly useful for studying the regulation of mRNA levels in specific regions of the brain. For example, the relative levels of a specific mRNA can be determined in specific loci of brain tissue by performing cytosolic hybridizations on areas dissected by the Palkovits punch microsampling technique (Palkovits and Brownstein, 1982). As described in Section 3.3, these filter-hybridization methods have been used to study the effects of various hormonal manipulations on steady-state POMC mRNA levels in the anterior and intermediate lobes of the rat (Chen et al., 1983; Schachter et al., 1982).

### 3.3. Characterization of Specific mRNAs

To ascertain whether a DNA probe is hybridizing to a discrete RNA population, further structural analysis of the RNA is required. Instead of directly binding an RNA sample to a single area of a filter, the RNA may first be size-fractionated by electrophoresis on an agarose gel (Potter et al., 1980) and then transferred and coupled to a filter (Seed, 1982; Thomas, 1980). If, however, the level of a specific RNA in a given tissue is predicted to be quite low, poly(A+)-containing RNA representing 1–5% of total RNA can be isolated by oligo(dT) cellulose chromatography (Aviv and Leder, 1972) prior to gel fractionation. At this point the RNA is of course ready for hybridization with a radiolabeled DNA probe. A homogeneous mRNA population complementary to the cDNA will appear as a single band on an autoradiogram. Standards of known molecular weights migrating in parallel lanes on the gel provide a size calibration of the hybridizing material. An example is shown in Figure 3.

The sizes and number of mRNA species may also be determined by Northern gel blot analysis. Again POMC mRNAs from rat pituitary and brain have been probed with POMC cDNA. The reactive species from AL and NIL of the pituitary and from hypothalamus and cerebral cortex were indistinguishable in size (approximately 1100 nucleotides) while those prepared from amygdala, cerebellum, and midbrain central gray matter were about 200 nucleotides shorter (Fig. 3; Civelli et al., 1982). Therefore, the presence of a single mRNA band in the two pituitary regions and in the hypothalamus constitutes strong evidence that the POMC-derived peptides in these tissues originate via transcription of the same gene. The structure and function of the POMC-related mRNA sequences in brain regions outside the hypothalamus await futher characterization. Whether the smaller RNAs actually represent functional (i.e., translatable) mRNAs must be similarly determined. It is noteworthy that, to date, no evidence has been presented for localization of any POMC-derived peptides within cell bodies of brain tissue outside the hypothalamus other than a single report of immunoreactive ACTH-like material being synthesized *in vitro* by dispersed amygdala cells (Pacold et al., 1978).

The importance of conducting RNA structural analysis in conjunction with quantitation studies is to confirm that cDNA probes are not hybridizing exclusively to sequences of partial homology rather than to those messages of interest. In an example of such a situation the cDNA probes for $\alpha$-fetoprotein and albumin were observed to hybridize to a continuum of mRNAs on a Northern blot. None of the mRNAs recognized by these cDNA probes corresponded to mRNAs for $\alpha$-fetoprotein or albumin, thereby indicating the partial homology of these probes to a large number of messages (Schachter and Toran-Allerand, 1982). An additional advantage is that these studies likewise reveal whether quantitation studies are being conducted under conditions of appropriate stringency.

There is yet another factor that must be considered in the preparation of mRNAs for hybridization studies. Characterization of mRNAs coding for specific proteins has revealed that, with few exceptions, the RNAs are polyadenylated. As described previously, there is a population of polysome-bound, non-polyadenylated mRNAs in the brain which represents a substantial portion (approximately half) of the total mRNA sequence complexity (Bantle and Hahn, 1976; Chaudhari and Hahn, 1983; Chikaraishi, 1979).

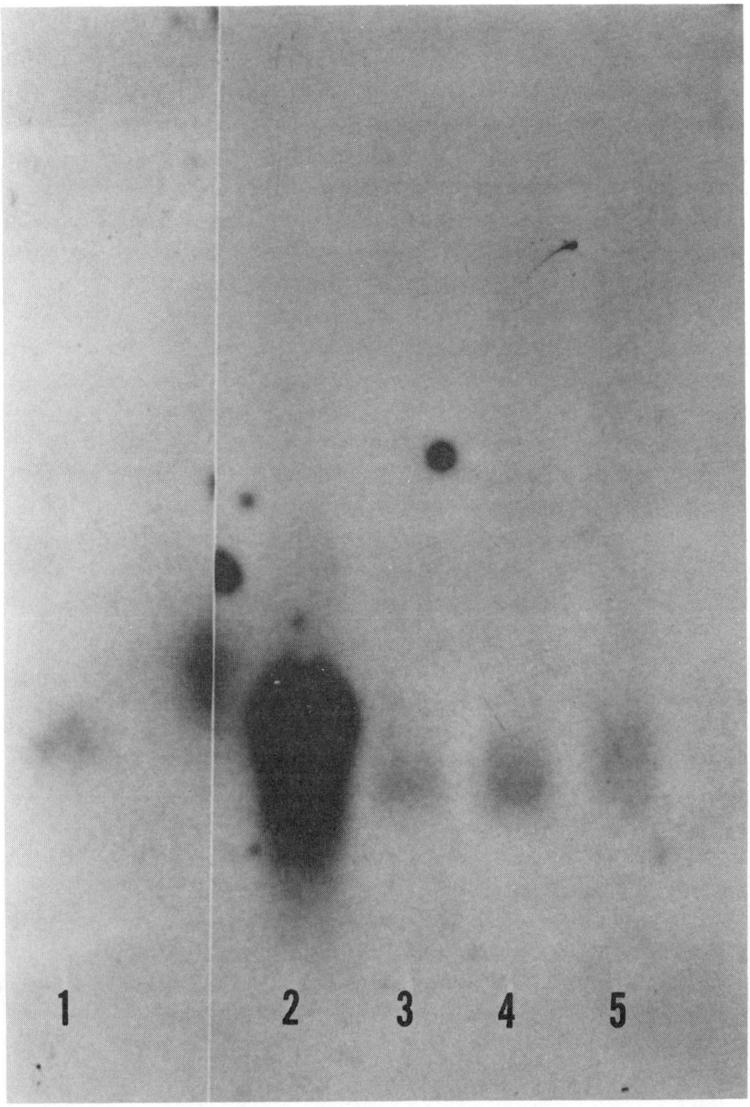


**Figure 3.** Northern blot hybridizations of POMC mRNAs prepared from rat pituitary and brain tissues. RNAs from pituitary and various brain regions of the rat were analyzed for their content of POMC mRNA. RNA was prepared by the procedure of Chirgwin et al. (1979) from tissues of adult female Sprague-Dawley rats. Total or poly(A+) RNAs were fractionated on a 1.2% agarose gel and then transferred by the procedure of Thomas (1980) to Gene Screen (New England Nuclear) filter support. The filter was incubated with a [$^{32}$P]-labeled POMC DNA probe, washed to remove nonhybridized probe, and then exposed for 5 days to X-ray film in the presence of an intensifying screen at −70°C. Lane 1, 30 μg pituitary total RNA; 2, 12.5 μg hypothalamic poly(A+) RNA; lane 3, 20 μg amygdala poly(A+); lane 4, 20 μg midbrain–central gray region poly(A+) RNA; lane 5, 20 μg cerebellar poly(A+) RNA.

**Figure 4.** POMC mRNA in hypothalamic RNA. Total RNA was extracted from female rat hypothalamus (Chirgwin et al., 1979). Thirty micrograms of total hypothalamic RNA were fractionated by agarose gel electrophoresis, transferred to Gene Screen, and hybridized to [$^{32}$P]-labeled POMC using the experimental conditions described in Figure 3. Only one species of POMC mRNA (1100 nucleotides) was detected in this preparation of hypothalamic RNA containing both poly(A+) and poly(A−) mRNAs.

Recently Herbert and co-workers (personal communication) have found two RNAs in the hypothalamus which anneal to an enkephalin cDNA probe. Although the two species are present in approximately equal concentrations, only one species is polyadenylated. The other form of enkephalin mRNA thus provides the first identification of a specific mRNA in brain which is not polyadenylated. The functional significance of non-polyadenylated enkephalin RNA in brain is not presently known. Interestingly, only poly (A+)-containing enkephalin RNA has been found in total RNA isolated from adrenal gland. Moreover, the extent to which neuropeptide hormones other than enkephalin are encoded by both polyadenylated and non-polyadenylated mRNAs has not been determined. In the case of POMC mRNA, only a single poly (A+)-containing species is observed in hypothalamic RNA (see Fig. 4), thus suggesting that all the POMC mRNA in this tissue is polyadenylated.

In this section we have discussed the methods available for examining the levels of specific mRNAs. The ability to localize the sites at which a peptide hormone is synthesized in the brain coupled with the ability to measure very low levels of its corresponding mRNA provides for the first time an effective means of investigating the regulation of specific genes in neural tissue. In Section 4 we will describe how these techniques have been successfully applied to the study of regulation for individual peptide hormone genes.

## 4. REGULATION OF PEPTIDE HORMONE GENE EXPRESSION IN THE BRAIN AND PITUITARY

Peptide hormone gene expression can be regulated at many steps from transcription of the mRNA to processing of the preprohormone (as illustrated in Fig. 5). In response to specific hormonal and environmental effectors, these genes are transcribed into heterogeneous nuclear RNAs (hnRNAs) which are then processed to mature mRNA species within the nucleus. After transport to the cytoplasm, such mRNAs are either translated into protein, stored, or degraded. In turn these proteins subsequently experience co- and post-translational modifications to yield the mature peptide hormones. The peptides are then stored in secretory granules to await the proper stimulus signaling secretion. Regulation, therefore, can occur at any of these steps.

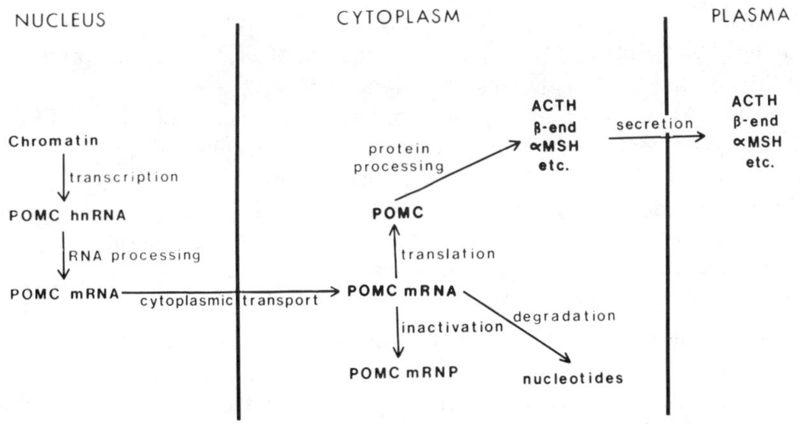


**Figure 5.** Pathway of expression of the POMC gene. Abbreviations: hnRNA, heteronuclear RNA; ACTH, adrenocorticotropic hormone; mRNP, ribonuclear particle; $\alpha$-MSH, melanocyte-stimulating hormone.

Although these steps have been examined individually in a variety of different systems, information on the regulation of brain peptide hormone gene expression is in many cases restricted to the levels of content and secretion. For example, the hypothalamic $\beta$-endorphin content in female rats responds to changing steroid levels during estrogen treatment, pregnancy, and parturition either by changing the synthesis or the release of this peptide (Wardlaw and Frantz, 1982). Moreover, the concentration of $\beta$-endorphin released from the hypothalamus into the hypophyseal portal blood of the female monkey varies throughout the menstrual cycle, again implying that the hypothalamic release of this peptide is dependent on the presence of ovarian steroids (Wardlaw et al., 1981). The possibility further exists that certain factors which modulate POMC peptide content and secretion from hypothalamic neurons may also affect earlier steps in its expression pathway. The regulation of POMC gene expression in the pituitary will be described as a model system for studying peptide hormone expression in the brain because POMC synthesis in the pituitary shares many aspects of its production in common with neural tissue.

### 4.1. POMC Gene Regulation in the Pituitary

A variety of hormonal and environmental factors selectively modulates the synthesis and secretion of POMC peptides in the anterior and neurointermediate lobes of the pituitary. Several of these modulators originate in the brain. For example, CRH (Vale et al., 1981; Yates and Maran, 1974) and vasopressin (Yates and Maran, 1974), two hypothalamic peptides, stimulate the synthesis and release of ACTH and $\beta$-endorphin in the AL. Neurotransmitters (e.g., norepinephrine, serotonin, and dopamine) which are transported to the pituitary either by the portal blood or via direct innervation similarly play important roles in the regulation of POMC peptide production in the AL and the NIL (Yates and Maran, 1974).

In the AL of the pituitary the synthesis and release of the ACTH and $\beta$-endorphin peptides are subject to both positive regulation by CRH (Gillies and Lowry, 1978; Vale et al., 1981) and negative-feedback regulation by glucocorticoids (Gillies and Lowry, 1978; Herbert et al., 1981; Roberts et al., 1979a; Schachter et al., 1982). However, in the NIL the secretion and synthesis of POMC peptides are regulated instead by catecholamines and not by CRH or glucocorticoids. Specifically, dopamine administered at physiological concentrations inhibits the release of the POMC-derived peptides (i.e., $\alpha$-MSH and $\beta$-endorphin) from the rat NIL (Farah et al., 1982; Lepine and Dupont, 1981; Vermes et al., 1980). Other factors such as circadian cycles, gonadal steroids, vasopressin, norepinephrine, serotonin, and mineralocorticoids have also been shown to regulate POMC peptide secretion differentially in the AL and NIL of the pituitary (Bruni et al., 1982; Giguere et al., 1981; Lim et al., 1982; Petraglia et al., 1982; Yates and Maran, 1974). A major point of interest then becomes a determination of whether expression of the POMC gene itself is influenced by any of those factors which modulate the secretion of its peptide. Several approaches previously described for the quantitation of mRNAs have been used to measure the levels of POMC mRNA in response to these effectors.

### 4.2. Effects of Dopaminergic Compounds on POMC Expression

The dopaminergic modulation of POMC mRNA levels in the rat NIL, AL, and hypothalamus has been investigated using the dot blot hybridization assay (Chen et al., 1983; see Fig. 6). Administration of a dopamine agonist, 2-bromo-$\alpha$-ergocryptine, to adult female rats significantly reduced the content of POMC mRNA in the NIL. In constrast treatment for 7 days with the dopamine antagonist haloperidol resulted in a 5- to 7-fold increase in POMC mRNA levels in the NIL. These dopaminergic effects on POMC mRNA levels in the NIL were both time and dose dependent with the initial elevation of POMC mRNA becoming detectable as early as 6 hr after administration of haloperidol. In agreement with the observations made by Hollt and colleagues using an indirect translation–immunoprecipitation scheme (Hollt et al., 1982), these results indicate that the effects of dopaminergic compounds on POMC mRNA levels in NIL parallel those of its effects on secretion. On the other hand neither haloperidol nor ergocryptine effected a change on POMC mRNA levels in the AL or the hypothalamus greater than 20% of the control values (Chen et al., 1983). Again, an absence of an effect on POMC mRNA levels reflected a corresponding absence of dopaminergic effects on POMC peptide secretion from the AL.

Although the mechanisms of dopaminergic inhibition of POMC mRNA levels in the rat NIL have not been determined, a possible mechanism has been described recently for PRL. The dopaminergic inhibition of PRL production is mediated by a decrease in PRL mRNA levels which is in turn controlled by a rapid inhibition of transcription of the prolactin gene (Ben-Jonathan et al., 1977; Maurer, 1980, 1981). This inhibition of gene transcription which can be mimicked by a rise in cellular cAMP concentration may mediate the dopaminergic inhibition of PRL mRNA levels (Hill et al., 1976; Maurer, 1982). It is conceivable that dopaminergic compounds may regulate POMC mRNA levels in rat NIL by a similar mechanism.

### 4.3. Effects of Glucocorticoids on POMC Expression

The modulation of POMC gene expression by glucocorticoids has been demonstrated both *in vivo* and *in vitro* (Chen et al., 1983; Nakanishi et al., 1977; Roberts et al., 1979a; Schachter et al., 1982). Using a coupled cell-free

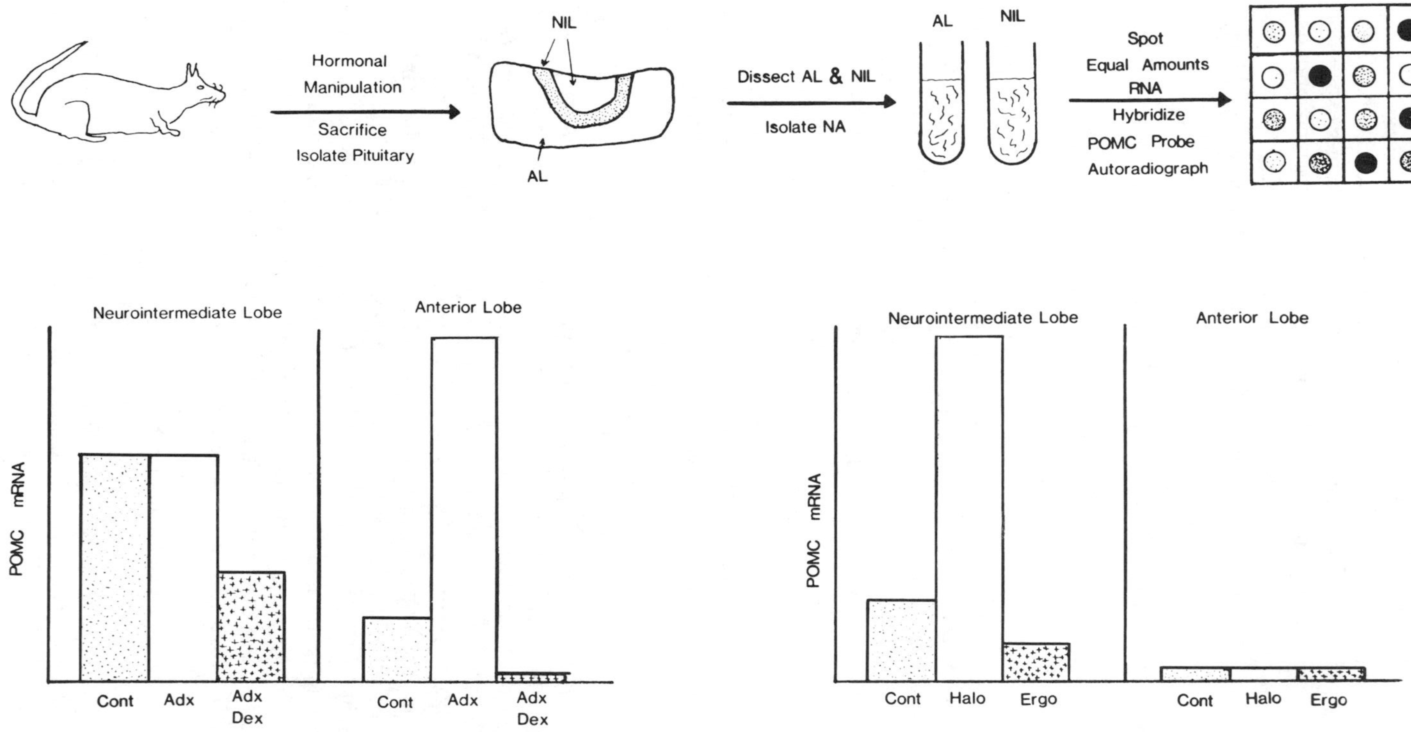


**Figure 6.** Summary diagram of POMC mRNA regulatory studies. Glucocorticoids exert regulatory effects on POMC mRNA in the AL of the pituitary. Conversely, the NIL is the site at which dopaminergic compounds affect the levels of POMC mRNA. See Sections 4.2 and 4.3 for a description of this differential regulation. For experimental details see Chen et al. (1983) and Schachter et al. (1982).

translation–immunoprecipitation system, Roberts et al. (1979a) have shown that the glucocorticoid-mediated decrease in POMC peptide synthesis in the AL-derived AtT20 tumor cell line is a direct result of a decrease in cytoplasmic POMC mRNA levels. Using similar methods Nakanishi et al. (1977) have described an increase in total pituitary POMC mRNA content after adrenalectomy and a decrease following subsequent dexamethasone treatment. Birnberg et al. (1982) and Herbert et al. (1981) similarly reported that POMC mRNA levels progressively increased 10-fold in the AL following adrenalectomy. Conversely, Schachter et al. (1982) (see Fig. 6) have shown that glucocorticoids dramatically decrease the levels of POMC mRNA by 30- to 50-fold in the anterior lobe of rats which have been adrenalectomized for 7 days. However, only a twofold suppression of POMC mRNA in the NIL was seen with this same treatment, an observation consistent with that of Herbert et al. (1981), who found no significant change in POMC mRNA levels of NIL.

### 4.4. Mechanisms for POMC Regulation

The possible mechanisms by which the cytoplasmic POMC mRNA levels might be depressed by glucocorticoids include (1) a decrease in the transcriptional activity of the POMC gene and (2) an increase in the degradation of POMC mRNA. There is presently no way to analyze the second possibility *in vivo* because such studies require radioactively labeling the RNA to high specific activity. It is possible, however, to test the first mechanism. An *in vitro* nuclear transcription assay can be used to examine the transcriptional activity of a specific gene by measuring the amount of radioactive precursors that are incorporated into mRNA by the RNA polymerase II molecules that have previously initiated transcription on the RNA. Using such a "runoff" assay, Eberwine and Roberts (1984) and Birnberg et al. (1982) have shown that transcriptional activity of the POMC gene rapidly decreases in the AL of rats following glucocorticoid treatment. As expected, this is in contrast to the increased rate of transcription of the POMC gene which is observed following adrenalectomy. Thus, the changes in POMC mRNA which occur in response to changing concentrations of glucocorticoids can be explained at least in part by changes in the levels of POMC gene transcription. Of course, the possibility that the glucocorticoids may also be modulating other aspects of RNA metabolism cannot be excluded at this point. For example, estradiol and progesterone have been shown to increase the stability of conalbumin mRNA from a half-life calculated to be 2.9 hr in the absence of the hormone to a minimum of 8 hr in their presence (Palmiter et al., 1981).

Several explanations must be considered for the differential regulation of POMC gene expression by glucocorticoids and dopaminergic compounds.

First, several similar nonallelic POMC genes may exist but only one may be expressed in an individual cell type. Conceivably such separate POMC genes would exhibit differing responses to glucocorticoids and to dopamine. Secondly, cells expressing a single POMC gene might possess characteristic sensitivities to certain hormones or neuromodulators as might be demonstrated, for example, by the expression of functionally distinct receptors.

It must be noted with regard to the possibility of multiple genes that only one POMC gene has been found in rat, human, and bovine species (reviewed by Eberwine and Roberts, 1983). It is unlikely that the existence of multiple POMC genes accounts for the differential regulation of POMC gene expression because extensive restriction endonuclease analyses of genomic DNA and sequence analysis of isolated POMC gene fragments (Eberwine and Roberts, 1983) have not revealed the presence of multiple POMC genes in these species. In the one exception Uhler (1982) has found two POMC-like genes in the mouse. Only one of these genes is physiologically functional; the other gene is a pseudo-gene containing terminators and mutations in critical amino acids.

Experimental evidence does support the possibility that functionally different receptors for glucocorticoids and dopaminergic compounds are expressed separately in the AL and NILs of the pituitary. Steroid autoradiographic studies have demonstrated that the POMC cells of the AL are able to concentrate glucocorticoids while the POMC cells of the NIL cannot (Rees et al., 1977; Warembourg, 1975). The absence of translocatable glucocorticoid receptors in the cells of the NIL in all likelihood accounts for the lack of POMC gene responsiveness to glucocorticoids in this tissue. On the other hand the well-characterized dopamine D2 receptor in the plasma membrane of the NIL cells appears to be of the same type as that found in the lactotrophs of the AL (Kebabian and Caline, 1979). Whether the POMC-producing corticotrophs of the AL possess D2 dopamine receptors is not clear. In pituitary cells fractionated by sucrose gradient sedimentation, dopamine inhibited cAMP accumulation in lactotroph-rich populations but not in the corticotroph population (Kebabian and Caline, 1979). Conceivably, the inability of dopamine to influence cAMP accumulation in the AL corticotrophs may account for its corresponding lack of influence on POMC mRNA levels and biosynthesis. Thus, regulation of POMC gene expression in the AL and NIL of the pituitary may depend primarily on the differential responses of the POMC-expressing cells to stimulatory and inhibitory signals.

This final point, suggesting that differential gene regulation may result from differential recognition of modulators, may become an important concept in the study of gene expression in the brain. In one example of differential regulation in neurons Morrell and colleagues have shown by immunohistochemistry and steroid autoradiography that estrogen is concentrated by only half of the neurophysin cells in the periventricular nucleus and by only 10% of the POMC neurons in the periarcuate region (Morrell and Pfaff, 1983; Rhodes et al., 1982). Thus, the value of the *in situ* hybridization histochemistry technique becomes apparent for resolving issues of this nature. This methodology will permit the study of specific mRNAs in individual neurons, a definite necessity for investigating a system as complex as the brain.

## 5. CONCLUDING REMARKS

In this chapter we have discussed the application of molecular biology to the study of gene expression in the brain. Prior to the development of recombinant DNA methodology, the examination of a specific neuropeptide hormone was limited to the measurement of its tissue levels and secretion. Peptide chemical analyses, precursor incorporation studies, and immunocytochemical localizations have proven highly valuable for identifying individual polypeptide species in neural tissue. However, these methods cannot distinguish between intercellular uptake and *de novo* synthesis of neurosecretory peptides. It is the ability to demonstrate the presence of a specific mRNA in a cell that convincingly establishes the capacity of that cell to synthesize its corresponding polypeptide.

Major advances in the study of gene expression in the brain have occurred as direct result of the refinement of experimental techniques, thereby enabling researchers to overcome the limitations of a system as complex as neural tissue. Accordingly, the degree of sensitivity and specificity necessary for detecting the mRNA of a single-copy gene can now be achieved for a polypeptide that is being expressed by only a few hundred neurons. The technique of hybridization histochemistry in particular holds great promise for investigating gene expression in individual neurons and glial cells. Furthermore, the ability to detect changing levels of mRNA in micropunch tissue samples likewise makes it possible to monitor the transcriptional activity of single genes in response to specific effectors. This is well illustrated by the fluctuations of POMC and mRNA levels observed in the pituitary following administration of dopamine or dexamethasone (Chen et al., 1983; Schachter et al., 1982).

cDNA probes are currently available for more than a dozen neuropeptide hormones. Of course this number is expected to increase steadily as researchers use many of the approaches outlined herein to obtain probes for additional hormones. Even as cDNA probes are being developed for known neuropolypeptides, there also exists the exciting possibility of identifying other as yet unknown polypeptides. The discoveries of $\gamma$-melanocyte-stimulating hormone in the POMC precursor (Nakanishi et al., 1977) and the calcitonin gene-related peptide in the calcitonin precursor (Rosenfeld et al., 1981) are two striking examples of the characterization of new genes before the isolation of the actual polypeptide product for the gene.

More importantly, it is now possible to examine all the stages of gene expression for a particular gene product ranging from the transcription of its hnRNA to the release of biologically active peptides from secretory granules. It is through the selective and judicious application of the techniques currently available that neurobiologists can now probe the behavioral and neuroendocrine responses of polypeptide hormone genes in the brain.

## ACKNOWLEDGMENTS

The authors wish to acknowledge the efforts of Edith Kupsaw, Vilma Lorenzo, and Pamela Ross in the preparation of this manuscript. The editorial assistance of James Eberwine, France Dionne, and Lynn Street is also greatly appreciated.

## REFERENCES

Advis, J. P., J. E. Krause, and J. F. McKelvy, *Endocrinology* **110,** 1238 (1982).

Atkinson, T. C., *Bio Techniques* **1,** 6 (1983).

Aviv, H. and P. Leder, *Proc. Natl. Acad. Sci. USA* **69,** 1408 (1972).

Baker, B. L. and D. S. Gross, *Amer. J. Anat.* **153,** 193 (1978).

Bantle, J. A. and W. E. Hahn, *Cell* **8,** 139 (1976).

Barchas, J., H. Akil, G. Elliott, R. Holman, and S. Watson, *Science* **200,** 964 (1978).

Ben-Jonathan, N., C. Oliver, H. J. Weiner, R. S. Mical, and J. C. Porter, *Endocrinology* **100,** 452 (1977).

Benton, W. D. and R. W. Davis, *Science* **196,** 180 (1977).

Birnberg, N. D., O. Civelli, J. C. Lissitzky, M. Hinman, and E. Herbert, *Endocrine Soc. Abst.* **220,** (1982).

Brahic, M. and A. T. Haase, *Proc. Natl. Acad. Sci. USA* **75,** 6125 (1978).

Britten, R. J. and D. E. Kohne, *Science* **161,** 529 (1968).

Bruni, J. F., R. L. Hawkins, and S. S. C. Yen, *Life Sci.* **30,** 1267 (1982).

Chatterjee, M., S. Baliga, and H. Munro, *J. Biol. Chem.* **251,** 2945 (1976).

Chaudhari, N. and W. E. Hahn, *Science* **220,** 925 (1983).

Chen, C. L. C., F. T. Dionne, and J. L. Roberts, *Proc. Natl. Acad. Sci. USA* **80,** 2211 (1983).

Chikaraishi, D., *Biochemistry* **18,** 3249 (1979).

Chirgwin, J. M., A. E. Przybala, R. J. McDonald, and W. J. Rutter, *Biochemistry* **18,** 5294 (1979).

Civelli, O., N. Birnberg, and E. Herbert, *J. Biol. Chem.* **247,** 6783 (1982).

Comb, M., P. H. Seeburg, J. Adelman, L. Eiden, and E. Herbert, *Nature (London)* **295,** 663 (1982).

Comstock, J. P., L. Chan, and A. R. Means, in W. T. Schrader and B. W. O'Malley, Eds., *Hormone Action and Molecular Endocrinology Workshop,* Baylor College of Medicine, Houston, TX, Chap. 8, p. 1 (1980).

Conner, B. J., A. A. Reyes, C. Morin, K. Itakura, R. L. Teplitz, and R. B. Wallace, *Proc. Natl. Acad. Sci. USA* **80,** 278 (1983).

Cooke, N. E., D. Coit, R. I. Weiner, J. D. Baxter, and J. A. Martial, *J. Biol. Chem.* **255,** 6502 (1980).

Crea, R., A. Kraszewski, T. Hirose, and K. Itakura, *Proc. Natl. Acad. Sci. USA* **75,** 5765 (1978).

Daniels-McQueen, S., D. R. McWilliams, S. Birken, R. Canfield, T. Landefeld, and I. Boime, *J. Biol. Chem.* **253,** 7109 (1978).

Das, H. K., P. A. Biro, S. N. Cohen, H. A. Erlich, A. von Gabain, S. K. Lawrance, P. G. Lemaux, H. O. McDevitt, B. M. Peterlin, M. F. Schulz, A. K. Sood, and S. Weissman, *Proc. Natl. Acad. Sci. USA* **80,** 1531 (1983).

Dokas, L. A., *Brain Res. Rev.* **5,** 177 (1983).

Eberwine, J. and J. L. Roberts, *DNA* **2,** 1 (1983).

Eberwine, J. H. and J. L. Roberts, *J. Biol. Chem.* **259,** 2166 (1984).

Edge, M. D., A. R. Greene, G. R. Heathcliffe, P. A. Meacock, W. Schuch, D. B. Scanlon, T. C. Atkinson, C. R. Newton, and A. F. Markham, *Nature* **292,** 756 (1981).

Eipper, B. A. and R. E. Mains, *Endo. Rev.* **1,** 1 (1980).

Farah, J. M., Jr., D. S. Malcolm, and G. P. Mueller, *Endocrinology* **110,** 657 (1982).

Fiddes, J. C., and H. M. Goodman, in G. Jagiello and H. J. Vogel, Eds., *Bioregulators of Reproduction,* Academic, New York, p. 279 (1981).

Furutani, Y., Y. Morimoto, S. Shibahara, M. Noda, H. Takahashi, T. Hirose, M. Asai, S. Inayama, H. Hayashida, T. Miyata, and S. Numa, *Nature* **301,**537 (1983).

Gee, C. and J. L. Roberts, *DNA* **2,** 155 (1983).

Gee, C. E., C. L. C. Chen, J. L. Roberts, R. Thompson, and S. J. Watson, *Nature,* **306,** 374-376 (1983).

Giguere, V., J. Cote, and R. Labrie, *Endocrinology* **109,** 757 (1981).

Gillam, S., K. Waterman, and M. Smith, *Nucleic Acids Res.* **2,** 625 (1975).

Gillies, G. and P. Lowry, *Endocrinology* **103,** 521 (1978).

Giudice, L. C. and I. M. Chaiken, *Proc. Natl. Acad. Sci. USA* **76,** 3800 (1979).

Godine, J. E., W. W. Chin, and J. F. Habener, *J. Biol. Chem.* **256,** 2475 (1981).

Grantham, R., C. Gautier, M. Guoy, M. Jacobzone, and R. Mercier, *Nucleic Acids Res.* **9,** 543 (1981).

Grunstein, M. and D. S. Hogness, *Proc. Natl. Acad. Sci. USA* **72,** 3961 (1975).

Gubler, U., D. L. Kilpatrick, P. H. Seeburg, P. L. Gage, and S. Udenfriend, *Proc. Natl. Acad. Sci. USA* **78,** 5484 (1981).

Gurr, J. A., J. F. Catterall, and I. A. Kourides, *Proc. Natl. Acad. Sci. USA* **80,** 2122 (1983).

Herbert, E., N. Birnberg, J. C. Lissitzky, O. Civelli, and M. Uhler, *Neuroscience Newsletter* **12,** 16 (1981).

Hill, M. K., R. M. MacLeod, and P. Orcut, *Endocrinology* **99,** 1612 (1976).

Hobart, P., R. Crawford, L. P. Shen, R. Pictet, and W. J. Rutter, *Nature* **288,** 137 (1980).

Hollt, V., I. Haarmann, B. R. Slizinger, and A. Herz, *Endocrinology* **110,** 1885 (1982).

Houghton, M., A. G. Stewart, J. S. Emtage, M. A. W. Eaton, J. C. Smith, T. P. Patel, H. M. Lewis, A. G. Porter, J. R. Birch, T. Cartwright, and N. H. Carey, *Nucleic Acids Res.* **8,** 1913 (1980).

Hudson, P., J. Haley, M. Cronk, J. Shine, and H. Niall, *Nature* **291,** 127 (1981).

Kebabian, J. W. and D. B. Caline, *Nature (London)* **277,** 93 (1979).

Khorana, H. G., K. L. Agarwal, P. Bresmer, H. Buchi, M. H. Caruthers, P. J. Cashion, M. Fridkin, E. Jay, K. Kleppe, R. Kleppe, A. Kumar, R. C. Loewen, R. C. Miller, K. Minamoto, A. Panel, U. L. Raybhandary, and J. H. Van de Sande, *J. Biol. Chem.* **251,** 565 (1976).

Kilpatrick, D. L., T. Taniguchi, B. N. Jones, A. S. Stern, J. E. Shively, J. Hullihan, S. Kimura, S. Stein, and S. Udenfriend, *Proc. Natl. Acad. Sci. USA* **78,** 3265 (1981).

Krause, J. E., J. P. Advis, and J. F. McKelvy, *Endocrinology* **111,** 344 (1982).

Krieger, D. T., A. S. Liotta, M. J. Brownstein, and E. A. Zimmerman, *Rec. Prog. Hormone Res.* **36,** 277 (1980).

Lauber, M., M. Camier, and P. Cohen, *Proc. Natl. Acad. Sci. USA* **76,** 6004 (1979).

Lepine, J. and A. Dupont, *Endocrine Soc. Abst.* **108,** 385 (1981).

Lim, A. T., B. A. K. Khalid, J. Clements, and J. W. Funder, *J. Clin. Investigations* **69,** 1191 (1982).

Liotta, A. and D. T. Krieger, *Endocrinology* **106,** 1504 (1980).

Liotta, A., D. Gildersleeve, M. J. Brownstein, and D. T. Krieger, *Proc. Natl. Acad. Sci. USA* **76,** 1448 (1979).

Liotta, A., C. Loudes, J. F. McKelvy, and D. T. Krieger, *Proc. Natl. Acad. Sci. USA* **77,** 1880 (1980).

Maniatis, T., R. C. Hardison, E. Lacy, J. Lauer, C. O'Connell, D. Quon, G. K. Sim, and A. Efstratiadis, *Cell* **15,** 687 (1978).

Maniatis, T., E. G. Fritsch, and J. Sambrook, Eds., *Molecular Cloning,* Cold Spring Harbor Laboratory, Cold Spring Harbor, NY, p. 269 (1982).

Markham, A. F., M. D. Edge, T. C. Atkinson, A. R. Greene, G. R. Heathcliffe, C. R. Newton, and D. Scanlon, *Nucleic Acids Res.* **8,** 5193 (1980).

Matteucci, M. D. and M. H. Caruthers, *J. Amer. Chem. Soc.* **103,** 3185 (1981).

Maurer, R. A., *J. Biol. Chem.* **255,** 8092 (1980).

Maurer, R. A., *Nature* **294,** 94 (1981).

Maurer, R. A., *Endocrinology* **110,** 1957 (1982).

McWilliams, D. and I. Boime, *Endocrinology* **107,** 761 (1980).

Merritt, S. C., C. Tschudi, W. H. Konigsberg, and F. F. Richards, *Proc. Natl. Acad. Sci. USA* **80,** 1536 (1983).

Millan, M. J., *Modern Problems in Pharmaco-Psychiatry* **17,** 49 (1981).

Moen, R. C. and R. D. Palmiter, *Devel. Biol.* **78,** 450 (1980).

Morrell, J. and D. J. Pfaff, *Histo. and Cyt. Abst.* **IA,** (1983).

Nagata, S., H. Taira, A. Hall, L. Johnsrud, M. Streuli, J. Ecsodi, W. Boll, K. Cantell, and C. Weissman, *Nature* **284,** 316 (1980).

Nakane, P. K., G. Setalo, and J. E. Mazurkiewicz, *Ann. NY Acad. Sci.* **297,** 201 (1972).

Nakanishi, S., T. Kita, S. Taii, H. Imura, and S. Numa, *Proc. Natl. Acad. Sci. USA* **74,** 3283 (1977).

Narang, S. A., H. M. Hsiung, and R. Brousseau, in R. Wu, Ed., *Methods in Enzymology,* Vol. 68, Academic, New York, p. 90 (1979).

Nilson, J. H., A. R. Thomason, M. T. Cserbak, C. L. Moncman, and R. P. Waychik, *J. Biol. Chem.* **25,** 4679 (1983).

Noda, M., Y. Furutani, H. Takahashi, M. Toyosato, T. Hirose, S. Inayama, S. Nakanishi, and S. Numa, *Nature* **295,** 202 (1982).

Noyes, B. E., M. Mevarech, R. Stein, and K. L. Agarwal, *Proc. Natl. Acad. Sci. USA* **76,** 1770 (1979).

O'Farrell, P. H., *J. Biol. Chem.* **250,** 4007 (1975).

O'Farrell, P. H., E. Kutter, and M. Nakanishi, *Mol. Gen. Genetics* **179,** 421 (1980).

Ohkubo, H., R. Kageyama, M. Ujihara, T. Hirose, S. Inayama, and S. Nakanishi, *Proc. Natl. Acad. Sci. USA* **80,** 2196 (1983).

O'Malley, K. L., A. Mauron, J. Raese, J. D. Barchas, and L. Kedes, *Proc. Natl. Acad. Sci. USA* **80,** 2161 (1983)

Pacold, S. T., L. Kirsteins, S. Hojvat, and A. M. Lawrence, *Science* **199,** 804 (1978).

Palkovits, M., and M. Brownstein, *Strategies for Studying the Roles of Peptides in Neuronal Function,* Society for Neuroscience, Bethesda, MD, p. 60 (1982).

Palmiter, R. D., E. R. Mulvihill, J. H. Shepherd, and G. S. McKnight, *J. Biol. Chem.* 256, 7910 (1981).

Petraglia, F., A. Penalva, V. Locatelli, D. Cocchi, A. E. Panerai, A. R. Genazzani, and E. E. Muller, *Endocrinology* **111,** 1224 (1982)

Potter, E., A. K. Nicolaisen, E. S. Ong, R. M. Evans, and G. M. Rosenfeld, *Proc. Natl. Acad. Sci. USA* **78,** 662 (1980).

Rees, H., W. Stumpf, M. Sar, and S. Z. Petra, *Cell Tissue Res.* **182,** 347 (1977).

Rhodes, C. H., J. I. Morrell, and D. W. Pfaff, *J. Neurosci.* **2,** 1718 (1982).

Roberts, J. L., M. Phillips, P. A. Rosa, and E. Herbert, *Biochemistry* **17,** 3609 (1978).

Roberts, J. L., M. Budarf, J. Baxter, and E. Herbert, *Biochemistry* **17,** 3609 (1979a).

Roberts, J. L., P. Seeburg, J. Shine, E. Herbert, J. D. Baxter, and H. M. Goodman, *Proc. Natl. Acad. Sci. USA* **76,** 2153 (1979b).

Roberts, J. L., C. L. C. Chen, J. H. Eberwine, M. J. Q. Evinger, C. Gee, E. Herbert, and B. S. Schachter, *Rec. Prog. Hormone Res.* **38,** 227 (1982).

Rosa, P., P. Policastro, and E. Herbert, *J. Exp. Biol.* **89,** 215 (1980).

Rosen, J. M. and J. Monahan, in B. W. O'Malley and W. J. Schraeder, Eds., *Hormone Action Workshop,* Baylor College of Medicine, Houston, TX, Chap. 3, p. 1 (1976).

Rosenfeld, M. G., S. G. Amara, B. A. Roos, E. S. Ong, and R. M. Evans, *Nature (London)* **290,** 63 (1981).

Russell, J. T., M. J. Brownstein, and H. Gainer, *Endocrinology* **107,** 1880 (1980).

Schacter, B. S. and C. D. Toran-Allerand, *Dev. Brain Res.* **5,** 93 (1982).

Schacter, B. S., L. K. Johnson, J. D. Baxter, and J. L. Roberts, *Endocrinology* **110,** 1442 (1982).

Schally, A. V., A. Arimura, Y. Baba, R. M. G. Naik, H. Matsuo, T. W. Redding, L. Debeljuk, and W. F. White, *Biochem. Biophys. Res. Commun.* **43,** 393 (1971).

Schmale, H. And D. Richter, *Proc. Natl. Acad. Sci. USA* **78,** 766 (1981).

Schultsberg, M., D. Dreyfus, M. Gershon, T. Hokfelt, R. Eldo, G. Nilson, S. Said, and M. Goldstein, *Brain Res.* **155,** 239 (1978).

Seeburg, P., *DNA* **1,** 239 (1982).

Seed, B., *Nucleic Acids Res.* **10,** 1799 (1982).

Sehgal, P. B., in S. Pestka, Ed., *Methods in Enzymology,* Vol. **79,** Part B, Academic, New York, P. 111 (1981)

Singer, R. H. and D. C. Ward, *Proc. Natl. Acad. Sci. USA* **79,** 7331 (1982).

Singer-Sam, J., R. L. Simmer, D. H. Keith, L. Shively, M. Teplitz, K. Itakura, S. M. Gartler, and A. D. Riggs, *Proc. Natl. Acad. Sci. USA* **80,** 802 (1983).

Smith, M., in S. M. Weissman, E., *Methods of RNA and DNA Sequencing,* Praeger Scientific, New York, p. 23 (1983).

Sood, A. K., D. Pereira, and S. M. Weissman, *Prox. Natl. Acad. Sci. USA* **78,** 616 (1981).

Steiner, D. F., W. Kemmaler, H. S. Tager, and J. D. Peterson, *Fed. Proc.* **33,** 2105 (1971).

Sternberger, L. A., P. A. Hardy, J. J. Culculis, and H. C. Meyer, *J. Histochem. Cytochem.* **18,** 315 (1970).

Suchanek, G. and G. Kreil, *Proc. Natl. Acad. Sci. USA* **74,** 975 (1977).

Taylor, W. W., K. J. Collier, R. J. Deschenes, H. L. Weith, and J. E. Dixon, *Proc. Natl. Acad. Sci. USA* **78,** 6694 (1981).

Thomas, P. S., *Proc. Natl. Acad. Sci. USA* **77,** 5201 (1980).

Toran-Allerand, C. D., in J. D. Vincent and C. Kordon, Eds., *Biologie Cellulaire des Processus Neurosecretoires Hypothalamique,* Colloques Internationaux du C.N.R.S., Paris, Chap. 45, p. 759 (1979).

Toran-Allernad, C. D., *Brain Res.* **189,** 143 (1980).

Uhler, M. D., *Structural Organization Sequences of Pro-opiomelanocortin Genes in the Mouse,* PH. D. dissertation, University of Oregon, Eugene, p. 94 (1982).

Unterman, R. D., K. R. Lynch, H. L. Nakhasi, K. P. Dolan, J. W. Hamilton, D. V. Cohn, and P. Feigelson, *Proc. Natl. Acad. Sci. USA* **78,** 3478 (1981).

Vale, W., J. Spiess, C. Rivier, and J. Rivier, *Science* **213,** 1394 (1981).

Vermes, I., G. H. Mulder, P. G. Smelik, and F. J. H. Tilder, *Life Sci.* **27,** 1761 (1980).

Wahl, G. M., M. Stern, and G. R. Stark, *Proc. Natl. Acad. Sci. USA* **76,** 3683 (1979).

Wallace, R. B., J. Shaffer, R. F. Murphy, J. Bonner, T. Hirose, and K. Itakura,*Nucleic Acids Res.* **6,** 3543 (1979).

Wardlaw, S. L. and A. G. Frantz, *Endocrine Soc. Abst.* **II,** 589 (1982).

Wardlaw, S. L., L. Thoron, W. B. Wehrenberg, M. Ferin, and A. G. Frantz, *Endocrine Soc. Abst.,* 49 (1981).

Warembourg, M., *Cell Tissue Res.* **161,** 183 (1975).

Watson, S. J., J. Barchas, and C. H. Li, *Proc. Natl. Acad. Sci. USA* **74,** 5155 (1977).

White, B. A. and F. C. Bancroft, *J. Biol. Chem.* **257,** 8569 (1982).

Yates, F. F. and J. W. Maran, in J. Field, Ed., *Handbook of Physiology,* Vol. 4, Williams and Wilkens, Baltimore, MD, p. 367 (1974).

Young, R. A. and R. W. Davis, *Proc. Natl. Acad. Sci. USA* **80,** 1194 (1983).

Zimmerman, E. A., A. Liotta, and D. T. Krieger, *Cell Tissure Res.* **186,** 393 (1978).

# 3

# *Molecular Biology of the Mammalian Brain*

***Marcelle R. Morrison and W. Sue T. Griffin***

The regulation of synthesis and utilization of messenger RNAs (mRNAs) within mammalian brain cells may be similar in many respects to that in other inherently simpler organs and non-mammalian systems. However, the number of different neuronal and glial cell types in the mammalian brain has until recently precluded analysis of such regulation in individual brain cells. Due to this heterogeneity of cell types, knowledge of the basic anatomy, physiology, and biochemistry of the brain is a necessary background for studying its molecular biology. Several features of neurons themselves are obviously unique. For example, they are highly polymorphic, each neuronal cell type having an extraordinary degree of topographical specialization, necessitating precise intracellular transport of proteins destined for axons, dendrites, or synapses. The specialized and intricate synaptic interconnections between neurons and their electrical activity are other distinctive features.

The transcription pattern in brain is more complex than in other organs, perhaps due to the heterogeneity of different cell types and/or to a high transcriptional diversity in individual neurons or glia. For example, whole brain poly(A+) RNA and polysomal poly(A+) RNA complexity is several-fold higher than in other organs (Bantle and Hahn, 1976; Chikaraishi et al., 1978; Kaplan et al., 1978), and a complex class of developmentally induced poly(A−) RNAs which is unique to brain has been described (Chaudhari and Hahn, 1983; Chikaraishi, 1979; Van Ness et al., 1979).

Several of the early experiments on mRNA translation *in vitro* used isolated brain mRNAs (Gilbert, 1973; Lim et al., 1974; Zomzely-Neurath et al., 1972), and this technique has since been used to compare specific brain mRNA populations of various types (Bigbee and Eng, 1982; Brown, 1983; Cosgrove and Brown, 1981; Cosgrove et al., 1983; Czosnek et al., 1980; Gilbert and Strocchi, 1983; Hall and Lim, 1981; Mahony and Brown, 1979; Matthees and Campagnoni, 1980; Strocchi et al., 1981a, 1981b). The molecular analysis of neuropeptide precursors has been rapid (Nakanishi et al., 1978, 1979; Roberts et al., 1979), but it is only recently that recombinant DNA techniques have been applied to the study of other nervous system mRNAs such as those encoding tyrosine hydroxylase (Lamouroux et al., 1982), dopamine $\beta$-hydroxylase (O'Malley et al., 1983), and the acetylcholine receptor subunits (Noda et al., 1983). Analysis of sequences in an adult rat brain complementary DNA (cDNA) library is beginning to elucidate some general features of mRNA synthesis in the brain (Sutcliffe et al., 1983).

## 1. MODEL SYSTEMS FOR STUDYING NEURONAL DEVELOPMENT AND FUNCTION

Various strategies have been used to circumvent the problem of cell heterogeneity in studying the brain at the molecular level. One is to study relatively simple nervous systems in which homogeneous cell populations are more readily accessible. Indeed, the potential of recombinant DNA technology for analyzing aspects of neuronal function is currently best illustrated by the recent studies of Scheller et al. (1982, 1983) on the synthesis of bag cell neuropeptides in *Aplysia*. In addition, the high concentration of cholinergic cells and synapses in the electric organs of the electric ray has facilitated the identification of mRNAs synthesizing proteins unique to cholinergic synapses (Schmid et al., 1982) as well as the cloning of the acetylcholine receptor subunits (Noda et al., 1983).

Controlled biochemical and pharmacological studies are possible in organotypic monolayer cultures in which specific cell and developmental interactions are preserved (Gähwiler, 1981). However, the synthesis and utilization of mRNAs in specific cell types can be most easily studied in homogeneous cell populations such as tissue culture lines derived from neuronal and glial cells (Amano et al., 1972; Gilman and Nirenberg, 1971; Greene and Tischler, 1976).

Hybridization analyses in neuroblastoma (Amano et al., 1972) have shown that its poly(A+) mRNA complexity is higher than that of non-neuronal cells (Kaplan et al., 1978; Schrier et al., 1978) suggesting that the high transcriptional diversity found in whole brain at least in part reflects that in individual cells. Comparison of *in vitro* translation products of abundant neuroblastoma mRNAs shows that similar to other cell types 70% of the translational activity is encoded by mRNAs with different lengths of poly(A) and that most of these mRNAs are also present in the poly(A−) mRNA fraction (Morrison et al., 1979). These results indicate that most of the abundant poly(A−) RNAs in neuroblastoma, like other cells, are probably derived from the poly(A+) mRNAs by poly(A) cleavage (Gorski et al., 1975; Wilson et al., 1978). However, we found that several of the less abundant *in vitro* translation products of the poly(A−) RNAs were not synthesized at detectable levels by the poly(A+) RNAs. Either these are present at very low levels in the poly(A+) RNA fraction or they are examples of mRNAs specific to the poly(A−) RNA population (Chaudhari and Hahn, 1983). Certain mRNAs such as those encoding actin and the $R_1$ cAMP binding protein (Morrison et al., 1980b) are relatively abundant in the poly(A−) RNA population whereas others, for example the tubulin mRNAs, are 90% polyadenylated indicating that the rate of poly(A+) processing and/or poly(A−) RNA stability may be different for different neuronal mRNAs (Morrison et al., 1979). Analysis of abundant, translationally active brain mRNAs shows that they have similar properties to those of neuroblastoma (see Section 4).

In many cells a variable proportion of the cytoplasmic mRNA population is found in non-translated messenger ribonucleoprotein (mRNP) particles, and the distribution of individual mRNAs between the polysome and mRNP compartments regulates the rate of synthesis of specific proteins (Lee and Engelhardt, 1979; Lodish, 1974; McMullen et al., 1979). Protein synthesis in the brain decreases during development and can be disrupted by various treatments (Cosgrove and Brown, 1981; Lajtha and Dunlop, 1981) suggesting that translational control can modulate neuronal protein synthesis *in vivo*. The technique of *in vitro, in situ* translation could be used to determine the cell types in which protein synthesis is compromised in neuropathology (Capco and Jäckle, 1982).

In neuroblastoma approximately 30% of the cytoplasmic poly(A+) mRNAs are present in the mRNP fraction and several mRNAs are unequally distributed between the polysomes and this fraction (Croall and Morrison, 1980; Fig. 1). Actin, the $R_1$ cAMP binding protein, and protein 2 mRNAs are among those enriched in the polysome fraction whereas the $\alpha$-tubulin mRNA and several of unknown function (e.g., X, Y, Z, 25, and 31) are enriched in the mRNP fractions. The cytoplasmic distribution of some mRNAs correlates with their relative translational efficiency *in vitro*. However, some mRNAs are relatively less efficient (e.g., protein 2) yet are predominantly polysomal whereas other relatively efficient mRNAs (e.g., $\alpha$-tubulin) are enriched in the mRNP fraction. The factors influencing such cytoplasmic distributions are as yet unknown. It has been suggested that mRNA adenylation regulates translational efficiency (Benoff and Nadal-Ginard, 1979), but in neuroblastoma individual mRNAs in mRNP particles are as adenylated as are their polysomal counterparts. Thus, mRNA adenylation is not regulating mRNA loading onto polysomes in neuroblastoma (Croall and Morrison, 1980).

Some markers of neuronal differentiation can be induced in neuroblastoma and other neuronal cell lines by elevation of intracellular cAMP (Prashad and Kumar, 1974; Szutowicz et al., 1983), a possible *in vivo* modulator of neuronal metabolism (Greengard, 1976). Neuroblastoma differentiation is accompanied neither by changes in the steady-state levels nor the adenylation of total poly(A+) mRNAs (Morrison et al., 1977, 1980a) nor by large changes in mRNA sequence complexity (Felsani et al., 1978; Grouse et al., 1980). However, there are changes in the steady-state levels of several specific mRNAs. For example, cAMP increases the levels of its $R_1$ regulatory subunit in neuroblastoma cells (Prashad et al., 1979) due to an increase in the steady-state levels of the mRNA encoding this protein (Morrison et al., 1980b; Table 1). The mRNA encoding a very rapidly turning over protein, protein 2, possibly another cAMP binding protein (Strocchi et al, 1981a), is also induced by cAMP treatment (Morrison et al., 1980b). The rapid turnover of this protein indicates that its levels are finely tuned, perhaps suggesting a role as a modulator of cellular events. In differentiated neuroblastoma there is a concomitant decrease in the levels of the mRNAs encoding actin and the $\alpha$- and $\beta$-tubulins similar to that seen during brain development (see Sections 2, 3, and 7).

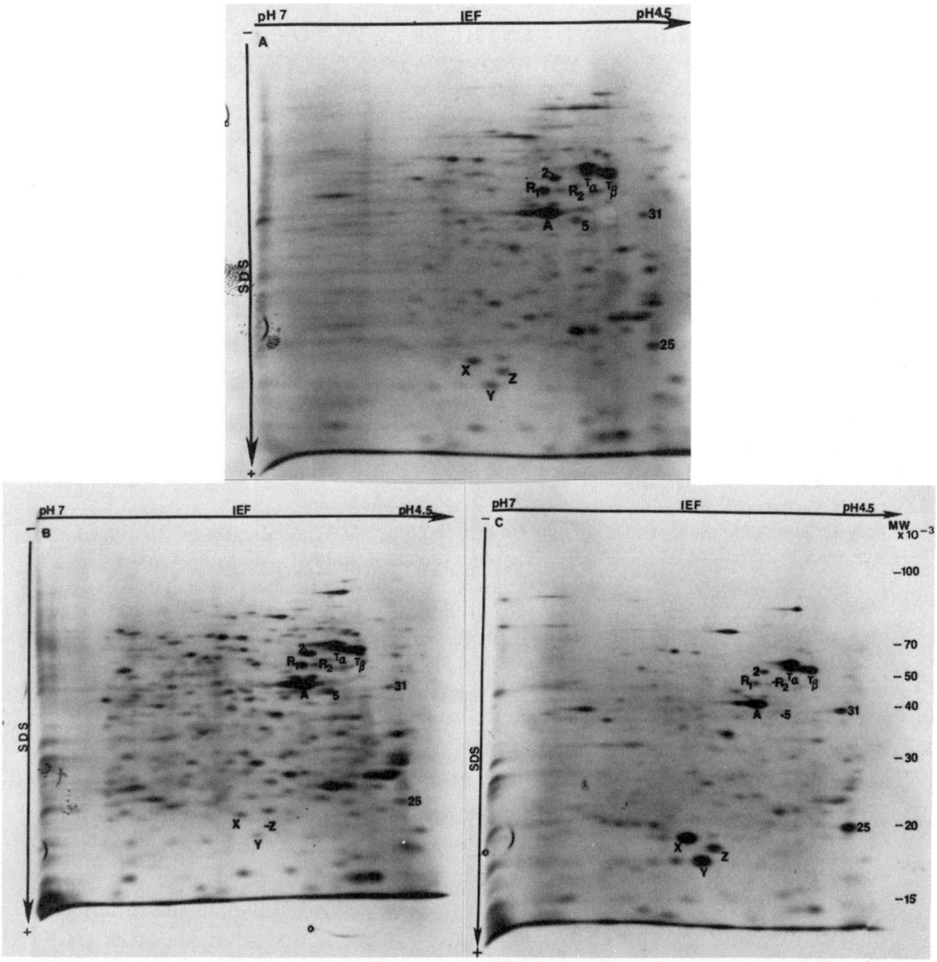


**Figure 1.** Patterns of *in vitro* translation products from neuroblastoma RNA: isoelectric focusing and SDS-gel analysis. Translation of RNA and gel analysis of [$^{35}S$]-methionine-labeled products were performed as described. Fluorographs are from gels containing 262,000 cts/min and exposed for 5 days at −70°C. Products from total cytoplasmic RNA (A), polysomal RNA (B), and non-polysomal RNA (C) were typical of four separate experiments and reproducible upon repetition of electrophoresis. A, actin; T, tubulin, $R_1$, $R_2$, cAMP-binding proteins. (From Croall and Morrison, 1980). With permission from *Journal of Molecular Biology*. Copyright: Academic Press Inc. (London) Ltd.

**TABLE 1. RELATIVE SYNTHESIS *IN VITRO* OF SELECTED PROTEINS BY POLYSOMAL RNAs FROM CONTROL AND $Bt_2$cAMP-TREATED CELLS**

| Protein | $[^{35}S]$ INCORPORATED IN $Bt_2$cAMP-TREATED CELLS RELATIVE TO CONTROL CELLS Experiment 1 | 2 | 3 | Average |
|---|---|---|---|---|
| $R_1$ | 1.74 | 1.44 | 1.41 | 1.53 |
| $R_2$ | 1.07 | 1.08 | 1.00 | 1.05 |
| Protein 2 | 1.38 | 1.20 | 1.04 | 1.20 |
| Actin | 0.79 | 0.96 | 0.80 | 0.85 |
| $\alpha$-Tubulin | 0.81 | 1.03 | 0.84 | 0.89 |
| $\beta$-Tubulin | 0.96 | 0.83 | 0.84 | 0.88 |

Polysomal RNAs were extracted from control and $Bt_2$cAMP-treated cells and translated at subsaturating inputs (4 $\mu$g) in 50 $\mu$L wheat germ protein-synthesizing systems *in vitro* (25 $\mu$Ci $[^{35}S]$-methionine input). The proteins synthesized *in vitro* were resolved by two-dimensional electrophoresis and selected proteins were quantified. The ratio of $[^{35}S]$ incorporated into proteins by mRNAs from $Bt_2$cAMP-treated cells relative to mRNAs from control cells was found in three separate experiments. The results for each experimental mRNA are the average of two separate translations *in vitro* and two-dimensional separations. (From Morrison et al., 1980b). With permission from *European Journal of Biochemistry*. Copyright: FEBS.

We have constructed a neuroblastoma cDNA library containing most of the RNA sequences transcribed in these cells (see Section 10) and plan to use sequences isolated from this library to determine (1) whether cAMP acts at a transcriptional or post-transcriptional level (Darnell, 1982); (2) the half-lives of specific mRNAs such as those encoding the actins and the tubulins; and (3) whether the synthesis of the mRNAs encoding the different $\alpha$- and $\beta$-tubulin subunits (see Sections 2, 3, and 7) is regulated by the pool of unpolymerized tubulin as has been proposed in other cell lines (Cleveland et al., 1981).

## 2. REGULATION OF mRNA LEVELS DURING RAT CEREBELLAR DEVELOPMENT

The rodent cerebellum has several characteristics which make it a suitable model system for analysis at the molecular level. Its postnatal development has been well characterized anatomically, electrophysiologically, and neurochemically (Altman, 1972; Gilad and Kopin, 1979; Woodward et al., 1969), thus laying the necessary foundation for analysis of its molecular biology. Furthermore, the cerebellum is less complex than other brain areas because the various cell types are interconnected in a regular geometry (Jacobson, 1978).

In the rodent neonate the major cell types in the cerebellum are the prenatally formed Purkinje and Golgi cells (15%), the glial cells (15%), and the germinal cells of the external granular layer (EGL, 70%) (Addison, 1911; Altman, 1972; Griffin et al., 1977, 1979; Zagon and McLaughlin, 1979). Granule cell proliferation is maximal at days 6–8 (Clark et al., 1978), and postmitotic granule cells are migrating into their final positions in the internal granular layer (IGL). By day 14 granule cell proliferation has declined by 60% (Clark et al., 1978), and the rate of synaptogenesis and myelination is high (Altman, 1972; Campagnoni et al., 1978). Cell division has virtually ceased by day 21 (Clark et al., 1978), but even after day 30 there is continued arborization of Purkinje cells (Addison, 1911). In the adult greater than 90% of cells are the granule cell interneurons of the IGL (Zagon and McLaughlin, 1979).

Hybridization experiments show that there is a reduction in the single-copy DNA transcribed into polysomal poly(A+) RNA during rat postnatal cerebellar development (Bernstein et al., in press). Gel analysis shows that the steady-state levels of several abundant proteins are developmentally regulated. For example, there is an age-related decrease in the steady-state levels of actin and the $\alpha$- and $\beta$-tubulins (Morrison et al., 1981) concomitant with a decrease in the *de novo* synthesis of these proteins (Morrison et al., 1981). Quantitation of translationally active mRNAs by two-dimensional gel analysis of their *in vitro* translation products shows that many of the abundant mRNAs are present at similar levels at different stages of cerebellar development (2, 6, 14, 30, and 90 days; Morrison et al., 1981). These abundant mRNAs likely encode "housekeeping" proteins necessary for basic cellular function in both immature and differentiated brain cells. Quantitation of developmentally regulated mRNAs shows that between postnatal days 2 and 6 there is a decline in the mRNAs encoding the $\alpha$-tubulins and those encoding unidentified proteins e and 5 (Morrison et al., 1981; Fig. 2), perhaps due to their relative abundance in prenatally formed cells whose numbers are decreasing proportional to the postnatally formed granule cells in the EGL. However, most of the decrease in the mRNAs encoding the tubulins and actin takes place between 14 and 30 days postnatal, a time when granule cell division is ending and synaptic connections are being formed. This decrease could therefore be a consequence of the cessation of neuronal cell division. The maintenance of high levels of all actin and tubulin mRNAs prior to 14 days may also reflect the high levels of synthesis of these proteins necessary for structuring the axons and dendrites (Olmstead and Borisy, 1973). Other groups have also documented a developmental decrease in these brain mRNAs (Gozes et al., 1980; Schmitt et al., 1977; Soreq et al., 1982).

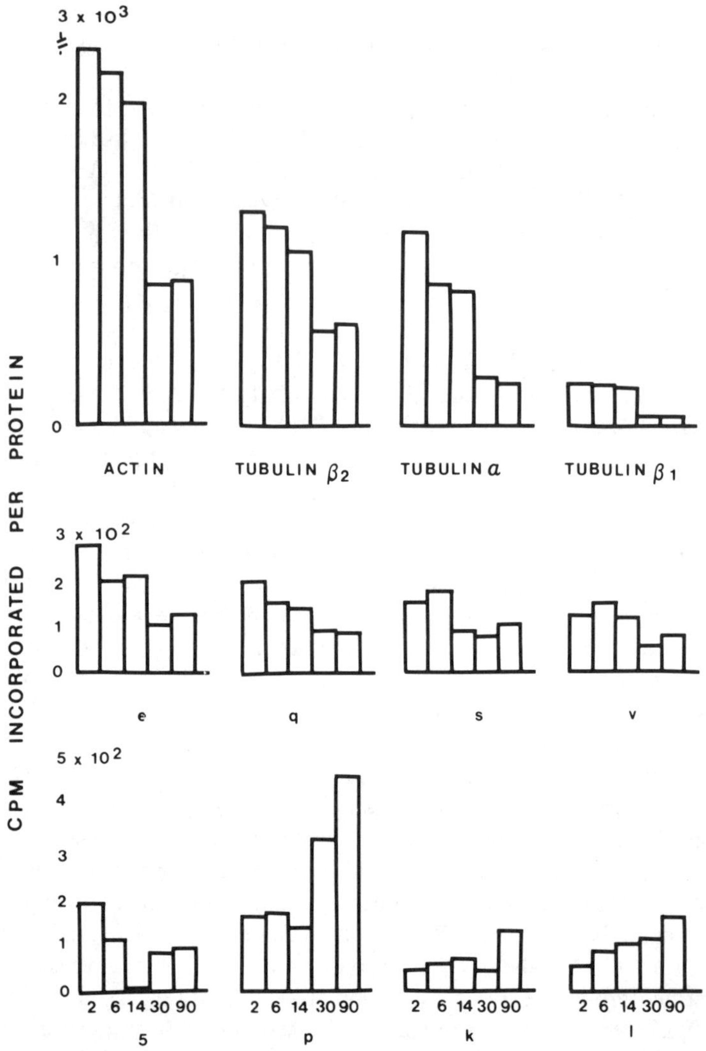


**Figure 2.** Histograms depicting the relative *in vitro* synthesis of certain cerebellar proteins by mRNAs isolated from cerebellum at various developmental stages (2, 6, 14, 30, and 90 days). Areas of gels corresponding to proteins of interest were excised and their content of radioactivity determined. The results are an average of results obtained from three RNA samples at each age, each analyzed in duplicate. The counts were corrected for 100,000 trichloroacetic-acid-precipitable counts loaded on each gel. (From Morrison et al., 1981.)

Interestingly, the ratios of the mRNAs encoding the different tubulin isoforms are also developmentally regulated. For example, there is a relatively greater decrease in the synthesis of the $\alpha$ subunits and in the minor $\beta$ subunit, $\beta_1$, relative to the synthesis of the major, more acidic $\beta$ subunit, $\beta_2$ (see Sections 3, 7, and 11).

mRNAs encoding several unidentified proteins are induced at different times during development. The levels of an mRNA encoding an acidic low-molecular-weight protein, P, also synthesized *in vivo*, show the greatest increase between 14 and 30 days. The levels of this mRNA and those encoding two other proteins of unknown function, k and l, are still increasing between 30 and 90 days postnatal suggesting the possibility of continued differentiation in mature brain.

## 3. REGULATION OF mRNA LEVELS DURING RAT CORTEX DEVELOPMENT

In contrast to the postnatally developing cerebellum, cell division in the rat cortex ceases perinatally. Changes in chromatin structure in these two brain areas parallel the onset of neuronal differentiation (Brown, 1978; Brown and Greenwood, 1982; Heizmann et al., 1980, 1982; Jaegar and Kuenzle, 1982). Comparison of the patterns of developmentally regulated mRNAs in the cortex and cerebellum would show whether changes in the relative steady-state levels of different mRNAs occur at equivalent developmental states in these two brain regions.

A number of the poly(A+) mRNAs isolated from 10-day and adult rat cortex are developmentally regulated (Fig. 3; Morrison and Griffin, in press-b) but many, including protein P mRNA, are identical to those in the cerebellum and the 10-day cortex (Morrison and Griffin, 1984). The relative amounts of the different tubulin mRNAs in adult cortex and cerebellum are similar. However, their levels in the 10-day cortex are much higher than in the 10-day cerebellum, particularly the $\alpha$- and $\beta_1$-tubulin mRNAs. The developmental events associated with these changes in the $\alpha$- and $\beta_1$-tubulin mRNA levels in the 10-day cortex are unknown. Interestingly, the $\beta_1$ subunit is synthesized at much higher levels in the fetal rat cortex than at any stage in cerebellar development (Morrison and Griffin, in press, 1984) illustrating that requirements for the $\beta_1$- and $\beta_2$-tubulin subunits differ not only developmentally but also between cortex and cerebellum. Analysis at the level of individual cells by *in situ* hybridization will identify the cell types expressing these mRNAs (see Section 11).

In cortex the different cell types have more dendritic and axonal arborization than does the major cerebellar cell population, the granule cell. Taken

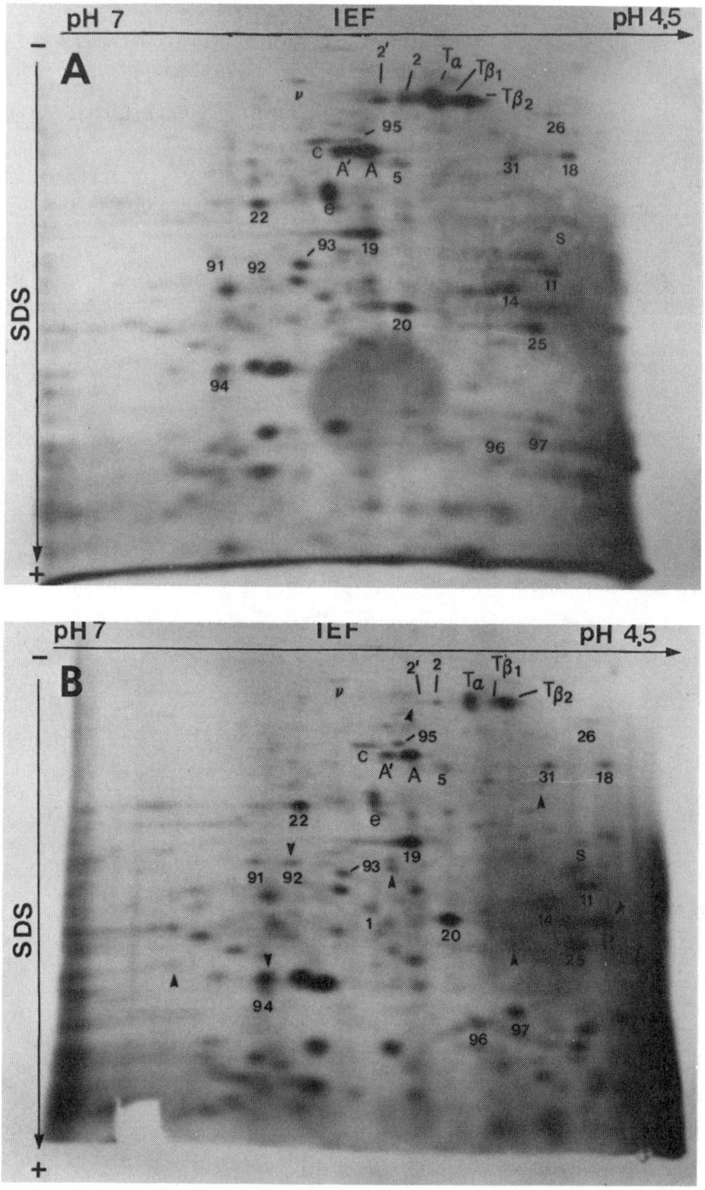


**Figure 3.** Two-dimensional gel electrophoresis of proteins synthesized *in vitro* by poly(A+) RNAs isolated from 10-day postnatal (A) and adult (B) rat cortex. 400,000 trichloroacetic-acid-precipitable counts were loaded on each gel. The gels were stained, fluorographed, dried, and exposed to X-ray film for 4 days. A = actin, $T_{\alpha}$, $T_{\beta_1}$, $T_{\beta_2}$ = $\alpha$-, $\beta_1$-, and $\beta_2$-tubulin. Numbered proteins correspond to those we previously identified in neuroblastoma.

together with the developmental patterns, it is therefore possible that the higher $\alpha$- and $\beta_1$-tubulin levels in immature cortex are related less to cell division *per se* than to the requirement for process formation. Alternatively, it is possible that in both cortex and cerebellum systemic factor(s) are directly influencing the levels of these mRNAs, even in cells at different developmental stages.

## 4. NON-ADENYLATED mRNAs IN BRAIN DEVELOPMENT

To determine whether or not brain also contains a neuroblastoma-like poly(A−) RNA population (Morrison et al., 1979; see Section 2), we have analyzed the abundant poly(A−) RNAs present in rat brain. In both the fetus and the adult we found that the poly(A−) RNAs contain between 30 and 40% of the total mRNA translational activity. Two-dimensional gel analysis of the translational products from adult rat cortex poly(A−) RNAs (Fig. 4) shows that as in neuroblastoma (Morrison et al., 1979), most of the poly(A−) mRNA translation products are similar to those encoded by the poly(A+) RNAs although several minor proteins may be uniquely encoded by the poly(A−) RNA population. Furthermore, the same mRNAs were relatively enriched in the non-adenylated fraction of brain as in neuroblastoma. Both in the adult (Fig. 4) and in the fetal brain (results not shown) there was a higher proportion of actin mRNAs in the poly(A−) RNA fraction and a lower proportion of tubulin mRNAs. The relative rates of poly(A) turnover and stability of the abundant mRNAs in both immature and mature brain must therefore be similar to those in neuroblastoma.

These results show that brain cells do contain a population of translationally active poly(A−) RNAs with properties similar to those described in neuroblastoma and other cell types which may be derived from the poly(A+) RNAs by poly(A) cleavage. However, this does not preclude the existence of another brain polysomal poly(A−) RNA class which is separate and unique from that described, above. This second class of poly(A−) RNAs, defined solely by hybridization analyses, comprises approximately 80,000 average-sized mRNAs, and in contrast to the abundant, translationally active poly(A−) mRNA class we have described, has no sequences in common with the poly(A+) mRNAs (Chaudhari and Hahn, 1983; Chikaraishi, 1979; Van Ness et al., 1979). Moreover, again in contrast to the abundant poly(A−) mRNAs, this poly(A−) RNA class is developmentally induced and, as suggested by Chaudhari and Hahn (1983), may play a crucial role in brain cell function.

We have shown by hybridization experiments with a recombinant DNA probe to U1 nuclear RNA (kindly provided by R. T. Manser and R. G. Gesteland; Manser and Gesteland, 1981) that there is appreciable leakage of

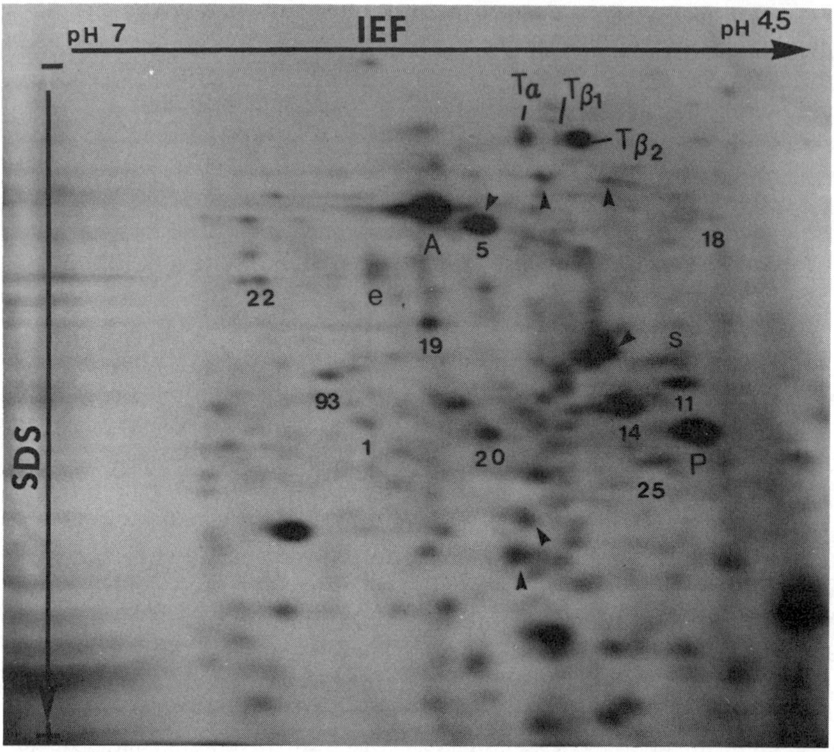


**Figure 4.** Two-dimensional gel electrophoretogram of proteins synthesized *in vitro* by poly(A−) RNAs isolated from adult rat cortex. 200,000 trichloroacetic-acid-precipitable counts were loaded on the gel. The gel was stained, fluorographed, dried, and exposed to X-ray film for 4 days. A = actin, $T_{\alpha_1}$, $\alpha_2$, $\beta_1$, $\beta_2$ = $\alpha_1$-, $\alpha_2$-, $\beta_1$-, and $\beta_2$-tubulin, P = protein P.

nuclear RNA into the cytoplasm during brain cell fractionation (Snider and Morrison, unpublished observations), possibly contributing some of the polysomal poly(A−) RNA complexity. Definitive proof that a biologically significant poly(A−) RNA class exists awaits the cloning and characterization of members of this class.

## 5. GRAFT VERSUS HOST DISEASE AS A MODULATOR OF DEVELOPMENTAL EVENTS

Genetic mutations have been used to modulate developmental events (for review see Hall et al., 1982). Ideally, such mutants would reveal how genes direct the formation of neurons, the specificity of neural circuits, and how mutations cause behavioral deficits (Quinn and Gould, 1979). Such mutants have been important in establishing neuroanatomical correlates of behavior although by their nature the perturbations in development caused by genetic mutations are irreversible. A number of mutations affect cerebellar development in mouse (Mullen, 1978; Sidman et al., 1965) with loss of specific cell populations (Fisher, 1983).

Environmental factors such as malnutrition (Griffin et al., 1977) and neonatal X-irradiation (Altman and Anderson, 1971; Griffin et al., 1978b), the latter specifically degranulating the cerebellum, have also been used to modulate developmental events in the cerebellum. For example, X-irradiation has been used to identify changes in the levels of cerebellar mRNAs associated with the loss of granule cells (Eliyahu and Soreq, 1982; Griffin and Morrison, unpublished observations).

We have found that graft versus host disease (GVHD) and its termination by alloantiserum treatment is also a modulator of molecular events in the developing rat cerebellum and, unlike the above examples, its effects are reversible. GVHD is a systemic immune disease induced by the grafting of fully or semi-allogeneic immunocompetent T lymphocytes into an immunoincompetent host which therefore does not recognize the grafted cells as foreign. The grafted lymphocytes can directly attack and kill host cells (Billingham, 1968) as well as indirectly affect tissues (Elson et al., 1977; Streilein, 1972) such as cerebellum which suffers as an innocent bystander of the graft versus host lymphocyte interactions. GVHD induced in human infants as a result of therapeutic measures (Grebe and Streilein, 1976; Pollack et al., 1982) shares many of the features that characterize experimentally induced GVHD in rodents. GVHD can be terminated in rodents by treatment with an alloantiserum specifically directed against the grafted lymphocytes (Griffin et al., 1981, 1982a; Silvers and Billingham, 1969).

Neonatally induced GVHD in humans and rodents occurs concomitant with cerebellar development. We therefore postulated that the altered gait of rats with GVHD might be causally related to altered cerebellar development. Indeed, by postnatal day 11 cerebellar development is seriously compromised by GVHD (Griffin et al., 1978a). DNA content and synthesis are less, and the size of the cerebellar layers is adversely affected (Griffin et al., 1980a, 1982b). The decrease in DNA synthesis is in part due to fewer total germinal cells, but of more interest from the molecular point of view, it is also due to a decrease in DNA synthesis in the remaining cells (Griffin et al., 1982b). Furthermore, progenitors of cell populations whose formation is greatly affected by the disease (e.g., the basket and stellate cells) respond promptly to alloantiserum treatment (Griffin et al., 1982b).

Accompanying the changes in acquisition of neuronal cells, the amount of rough endoplasmic reticulum is much less in cerebellar cells from 14-day-old

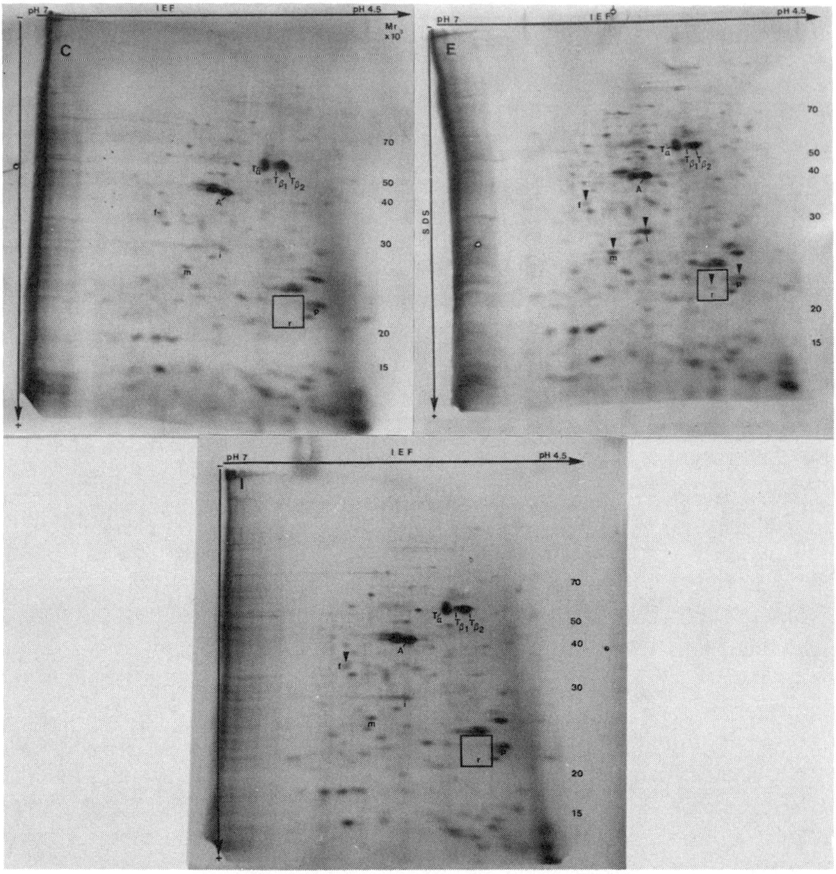


**Figure 5.** Two-dimensional gel electrophoretograms of [$^{32}$S]-methionine-labeled proteins synthesized *in vitro* by cytoplasmic RNAs isolated from cerebellum of 14-day-old animals with GVHD(E), diseased animals treated with alloantiserum (I), and their control littermates (C). Samples of 200,000 trichloroacetic-acid-precipitable cpm in 49 $\mu$L of the wheat germ lysate were loaded on each gel. The gels were stained, fluorographed, dried, and exposed to X-ray film for 8 days. A = actin; T = tubulin, r = protein r. Arrows identify altered protein levels. (From Griffin et al., 1982c.) With permission from *Journal of Neurochemistry*. Copyright: Raven Press.

animals with GVHD, Purkinje cells being particularly affected (Griffin et al., 1980a). Total cerebellar RNA as well as RNA synthesis are significantly less in diseased animals, and cytoplasmic RNA isolates are less translationally active *in vitro*. There are also alterations in the relative abundance of certain mRNAs in cerebella of affected animals (Fig. 5), including the induction of an mRNA encoding a protein, r, that is not present in either control cerebella, cerebral hemispheres, or in cerebella of starved or X-irradiated rats (Griffin et al., 1980b, 1982c). Alloantiserum treatment normalizes mRNA levels (Fig. 5; Griffin et al., 1982c).

We are particularly interested in the possible regulation of mRNAs encoding proteins associated with the decreased mitosis and cell division found in GVHD. Application of immunohistochemical and *in situ* hybridization techniques (see Section 11) to cerebella of rats with GVHD will unequivocally identify the affected cell types. In particular it should be possible to determine whether, as we postulate (Griffin, Alejos, and Morrison, unpublished results), the levels of mRNAs encoding the tubulins and actin fluctuate concomitant with cell division in the EGL in rats with decreased (GVHD), increased (alloantiserum treatment), and normal (control) DNA synthesis.

## 6. ISOLATION OF UNDEGRADED mRNAs FROM HUMAN POSTMORTEM BRAIN

Resolution of the molecular basis for the hemoglobinopathies was made possible by the isolation and characterization of human globin mRNAs (Forget, 1982). Similarly, the isolation and analysis of mRNAs from affected regions of diseased brain would be an important advance in the elucidation of the molecular changes accompanying and underlying basic pathology. Furthermore, molecular analysis of cells in relatively unaffected areas of diseased brain may reveal previously unrecognized targets of disease.

Our initial bias was that mRNAs would be rapidly degraded postmortem, but surprisingly total rat cerebellar RNAs isolated 16 hours postmortem are only negligibly degraded if cerebella are stored at 4°C (Morrison and Griffin, 1981a) and are 70% of control values when cerebella are stored at room temperature. Furthermore, the biological specific activity of RNAs isolated from cerebella kept at 4°C for 16 hr is not significantly less than RNAs isolated from freshly killed controls, and the RNAs isolated from cerebella kept at 25°C are 70% as active per microgram of RNA. Thus, RNA isolates from cerebella maintained at 4°C retain 85% of their original translational capacity while isolates from cerebella maintained at 25°C retain approximately 50%.

Human postmortem cerebella maintained at room temperature for either 4 or 16 hr postmortem have less recoverable RNA per gram wet weight than do postmortem rat cerebella (Table 2; Morrison and Griffin, 1981a), perhaps due to the higher lipid and protein content of human brain. The total RNA recovery per gram wet weight of human cerebella, 4 hr postmortem, is 70% of that from freshly killed rat and is approximately 50% after 16 hr at room

**TABLE 2. RECOVERY AND BIOLOGICAL SPECIFIC ACTIVITIES OF RNAs ISOLATED FROM RAT AND HUMAN CEREBELLA[a]**

| Postmortem Storage Time and Temperature | RNA Recovery ($\mu$g/g wet wt.) | Biological Specific Activity (cpm/$\mu$g RNA) |
|---|---|---|
| A. Rat, 0 hr | 634 | 7918 |
| Rat, 4 hr, RT[b] | 606 | 6328 |
| Rat, 16 hr, 4°C | 573 | 7425 |
| Rat, 16 hr, RT | 529 | 5605 |
| Human, 4 hr, RT | 344 | 5432 |
| Human, 16 hr, RT | 243 | 2958 |
| B. Rat, 0 hr | 791 | 6228 |
| Human, 4 hr, RT | 576 | 4181 |
| Human, 16 hr, RT | 417 | 3616 |

*Source:* Morrison and Griffin, 1981*a*. With permission from *Anal. Biochem.* Copyright: Academic Press, Inc.

[a]RNAs were isolated (A) from the cytoplasm of cells lysed in aqueous extraction buffer or (B) from whole cerebellum by homogenization in guanidinium chloride followed by centrifugation of the RNA through cesium chloride. The biological specific activities were assayed by translation of the RNAs at subsaturating (2 $\mu$g) inputs in a wheat germ cell-free system at 2.5 $\mu$Ci [$^{35}$S]-methionine/50 $\mu$L.

[b]RT designates room or ambient temperature.

temperature postmortem. The human RNAs are still translationally active with the 16-hr sample retaining 50% of translational activity compared to the freshly killed rat.

If these lower RNA recoveries and translational activities were indicative of selective mRNA degradation postmortem, such preparations would be of little use in the study of relative mRNA levels, but on the contrary the translation products of RNAs isolated from cerebella of rats 16 hr postmortem are virtually identical to those from the freshly killed rat control showing the usual spectrum of brain proteins, many of high molecular weight (Fig. 6A and B). Similarly, the translation products of the human samples are virtually identical both to each other and to that of the rat control (Fig. 7A and B); the relative amount of radiolabeled amino acid incorporated into different molecular weight proteins such as actin, the tubulins, and protein P being unchanged by postmortem storage (Morrison and Griffin, 1981a). Therefore, since the abundant brain mRNAs are degraded randomly and at similar rates for at least up to 16 hr postmortem, postmortem brain can be used to study alterations of steady-state mRNA levels in control and diseased human brain. We have now compared various methods of storing brain and find that similarly undegraded mRNAs can be isolated from specimens stored for at least 1 year at −70°C. Others (Gilbert et al., 1981) have isolated human brain polysomal poly(A+) RNAs which, although in very low yields, were translatable and not preferentially degraded. More recently Sadjel-Suklowska et al. (1983) have also obtained high yields of undegraded total mRNAs from human postmortem brain.

It is theoretically possible that less abundant mRNAs, not analyzed by our procedure, are selectively degraded postmorem. However, cDNAs reverse-transcribed from adult human cortex poly(A+) RNAs have the expected size distribution on denaturing gels, thus showing that the bulk of poly(A+) mRNAs are still intact at 16 hr postmortem (Michael et al., 1982; Morrison, 1983; Morrison and Sparkman, 1984; Sparkman et al., unpublished results; see Section 10). In order to determine whether the bulk of the complex class poly(A+) mRNAs are still present, it will be necessary to compare the hybridization kinetics of postmortem rat and human poly(A+) mRNA preparations to those of freshly killed rats. DNA probes to relatively rare mRNAs could be used to determine whether individual mRNAs are still present in these postmortem preparations at their *in vivo* levels.

## 7. CHANGES IN TUBULIN mRNAs DURING DEVELOPMENT OF THE HUMAN CORTEX

Direct quantitation of the steady-state levels of adult human brain tubulins is hindered because tubulins migrate anomolously on two-dimensional gels (Comings et al., 1982; Morrison and Rosenberg, 1983). However, the relative synthesis of the different tubulin isoforms can be quantitated by *in vitro* translation of RNA isolates. In order to determine whether the developmentally regulated changes in the tubulin isotypes in rat brain (Morrison and Griffin, 1984; Morrison et al., 1981) could be demonstrated in human cortex, we isolated total RNAs from fetal and adult human cortex using our standard procedure (Morrison and Griffin, 1981a) and purified poly(A+) RNAs by two cycles of oligo(dT)-cellulose hybridization (Morrison et al., 1977). As in the cerebellum, cortex RNA recoveries are between 250 and 300 μg/g wet weight. As expected, the poly(A+) RNAs constitute 1.5–2% of the total RNA, thus demonstrating that the poly(A+) RNAs are not deadenylated postmortem (Morrison and Griffin, 1984; Morrison et al., 1983). Profiles of their *in vitro* translation products (Fig. 8A and B) are in most respects similar to those at similar developmental stages in the rat (Morrison and Griffin, 1981b, 1984; Morrison et al., 1981; see Sections 2 and 3). For example, the mRNA encoding protein P is developmentally induced, and the relative levels of the mRNAs encoding the tubulins and actin decrease in amount during development.

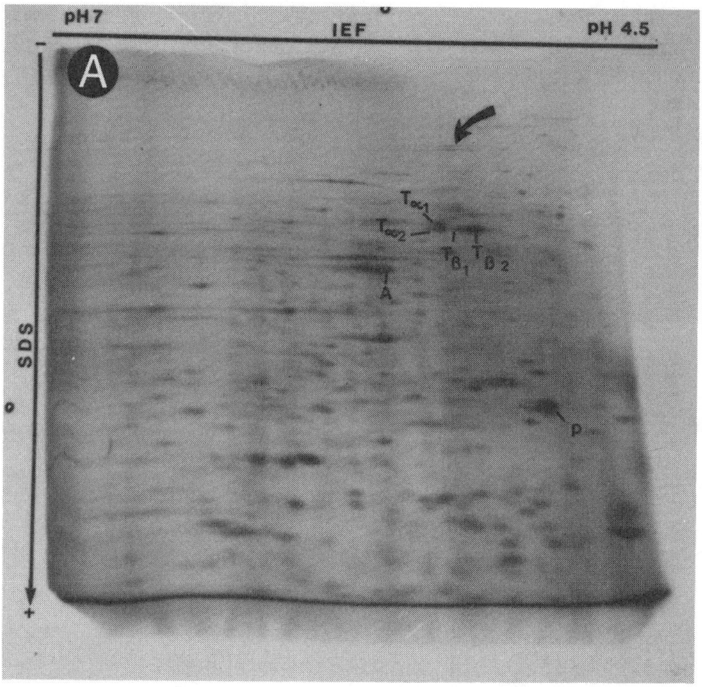


**Figure 6.** Two-dimensional gel analysis of [$^{35}$S]-methionine-labeled proteins synthesized *in vitro* by total cytoplasmic RNAs isolated from (A) fresh rat cerebellum and (B) rat cerebellum kept for 16 hr postmortem at room temperature. RNAs were isolated from the postnuclear supernatant fraction by extraction with guanidinium chloride. 623,000 trichloroacetic-acid-precipitable counts were loaded on each gel. The gels were stained, dried, and exposed to X-ray film for 13 days. A, actin; $T_{\beta_1}$, $\beta_1$-tubulin; $T_{\beta_2}$, $\beta_2$-tubulin; $T_{\alpha_1}$, $\alpha_1$-tubulin; $T_{\alpha_2}$, $\alpha_2$-tubulin; P, protein P. (From Morrison and Griffin, 1981a.) With permission from *Anal. Biochem.* Copyright: Academic Press, Inc.

The relative synthesis of both $\alpha$-tubulin subunits falls by more than three-fold during development. In the fetus the relative synthesis of the $\beta_1$ subunit is higher than that of the $\beta_2$ subunit whereas it has fallen to less than 25% of that level in the adult. In contrast, the relative synthesis of the $\beta_2$ subunit in the adult is unaltered during development. Changes in total $\alpha$- and $\beta$-tubulin levels have been confirmed (Morrison, Wilcutts, and Jamison, unpublished results) by dot blot hybridization (Beale et al., 1982; Thomas, 1980) of total $\beta$-tubulin probes, kindly donated by D. W. Cleveland (Cleveland et al., 1980).

Several different $\beta$-tubulin mRNAs are transcribed in the human brain (Hall et al., 1983; Lopata et al., 1983), and these mRNAs are present at different steady-state levels in a variety of cell types (Lopata et al., 1983).

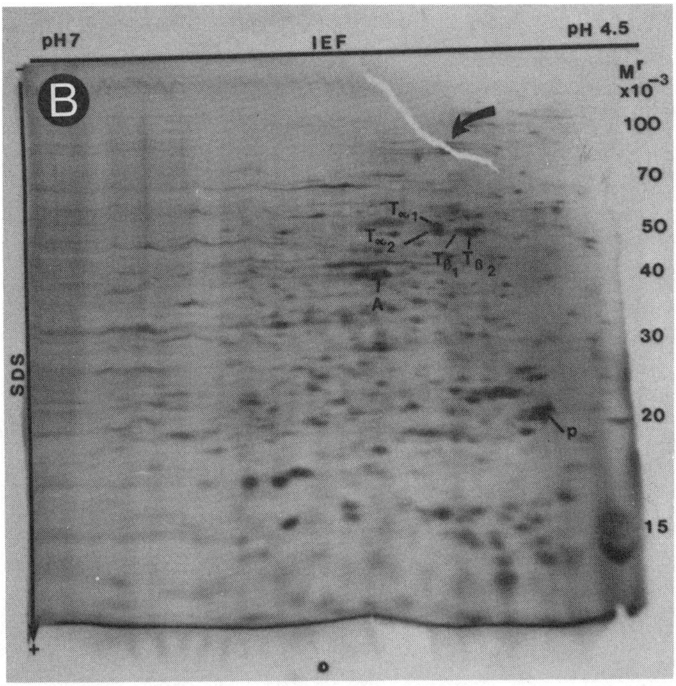


**Figure 6 (continued)**

Our results unequivocally demonstrate that the synthesis of two $\beta$-tubulin isotypes are developmentally regulated in human brain, indicative of unique cellular functions for the $\beta_1$- and $\beta_2$-tubulins. The distribution of the individual tubulin mRNAs in different cell types during development can now be analyzed using the technique of *in situ* hybridization (Griffin et al., 1983; 1984; see Section 11).

## 8. CHARACTERIZATION OF mRNAs PRESENT IN NON-NEURONAL BRAIN CELLS

One of the major questions in neuromolecular biology is to what extent transcription patterns differ in individual neuronal and glial cell types. Methods now exist for isolating relatively pure cell populations by physical means (for review see Althaus and Neuhoff, 1982), and the characterization of antibodies specific to individual cell types will doubtless result in the development of new methods of cell fractionation. Analysis of specific mRNAs in such cell isolates has not yet been attempted.

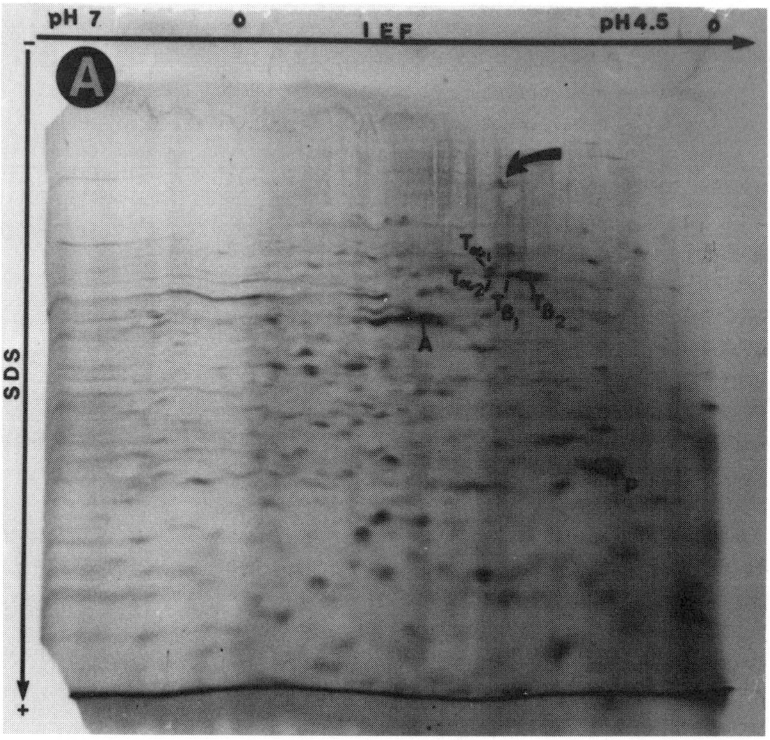


**Figure 7.** Two-dimensional gel analysis of [$^{35}$S]-methionine-labeled proteins synthesized *in vitro* by total cytoplasmic RNAs isolated from human cerebellum which was stored at room temperature for 4 hr (A) and 16 hr (B) postmortem. RNA was isolated from the postnuclear supernatant fraction by extraction with guanidinium chloride. 623,000 trichloroacetic-acid-precipitable counts were loaded on each gel. The gels were stained, dried, and exposed to X-ray film for 13 days. A, actin; $T_{\beta_1}$, $\beta_1$-tubulin; $T_{\beta_2}$, $\beta_2$-tubulin; $T_{\alpha_1}$, $\alpha_1$-tubulin; $T_{\alpha_2}$, $\alpha_2$-tubulin; P, protein P. (From Morrison and Griffin, 1981a.) With permission from *Anal. Biochem.* Copyright: Academic Press, Inc.

The optic nerve is composed of retinal ganglion cell axons ensheathed in oligodendroglial processes. Hence, as it contains no neuronal cell bodies, mRNAs isolated from it should be exclusively non-neuronal. Two-dimensional gel analysis of the *in vitro* translation products of total RNAs isolated from the human optic nerve show that the major translationally active mRNAs present are those encoding actin and the astrocytic (Bignami et al., 1973) glial fibrillary acidic proteins (Fig. 9; Morrison, 1983). The levels of all tubulin mRNAs are low, but as in other brain areas in the adult the $\beta_2$-subunit predominates. In addition, the levels of most mRNAs prominent in other

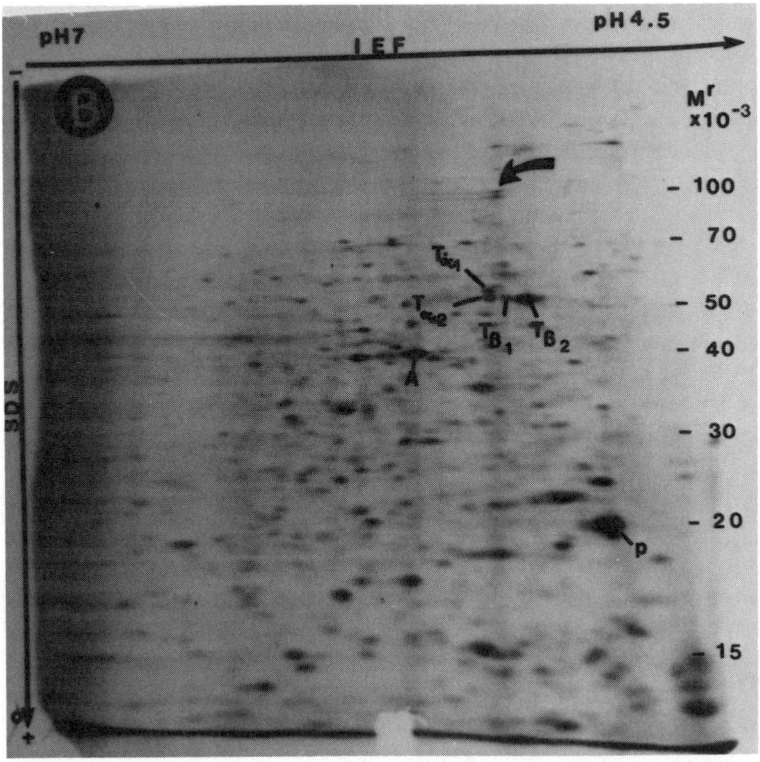


**Figure 7 (continued)**

brain areas are reduced. Since protein P was not synthesized at detectable levels, we conclude that the mRNA encoding this protein is uniquely neuronal. In contrast, several proteins are synthesized exclusively by the optic nerve mRNAs (indicated by arrows in Fig. 9) suggesting that these mRNAS are predominantly non-neuronal.

These results are the first proof at the mRNA level of transcriptional diversity in a particular subpopulation of mammalian brain cells. Construction of a recombinant cDNA library from this mRNA population would allow further characterization of the abundant and medium-abundant mRNAs synthesized by these cells (Norgard et al., 1980: Sutcliffe et al., 1983). Moreover, it should be relatively easy to isolate from such a library clones for non-neuronal mRNAs such as those encoding the glial fibrillary acidic protein.

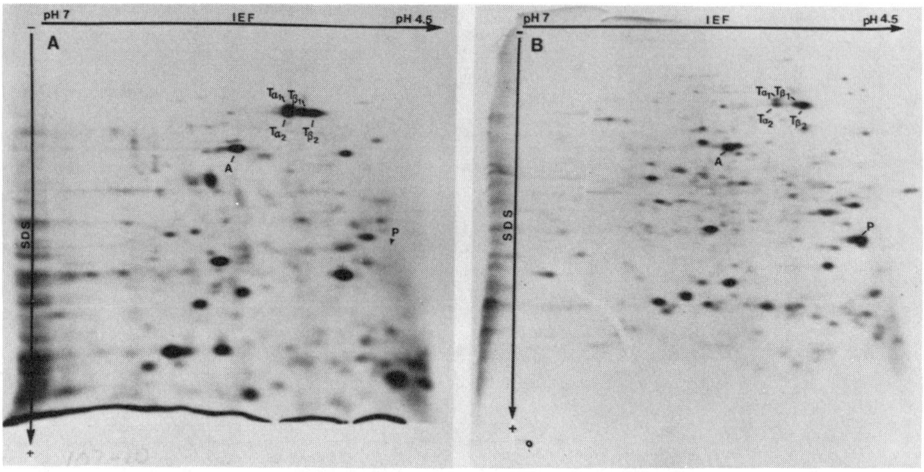


**Figure 8.** Two-dimensional gel electrophoretograms of *in vitro* translation products of poly(A+) RNA isolated from human fetal cortex (A) and human adult cortex (B). Postmortem fetal cortex was obtained through the Department of Obstetrics and Gynecology, Southwestern Medical School. Consent forms and protocols were approved by the Human Research Review Committee. Postmortem adult cortex was obtained through the medical examiner's office. Total cellular RNAs were obtained by homogenization of tissue in guanidinium chloride followed by pelleting the RNA through a cesium chloride cushion. After a final guanidinium chloride extraction total RNAs were ethanol-precipitated and the poly (A+) RNAs were isolated by two consecutive passages over oligo(dT)-cellulose. Each RNA fraction was translated at subsaturating RNA inputs in an *in vitro* wheat germ protein-synthesizing system in the presence of 25 $\mu$Ci of [$^{35}$S]-methionine. Two-dimensional gel electrophoresis was carried out according to O'Farrell. 350,000 trichloroacetic-acid-precipitable counts in 45 $\mu$L of wheat germ cell-free system were loaded on each gel. After electrophoresis gels were fluorographed and exposed to Kodak XR2 film for 7 days at $-70$°C. A = actin; $T_{\alpha_1}$, $T_{\alpha_2}$, $T_{\beta_1}$, $T_{\beta_2}$ = $\alpha_1$-, $\alpha_2$-, $\beta_1$-, and $\beta_2$-tubulin subunits, respectively; P = protein P. (From Morrison et al., 1983.) With permission from *J. Neurogenetics*. Copyright: Elsevier.

## 9. ALTERED mRNA LEVELS IN JOSEPH DISEASE

Joseph disease is an autosomal dominant, spinocerebellar degeneration of unknown etiology (Rosenberg et al., 1976). Attempts have been made to identify disease-specific protein changes in Joseph disease and other neurogenetic diseases by two-dimensional gel electrophoresis of abundant brain proteins (Comings, 1982; Rosenberg et al., 1981). The steady-state levels of several cerebellar proteins are higher in Joseph disease than control brain. However, increases in some of these proteins may be a consequence of

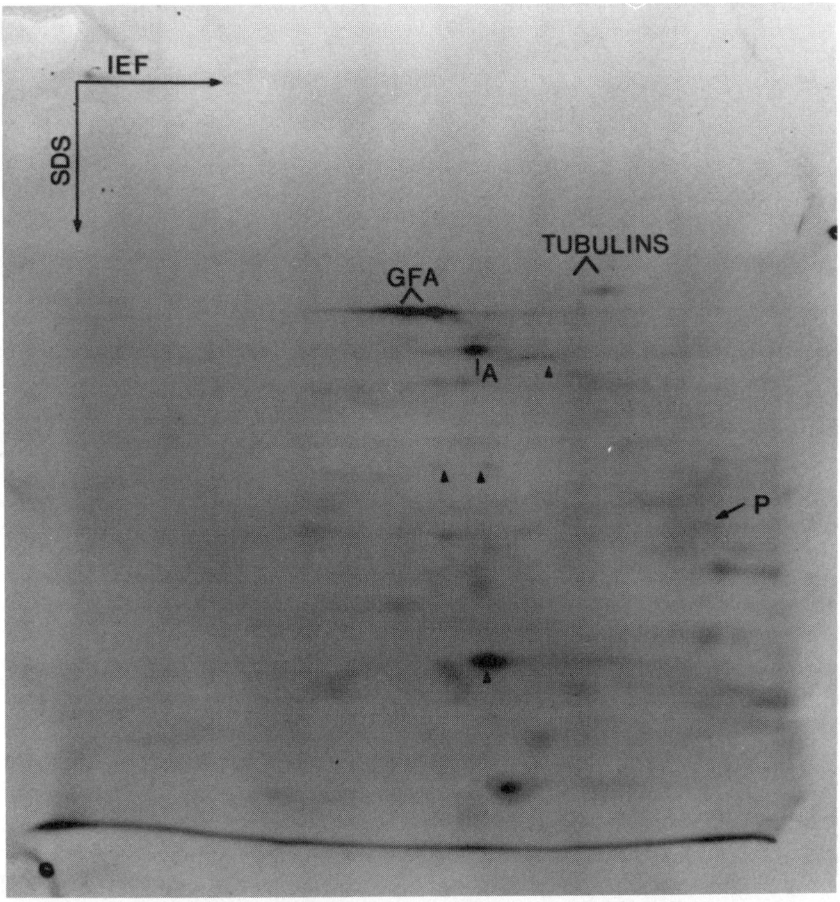


**Figure 9.** Two-dimensional gel electrophoretogram of proteins synthesized *in vitro* by mRNAs isolated from human optic nerve. 900,000 trichloroacetic-acid-precipitable counts were loaded on the gel. It was stained, fluorographed, dried, and exposed to X-ray film for 4 days. A = actin, GFA = glial fibrillary acidic protein, P = protein P.

post-translational protein modification or degradation. Direct comparison of *in vitro* translation products discriminates between post-translational alterations and those caused by changes in the levels of abundant mRNAs. For example, specific low-molecular-weight proteins which are present at relatively high levels in diseased tissue are not found among the *in vitro* translation products of either control or Joseph mRNAs, suggesting that they are the products of protein degradation (Fig. 10A and B; Morrison and Rosen-

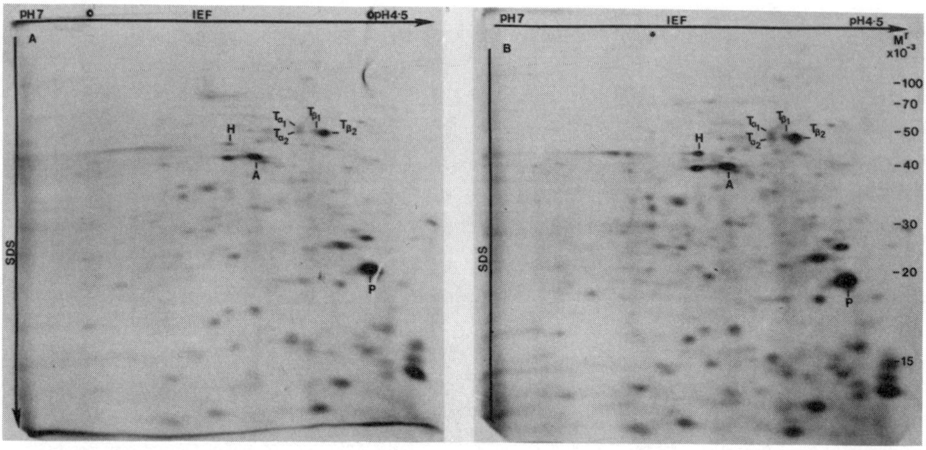


**Figure 10.** Two-dimensional gel electrophoretograms of proteins synthesized *in vitro* by mRNAs isolated from control (A) and Joseph disease type I cerebellum (B). 350,000 trichloroacetic-acid-precipitable counts were loaded on each gel and the fluorographed gels were exposed to X-ray film for 10 days. A = actin, H = glial fibrillary acid protein; $T_{\alpha_1}$ and $T_{\alpha_2}$ = $\alpha_1$- and $\alpha_2$-tubulin; $T_{\beta_1}$ and $T_{\beta_2}$ = $\beta_1$- and $\beta_2$-tubulin. (From Morrison and Rosenberg, 1983.) With permission from *Annals Neurology*. Copyright: Little Brown.

berg, 1983). In contrast, the increase in steady-state levels of glial fibrillary acidic protein in Joseph disease cerebellum (Rosenberg et al., 1981) is reflected in an increase in the levels of its mRNA.

The alterations in mRNA levels for glial fibrillary acidic protein are most probably secondary to cerebellar gliosis rather than being causal to the disease process. However, Comings (1982) has identified a possible polymorphism in a protein of low molecular weight in Joseph disease brain which may be more specific to this disease. Isolation of poly(A+) RNAs from Joseph cerebella and their translation into proteins of higher specific activity than those synthesized *in vitro* from total RNAs (Morrison et al., 1979) should allow us to determine whether this polymorphism can be detected among the *in vitro* translation products. If, as is likely, the genetic defect in Joseph disease cannot be identified by this approach, it will be necessary to identify the chromosomal region containing the mutation by the analysis of linkage relationships between polymorphic markers of known genomic location and the Joseph disease locus (Botstein et al., 1980).

## 10. CONSTRUCTION OF RECOMBINANT DNA LIBRARIES CONTAINING SEQUENCES EXPRESSED IN BRAIN

The techniques of recombinant DNA technology have been used in several ways to analyze gene expression in the brain. Specific sequences encoding several brain peptides, including those for pro-opiomelanocortin (Nakanishi et al., 1978: Roberts et al., 1979), preprolactin and preprogrowth hormone (Gubbins et al., 1979), the arginine-vasopressin-neurophysin II precursor (Land et al., 1982), proenkephalin (Comb et al., 1982), and somatostatin (Naylor et al., 1983), have been cloned and in some cases sequenced (Nakaishi et al., 1979).

The sequences encoding two mammalian neurotransmitter enzymes, tyrosine hydroxylase and dopamine $\beta$-hydroxylase, have also been cloned as have those encoding the electric ray acetylcholine receptor subunits and myelin basic protein (Lamouroux et al., 1982; Noda et al., 1983; O'Malley et al., 1983; Zeller et al., 1983). With the exception of pro-opiomelanocortin (Chen et al., 1983; Roberts et al., 1982), these cloned sequences have not yet been used to analyze their corresponding mRNA levels in brain.

Analysis of recombinant DNA clones picked at random from an adult rat brain cDNA library of 1629 clones (Milner and Sutcliffe, 1983; Sutcliffe et al., 1982, 1983) is defining general aspects of transcription in brain. By Northern analysis (Thomas, 1980) with different cloned sequences, it was shown that many relatively abundant poly(A+) RNAs are brain-specific (Milner and Sutcliffe, 1983). However, this method is not sensitive enough to detect these mRNAs if they are transcribed at very low levels in non-neuronal tissue. The clones corresponding to several of these mRNAs have been sequenced, short peptides they encode have been synthesized, and specific antibodies have been produced and used to determine the cell types within brain in which these proteins are localized (Sutcliffe et al., 1983).

In order to study not only the more abundant brain mRNAs but also those which are expressed at much lower levels, we have refined the conventional dG:dC cloning procedure of Villa-Komaroff et al. (1978) so as to efficiently construct large cDNA libraries from small amounts of template poly(A+) RNA from mouse neuroblastoma, rat cerebellum, and adult human cortex (Michael et al., 1982: Morrison, in press; Sparkman et al., unpublished results). We find that first-strand cDNAs can be synthesized (Buell et al., 1978; Land et al., 1981) as efficiently (20–40%) with adult human cortex poly(A+) RNAs as templates as with rat brain or mouse neuroblastoma templates. Moreover, denaturing gel electrophoresis shows that the cDNAs synthesized from each of these templates range in size from 500 to 10,000 nucleotide pairs with an average size of 1800 base pairs (Fig. 11; Morrison, 1983; Morrison and Sparkman, 1984; Sparkman et al., unpublished results). Therefore, the poly(A+) RNAs isolated from postmortem human brain are of the expected size range (i.e., not appreciably degraded).

In order to eliminate loss of 5′ mRNA sequences, we add 20–30 dA residues to the first cDNA strand using terminal transferase (TdT) (Deng and Wu, 1981) and prime second-strand reverse transcription with oligo(dT) (Land et al., 1981). By passing the tailed cDNA over a spun column (Mani-

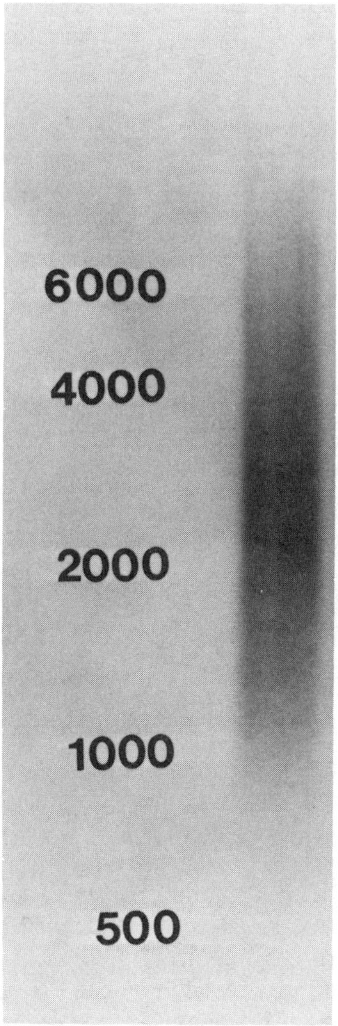


**Figure 11.** Denaturing agarose gel electrophoresis of [$^{32}$P] cDNAs reverse-transcribed from mouse neuroblastoma poly(A+) RNAs.

atis et al., 1982), equilibrated in water, we find that the percentage of double stranding is routinely better than 90%. Use of this column between each step of the procedure leads to DNA recoveries of greater than 90% at each stage (Sparkman et al., unpublished results).

The double-stranded DNA is size-fractionated on a column of Sepharose CL-4B and the kinetics of tailing with dC residues is followed for each fraction. The times for optimum transformation are determined by annealing aliquots of each time point of the dC-tailing reaction to dG-tailed pBR322 and using the recombinant plasmids to transform $CaCl_2$-treated *E. coli* K12 RR1 (Dagert and Ehrlich, 1979). The high-molecular-weight fraction from neuroblastoma gives a transformation efficiency of 2500 colonies/ng double-stranded DNA and the low-molecular-weight fraction gives a transformation efficiency of 5000 colonies/ng double-stranded DNA. Under these conditions 95% of the clones are ampicillin sensitive, indicating the presence of inserts in the *Pst 1* site of pBR322.

Although our transformation efficiencies were already high, we reasoned that the efficient tailing of both 3′ ends of the double-stranded DNA molecules might be hindered by possible recessing of some second-strand 3′ ends (Michelson and Orkin, 1982) thus lowering transformation efficiencies. Indeed, blunt ending the size-fractionated double-stranded DNAs by mild S1 nuclease digestion increases the transformation efficiencies several-fold without decreasing the average insert size (Sparkman et al., unpublished results).

Using this methodology we have now constructed a library containing approximately 100,000 clones containing sequences complementary to mouse neuroblastoma poly(A+) RNAs and are in the process of setting up libraries of similar size for rat cerebellum and adult human cortex. A large proportion of the libraries will be stored on nitrocellulose filters (Hanahan and Meselson, 1980) for ease of duplication and screening. Importantly, screening the neuroblastoma library with the chicken actin as well as $\alpha$- and $\beta$-tubulin recombinant probes shows that recombinants containing these sequences can be recovered by screening as few as several hundred colonies. Thus, these libraries should contain sequences encoding even relatively rare brain or neuroblastoma mRNAs.

With the high efficiencies of first- and second-strand synthesis and with the high DNA recoveries at each step of the procedure, we can now construct libraries containing $1 \times 10^6$ colonies from less than 1 $\mu$g of poly(A+) RNA. This now makes it feasible to construct cDNA libraries from sequences transcribed in discrete brain regions or in purified cell populations, thus enabling us to more easily isolate sequences unique to these RNA populations. cDNA libraries of this size can also be used to determine the sequence complexity of the template RNAs (Norgard et al., 1980; Van Ness and Hahn, 1980; Woods et al., 1980) and could enable us to directly determine whether the transcriptional diversity of a single neuronal population is indeed higher than that of non-neuronal cells (Kaplan and Finch, 1982). Moreover, sequences from the brain libraries can now be used to analyze their cellular distribution using the *in situ* hybridization techniques described in Section 11.

## 11. ANALYSIS OF mRNAs IN INDIVIDUAL CELLS BY *IN SITU* HYBRIDIZATION

Studies on the developmental regulation of abundant mRNAs in whole brain and/or specific brain areas demonstrate that the heterogeneity of cell types in brain preclude the detection of subtle changes in mRNA levels that likely accompany developmental events in individual cells and cell types. For example, since neonatal rat cerebellum is primarily committed to proliferation whereas the cerebral hemispheres are primarily committed postnatally to neuronal differentiation (Blue and Parnavelas, in press), it is surprising that the levels of many *in vitro* translation products of cerebellar RNAs are remarkably similar to those of cerebral hemisphere RNAs (Morrison and Griffin, 1984; Morrison et al., 1981, 1983).

Although we have identified a number of specific mRNAs which are developmentally regulated in different brain regions (Morrison et al., 1981, 1983), it is not possible to identify precisely the cell populations in which these mRNA changes take place. However, using the technique of *in situ* hybridization, precise localization and quantitation of specific mRNAs within individual cell populations is possible. For example, an mRNA encoding pro-opiomelanocortin has been detected in individual pituitary cell populations by *in situ* hybridization (Roberts et al., 1982). Again, using *in situ* hybridization, cell differentiation in tissues other than brain has been shown to be associated with the induction of a number of specific mRNA sequences in defined cell populations (Hafen et al., 1983; Harrison et al., 1974; Jakowlew et al., 1982; Lynn et al., 1983; MacDonald et al., 1982; Singer and Ward, 1982).

Hybridization specificity has been demonstrated in several systems (Brahic and Haase, 1978; MacDonald et al., 1982; Szabo et al., 1977). Furthermore, using saturating levels of radiolabeled probe and appropriate controls, *in situ* hybridization is quantitative (Angerer and Angerer, 1981; Brahic and Haase, 1978; Capco and Jeffery, 1978; Haase et al., 1981, 1982; Jeffery, 1982; Szabo et al., 1977; Venezky et al., 1981). Furthermore, although hybridization of total RNAs isolated from whole brain to a specific DNA probe is sometimes too insensitive to detect the levels of specific mRNAs, *in situ* hybridization has been used to detect as few as 1–10 copies of a specific mRNA present in a small population of cells (Haase et al., 1981). Such studies demonstrate the power of *in situ* hybridization to detect changes in the levels of relatively rare, developmentally regulated mRNAs present in select brain cell populations, perhaps in only a few cells (Beckmann et al., 1981; Kaplan and Finch, 1982).

It is as yet unknown what transcriptional changes are necessary for select brain cell populations to perform the tasks associated with a particular developmental stage. Analysis of brain homogenates indicates that cells performing developmental tasks correspondingly alter the levels of specific mRNAs encoding proteins necessary for these tasks. For example, the levels of the abundant mRNAs encoding the $\alpha$- and $\beta$-tubulin subunits are developmentally regulated in rat brain, being present at high levels during fetal and postnatal brain growth spurts and at relatively low levels in the adult (Gozes et al., 1980, Marotta et al., 1979; Morrison et al., 1981, 1984; see Sections 2, 3, and 7). We therefore postulated that cerebellar EGL cells engaged either in mitosis or in preparation for migration would have relatively higher levels of the tubulin mRNAs than would non-mitotic, postmigratory cells such as Purkinje cells and granule cells in the IGL.

By *in situ* hybridization to the corresponding DNA probes, we compared the relative levels of total poly(A+) mRNAs and of the specific mRNAs encoding the $\beta$-tubulins in the various cerebellar cell types of the 10-day-old rat (Griffin et al., 1983; 1984). The distribution of the mRNAs was estimated by counting autoradiographic grains over the various cell types. Background counts on sections either treated with RNase prior to hybridization with cDNA or hybridized with nick-translated plasmid DNA were subtracted to account for non-specific binding (Griffin et al., 1984). Because background grains are uniformly low (further demonstrated by the paucity of grains over the glomerulus in Fig. 12C) and grain counts over individual cells of a population are highly reproducible with standard errors (SEM) averaging less than 10%, it is possible to quantitate relative levels of mRNAs and thus compare their levels in different cell types (Griffin et al., unpublished results).

The large Purkinje cells contain the greatest amount of poly(A+) RNA, the levels decreasing with cell size to the granule cells in the IGL. However, cell size was not the only determinant of poly(A+) RNA levels since migrating granule cells had only half the amount of poly(A+) RNA as did similar size granule cells in either the EGL or the IGL (Fig. 12).

When compared to the total mRNA distribution, there is an uneven distribution of the $\beta$-tubulin mRNAs in the various cerebellar cell types in 10-day-old rat. Although the amount of total poly(A+) mRNA is similar in premigratory cells in the mantle zone of the EGL, germinal cells in the matrix zone of the EGL, and granule cells in the IGL, the $\beta$-tubulin mRNA levels are twice as high in the EGL cells as in the IGL cells. As with total poly(A+) RNA, the amount of $\beta$-tubulin mRNA is significantly less in migrating cells than in granule cells in either EGL or IGL (Griffin et al., 1984). The relatively greater abundance of $\beta$-tubulin mRNA in the premigratory mantle zone of the EGL compared to the germinal cells of the matrix zone may reflect the necessity for tubulin synthesis prior to migration. The paucity of tubulin and total mRNA in migrating cells may be indicative of a commitment of cellular function to migration rather than to protein synthesis. These results indicate that the relatively high levels of $\beta$-tubulin mRNAs in the neonatal cerebellum are at least in part due to their higher levels in the EGL. A similar analysis of $\beta$-tubulin mRNA levels in the adult cerebellum would determine whether the overall content of these mRNAs also decreases developmentally in the other cell types.

Since the $\beta$-tubulin probe used in these studies contains all of the coding

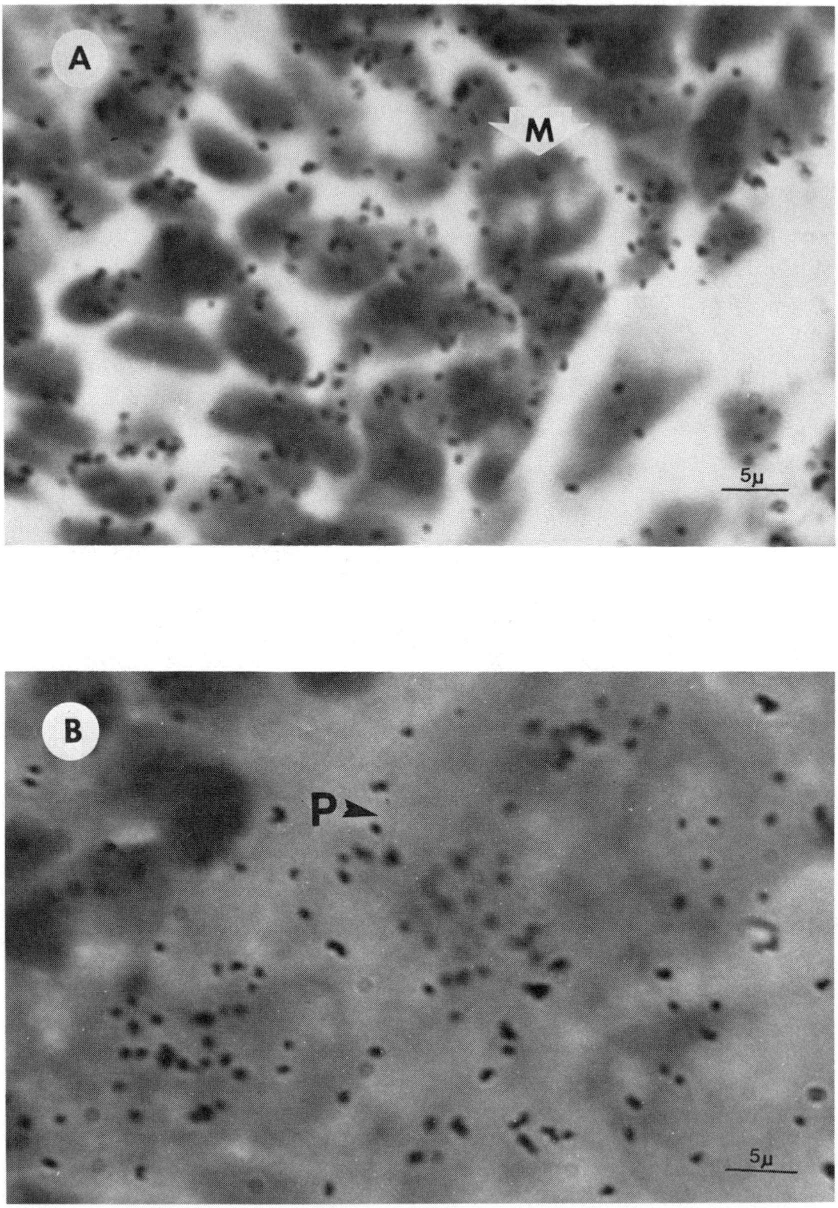


**Figure 12.** *In situ* hybridization to specific cell types. Photomicrographs at 1000× of (A) EGL showing mitotic figures (m = metaphase); (B) Purkinje (P) cells; and (C) IGL cells surrounding mossy fiber glomeruli (glom). These photomicrographs are of 8-μm sections with [$^3$H] cDNAs *in situ* hybridizated to mRNAs.

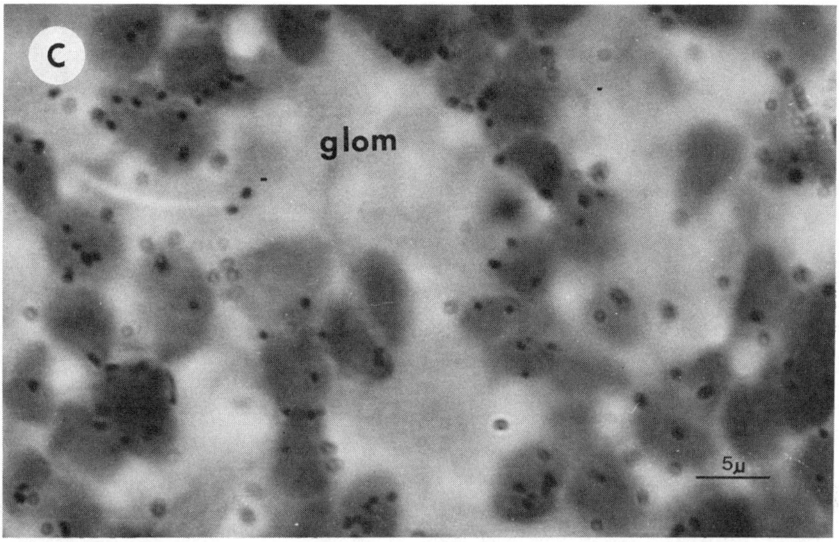


**Figure 12 (continued)**

sequence (Cleveland et al., 1980), it will hybridize to both the $\beta_1$- and $\beta_2$-tubulin mRNAs. However, *in situ* hybridization with recombinant DNA probes specific for each tubulin isotype will reveal any differential distribution of these mRNAs in different cell types and identify the cell types with the dramatic changes in $\beta_1$-tubulin mRNA levels during development.

Because neurons synthesize proteins that are transported great distances along neuronal processes, it is important to identify cells that contain both a specific protein and its encoding mRNA. A neuronal cell population inferentially associated with a neuropeptide has been identified by histochemistry, and on an adjacent section the mRNA encoding the peptide has been localized by *in situ* hybridization to morphologically similar cell types (Pochet et al., 1981). However, combining immunohistochemistry and *in situ* hybridization on the same histological section would unequivocally distinguish between cells that merely store a protein and those with the capacity to synthesize the protein. In this way cells of origin of a protein can be positively identified in brain areas that might be far removed from cell terminals and cells that take up the protein.

We have successfully combined the immunohistochemistry and *in situ* hybridization techniques in sections of cerebellum from 10-day-old rat. Using an antibody to glial fibrillary acidic protein (kindly given us by Dr. Gaj Nilaver) and cDNA reverse-transcribed from poly(A+) RNA isolated from 10-day cerebellum, we show (Fig. 13) that glial cells in white matter can be

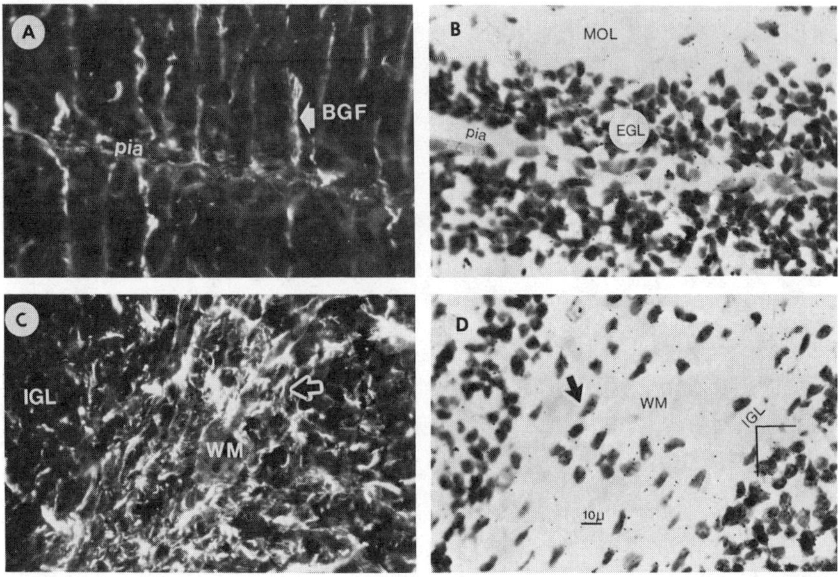


**Figure 13.** These photomicrographs (1000×) are of a midsagittal section of the secondary fissure of the cerebellum from a 10-day-old Fischer rat. A and B depict an area surrounding the pial surface of the fissure including the external granular layer (EGL) on either side of the fissure and the molecular layer on the upper side. In A the Bergmann glial fibers (BGF) and the *Glia limitans* at the pial surface of the fissure (pia) were fluorescently labeled by reacting fixed frozen 8-$\mu$m sections with an antibody to glial fibrillary acidic protein and fluoresceinated immunoglobulin. In B the same section shown in A was incubated with [$^3$H] cDNA reverse-transcribed from mRNAs isolated from 10-day-old rat cerebellum. The same area is shown in A and B. C and D were stained similarly to A and B, but the area shown is in the white matter near the secondary fissure. In C the fluorescence is associated with fibrous astrocytes in the white matter (WM) and with velate (less fibrous) astrocytes in the internal granular layer (IGL). In D [$^3$H] DNA labeling is associated with cells in WM and IGL. A–D show areas within a single section of cerebellum to illustrate the clarity of immunohistochemical and hybridization histochemical localization of protein and mRNAs on the same section. Black arrows denote two çells that appear fluorescent in C and have grains above them in D. (From Griffin et al., 1983.) With permission from *Brain Research Bulletin*. Copyright: Ankho International, Inc.

identified by immunofluorescence of the antigen–antibody complex and that the same cells also have autoradiographic grains, indicative of hybridization of cDNA to poly(A+) RNA (Griffin et al., 1983).

Combining these techniques allows us to address a number of other interesting developmental questions, for example, the role of actin in migration. Sadler et al. (1982) have shown an uneven distribution of actin over motile cells during neurulation. Similarly, we have shown actin-like immunoreactivity over the premigratory cells in the EGL of 10-day rat cerebellum (Fig. 14). (The rabbit–anti–chicken non-muscle actin antibody used was

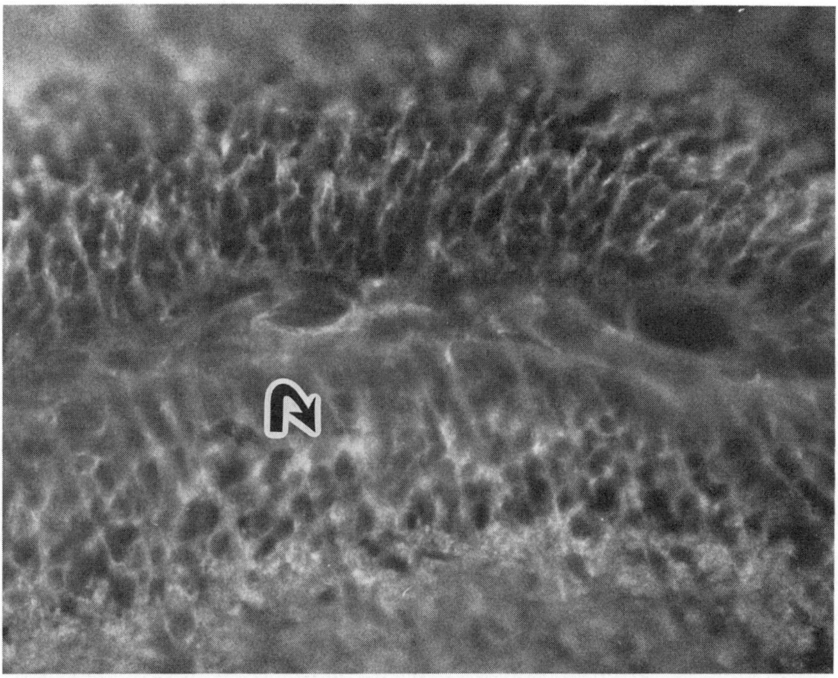

**Figure 14.** Actin-like immunoreactivity in cerebellum from 10-day-old rat. Rabbit–anti-chicken non-muscle actin antiserum diluted 1:400 was applied to 10-$\mu$m cryostat sections from formalin-fixed cerebellum frozen in isopentane and liquid nitrogen, incubated overnight at 4°C, incubated 30 min with fluoresceinated (FITC) anti-rabbit immunoglobulin, and stained with toluidine blue. Note the intensity of fluorescence associated with the premigratory layer of the EGL (arrow).

kindly given us by Dr. A. Bretscher.) Using *in situ* hybridization we can now determine whether such localization of actin is related to correspondingly higher levels of actin mRNA in the premigratory cells and discern the time course of the acquisition of this mRNA by cells as they "prepare" for a specific developmental task.

## 12. CONCLUDING REMARKS

Our characterization of abundant mRNAs by analysis of their *in vitro* translation products has provided detailed information on the changes in steady-state mRNA levels taking place during brain and neuroblastoma differentiation as well as on more general aspects of mRNA structure and utilization in the nervous system. Quantitation of specific mRNAs using radiolabeled recombinant DNA probes has confirmed that the measurements of translationally active tubulin and actin mRNAs by this method are indeed an accurate indication of their steady-state levels.

In order to analyze more completely transcriptional patterns in brain and to determine the control mechanisms operative at specific developmental stages, recombinant cDNA libraries are now essential. Such a library has already been established from adult rat brain poly(A+) RNAs (Sutcliffe et al., 1983) and used to determine the relative abundance of specific mRNAs and to show that many such sequences are unique to brain. Our cDNA library from rat cerebellum will be similarly useful in characterizing mRNA populations in the adult rat cerebellum and in identifying sequences that are not only brain-specific or region-specific but also developmentally regulated. Comparison of the genomic sequences from which such developmentally regulated mRNAs are transcribed will begin to elucidate the transcriptional and post-transcriptional controls operative in brain development.

Construction of recombinant DNA libraries will be essential not only for characterizing the brain poly(A+) mRNA population but also the poly(A−) mRNA class. The characteristics of the poly(A−) mRNAs are tantalizing in that they imply an important functional role in the mature brain. Analysis of this class may well lead to unexpected insights into metabolic regulation in brain. A functional role for individual members of this poly(A−) class has not, except for the histone mRNAs, been confirmed. Sequence analysis of cloned members of this population and analysis of any translation products would begin to elucidate their function(s).

There are now two ways to approach the fundamental question of mRNA transcription in individual cell types in brain. First, purified populations of specific brain cells can be isolated, and using our efficient cDNA cloning procedure, cDNA libraries can be generated from cell isolates, tissue sections, and regional punches. Such cDNA libraries could be used to determine whether the known high level of transcription in brain is due to diversity within each cell type and/or whether it merely reflects the large number of brain cell types. Such a result would have important implications for the specialization necessary for efficient functioning of brain cells.

Monoclonal antibodies show directly (McKay et al., 1981) and RNA complexity studies show indirectly (Beckmann et al., 1981; Schrier et al., 1978) that some mRNAs are present in only one or a few brain cell types. Therefore, many of these cell-specific mRNAs may be present at undetectable levels in whole brain homogenates. Clones containing such sequences should thus be more easily isolated from cell-specific cDNA libraries.

The second technique for localizing and subsequent quantitation of mRNAs in individual brain cell types is that of *in situ* hybridization. Biotinylated DNA probes from recombinant cDNA libraries could be rapidly screened for sequences expressed in particular cell types (Brigati et al., 1983; Singer and Ward, 1982). Such *in situ* hybridization should document with greater assurance the presence or absence of a specific mRNA encoding a protein in a particular cell type than would the presently available method of immunohistochemical analysis. This technique can also be used to demonstrate the extent of cellular specialization in brain and also the degree to which neurons defined morphologically and physiologically are biochemically similar (Chan-Palay et al., 1981).

Of equal importance, *in situ* hybridization using radiolabeled DNA probes can quantitate mRNA levels in individual cells. Indeed, we have for the first time documented the relative levels of mRNAs, those encoding the $\beta$-tubulins, in individual populations of brain cells and have shown that $\beta$-tubulin mRNA levels are higher in the dividing and premigratory cells of the developing cerebellum (Griffin et al., 1984). With the availability of suitable probes we can precisely delineate the stage in development and the cell types in which mRNAs fluctuate developmentally. Similarly, by analysis of brains whose development is modulated by GVHD, we can address the question of whether the levels of mRNAs encoding proteins necessary for cell division fluctuate depending on the level of DNA synthesis. Other cellular functions altered by GVHD can be similarly investigated. Furthermore, it now becomes possible to assess changes in gene expression in cells whose synaptic input is modulated by various means, including X-irradiation, drug or alcohol treatment, and genetic mutations.

The demonstration that representative populations of brain mRNAs survive for considerable periods postmortem (Morrison et al., 1981) is an important breakthrough in the characterization of diseases of the nervous system. We have already analyzed developmental and disease-related changes in abundant mRNA levels in human postmortem tissue (Morrison and Rosenberg, 1983; Morrison et al., 1984), and the construction of recombinant DNA libraries containing sequences expressed in human brain (Sparkman et al., unpublished results) will now allow us to analyze changes in less abundant mRNAs. As undegraded mRNAs can be isolated from postmortem brains stored frozen (Morrison et al., 1981), analysis of mRNA levels in autopsy material, for example that stored in brain banks, now becomes feasible. Recent positive *in situ* hybridization results on formalin-fixed tissue (Brigati et al., 1983; Griffin et al., 1983; 1984) suggest that this too may be a possible source of material for such analyses.

Thus, the technology is now available to characterize neuropathology at the cellular level. Analysis of mRNA changes in diseased brain are of obvious relevance in documenting gross pathological changes in transcription patterns. *In situ* hybridization and immunohistochemistry can now be used, perhaps even in combination with computer reconstruction (German et al., in press), to investigate more critically the specific cell losses so characteristic of diseases such as Huntington's, Parkinson's, amyotrophic lateral sclerosis, multiple sclerosis, and Alzheimer's. *In situ* hybridization of probes to mRNAs encoding specific neurotransmitter enzymes and abundant "housekeeping" proteins can now be used to determine whether the remaining cells in affected brain areas are transcriptionally normal. Furthermore, this technique can also be used to document the transcriptional changes in cell types not presently identified as compromised and thus will pinpoint more precisely the initial cell targets of disease.

A large proportion of inherited diseases in man have prominent neurological features, but due to the dependence of neural function on metabolic homeostasis, many genes affecting neural function may not themselves be brain-specific. Examples are hypoxanthine phosphoribosyltransferase, the enzyme deficient in Lesch-Nyhan syndrome, and phenylalanine hydroxylase, the enzyme deficient in phenylketonuria. The cDNAs complementary to these mRNAs have been cloned (Jolly et al., 1983; Nussbaum et al., 1983; Robson et al., 1982), thus making possible identification of the mutations that initiate the cascade of events in these diseases. Furthermore, recombinant DNA probes isolated from genomic or cDNA libraries (Shows et al., 1982) are now being used to identify polymorphisms in genomic DNA. These polymorphisms are being used in family studies to localize the mutant genes responsible for neurogenetic diseases of unknown etiology to specific chromosome regions. Recently this technique has been used to show that the mutant gene associated with Duchenne muscular dystrophy lies on the short arm of the X chromosome (Murray et al., 1982). The recombinant DNA clones encoding brain-specific proteins including the neurotransmitter enzymes (Lamouroux et al., 1982; O'Malley et al., 1983) can now be used to map their genes and thus to define possible relationships to neurogenetic disease.

## ACKNOWLEDGMENTS

This work was supported by Grants HD 14886 and GM 27214 to Marcelle R. Morrison, by Grant AI 14663 to W. Sue T. Griffin, and by the Leland Fikes Foundation. We acknowledge the invaluable contributions of D. Sparkman, S. Pardue, J. Snider, D. Croall, M. Alejos, R. Brodeur, K. Hall, E. Crom, and L. Williams to these projects, the helpful comments of F. Baskin, R. Rosenberg, and D. Woodward, and the excellent secretarial assistance of C. Beisert.

## REFERENCES


Addison, W. H. F., *J. Comp. Neurol.* **21,** 459 (1911).

Althaus, H. H. and V. Neuhoff, in I. R. Brown, Ed., *Molecular Approaches to Neurobiology,* Academic, New York, p. 342 (1982).

Altman, J., *J. Comp. Neurol.* **145,** 353 (1972).

Altman, J. and W. J. Anderson, *Exp. Neurol.* **30,** 492 (1971).

Amano, T., E. Richelson, and M. Nirenberg, *Proc. Natl. Acad. Sci. USA* **69,** 258 (1972).

Angerer, L. M. and R. C. Angerer, *Nucleic Acids Res.* **9,** 2819 (1981).

Bantle, J. A. and W. E. Hahn, *Cell* **8,** 139 (1976).

Beale, E. G., J. L. Hartley, and D. K. Granner, *J. Biol. Chem.* **257,** 2022 (1982).

Beckmann, S. L., D. M. Chikaraishi, S. S. Deeb, and N. Sueoka, *Biochemistry* **20,** 2684 (1981).

Benoff, S. and B. Nadal-Ginard, *Biochemistry* **18,** 494 (1979).

Bernstein, S. L., A. E. Gioio, and B. B. Kaplan, *J. Neurogenet.* **1,** 1 (1983).

Bigbee, J. W. and L. F. Eng, *J. Neurochem.* **38,** 130 (1982).

Bignami, A., L. F. Eng, D. Dahl, and C. T. Uyeda, *Brain Res.* **43,** 429 (1973).

Billingham, R., *Harvey Lecture Series,* Vol. 62, Academic, New York, p. 21 (1968).

Blue, M. E. and J. G. Parnavelas, *J. Neurocytol.* **12** (in press).

Botstein, D., R. L. White, M. Skolnick, and R. W. Davis, *Am. J. Hum. Genet.* **32,** 314 (1980).

Brahic, M. and A. T. Haase, *Proc. Natl. Acad. Sci. USA* **75,** 6125 (1978).

Brigati, D. J., D. Myerson, J. J. Leary, B. Spalholz, S. Z. Travis, C. K. Y. Fong, G. D. Hsiung, and D. C. Ward, *Virology* **126,** 32 (1983).

Brown, I. R., *Biochem. Biophys. Res. Comm.* **84,** 285 (1978).

Brown, I. R., *J. Neurochem.* **40,** 1490 (1983).

Brown, I. R. and P. Greenwood, in I. R. Brown, Ed., *Molecular Approaches to Neurobiology,* Academic, New York, p. 41 (1982).

Buell, G. N., M. P. Wickens, F. Payvar, and R. T. Schimke, *J. Biol. Chem.* **253,** 2471 (1978).

Campagnoni, C. W., G. D. Carey, and A. T. Campagnoni, *Arch. Biochem. Biophys.* **190,** 118 (1978).

Capco, D. G. and H. Jäckle, *Dev. Biol.* **94,** 41 (1982).

Capco, D. G. and W. R. Jeffery, *Dev. Biol.* **67,** 137 (1978).

Chan-Palay, V., G. Nilaver, S. L. Palay, M. C. Beinfeld, E. A. Zimmerman, J. Y. Wu, and T. L. O'Donohue, *Proc. Natl. Acad. Sci. USA* **78,** 7787 (1981).

Chaudhari, N. and W. E. Hahn, *Science* **220,** 924 (1983).

Chen, C. L. C., F. T. Dionne, and J. L. Roberts, *Proc. Natl. Acad. Sci. USA* **80,** 2211 (1983).

Chikaraishi, D. M., *Biochemistry* **18,** 3249 (1979).

Chikaraishi, D. M., S. S. Deeb, and N. Sueoka, *Cell* **13,** 111 (1978).

Clark, B. R., M. E. Weichsel, Jr., and R. E. Poland, *Biol. Neonate* **34,** 209 (1978).

Cleveland, D. W., M. A. Lopata, R. J. MacDonald, N. J. Cowan, W. J. Rutter, and M. W. Kirschner, *Cell* **20,** 95 (1980).

Cleveland, D. W., M. A. Lopata, P. Sherline, and M. W. Kirschner, *Cell* **25,** 537 (1981).

Comb, M., P. H. Seeburg, J. Adelman, L. Eiden, and E. Herbert, *Nature* **295,** 663 (1982).
Comings, D. E., *Clin. Chem.* **28,** 798 (1982).
Comings, D. E., N. G. Carraway, and A. Pekkula-Flagan, *Clin. Chem.* **28,** 790 (1982).
Cosgrove, J. W. and I. R. Brown, *J. Neurochem.* **36,** 1026 (1981).
Cosgrove, J. W., J. J. Heikkila, A. Marks, and I. R. Brown, *J. Neurochem.* **40,** 306 (1983).
Croall, D. E. and M. R. Morrison, *J. Mol. Biol.* **140,** 549 (1980).
Czosnek, H., D. Soifer, and H. M. Wisniewski, *J. Cell Biol.* **85,** 726 (1980).
Dagert, M. and S. D. Ehrlich, *Gene* **6,** 23 (1979).
Darnell, J. E., Jr., *Nature* **297,** 365 (1982).
Deng, G. and R. Wu, *Nucleic Acids Res.* **9,** 4173 (1981).
Eliyahu, D. and H. Soreq, *J. Neurochem,* **38,** 313 (1982).
Elson, C. O., R. W. Reilly, and I. H. Rosenberg, *Gastroenterology* **72,** 886 (1977).
Felsani, A., F. Berthelot, F. Gros, and B. Croizat, *Eur. J. Biochem.* **92,** 569 (1978).
Fisher, M., *Int. Soc. Dev. Neurol. Abst.* **4,** 2 (1983).
Forget, B. G., *Rec. Prog. in Hormone Res.* **38,** 257 (1982).
Gähwiler, B. H., *J. Neurosci. Meth.* **4,** 329 (1981).
German, D. C., D. S. Schlusselberg, and D. J. Woodward, *J. Neurol. Trans.* (in press).
Gilad, G. M. and I. J. Kopin, *J. Neurochem.* **33,** 1195 (1979).
Gilbert, J. M., *Biochem. Biophys. Res. Comm.* **52,** 79 (1973).
Gilbert, J. M. and P. Strocchi, *J. Neurochem.* **40,** 153 (1983).
Gilbert, J. M., B. A. Brown, P. Strocchi, E. D. Bird, and C. A. Marotta, *J. Neurochem.* **36,** 976 (1981).
Gilman, A. G. and M. Nirenberg, *Proc. Natl. Acad. Sci. USA* **68,** 2165 (1971).
Gorski, J., M. R. Morrison, C. G. Merkel, and J. B. Lingrel, *Nature* **253,** 749 (1975).
Gozes, I., A. de Baetselier, and U. Z. Littauer, *Eur. J. Biochem.* **103,** 13 (1980).
Grebe, S. and W. Streilein, *Adv. Immun.* **22,** 119 (1976).
Greene, L. A. and A. S. Tischler, *Proc. Natl. Acad. Sci. USA* **73,** 2424 (1976).
Greengard, P., *Nature* **260,** 101 (1976).
Griffin, W. S. T., D. J. Woodward, and R. Chanda, *J. Neurochem.* **28,** 1269 (1977).
Griffin, W. S. T., J. R. Head, D. J. Woodward, and C. Carrol, *Nature* **275,** 315 (1978a).
Griffin, W. S. T., D. J. Woodward, and R. Chanda, *Brain Res. Bull.* **3,** 369 (1978b).
Griffin, W. S. T., J. R. Head, and M. F. Pacheco, *Brain Res. Bull.* **4,** 313 (1979).
Griffin, W. S. T., M. A. E. Eriksson, E. N. Crom, and J. R. Head, *Brain Res. Bull.* **5,** 673 (1980a).
Griffin, W. S. T., J. R. Head, S. Pardue, and M. R. Morrison, *J. Neurochem.* **35,** 880 (1980b).
Griffin, W. S. T., J. R. Head, and M. R. Morrison, *Transplant. Proc.* **13,** 1237 (1981).
Griffin, W. S. T., E. N. Crom, G. M. Cowan, and J. R. Head, *Proc. Roy. Soc. Lond.* **215,** 289 (1982a).
Griffin, W. S. T., E. N. Crom, and J. R. Head, *Proc. Natl. Acad. Sci. USA* **79,** 4783 (1982b).
Griffin, W. S. T., B. J. Snider, and M. R. Morrison, *J. Neurochem.* **39,** 1412 (1982c).
Griffin, W. S. T., M. Alejos, G. Nilaver, and M. R. Morrison, *Brain Res. Bull.* **10,** 597 (1983).
Griffin, W. S. T., M. Alejos, and M. R. Morrison, *J. Cell. Biochem., Supp.* **8B,** 115 (1984).

Grouse, L. D., B. K. Schrier, C. H. Letendre, M. Y. Zubairi, and P. G. Nelson, *J. Biol. Chem.* **255,** 3871 (1980).

Gubbins, E. J., R. A. Maurer, J. L. Hartley, and J. E. Donelson, *Nucleic Acids Res.* **6,** 915 (1979).

Haase, A. T., P. Ventura, C. J. Gibbs, Jr., and W. W. Tourtellotte, *Science* **212,** 672 (1981).

Haase, A. T., I. Stowring, J. D. Harris, B. Traynor, P. Ventura, R. Peluso, and M. Brahic, *Virology* **119,** 399 (1982).

Hafen, E., M. Levine, R. L. Garber, and W. J. Gehring, *The EMBO J.* **2,** 617 (1983).

Hall, C. and L. Lim, *Biochem. J.* **196,** 327 (1981).

Hall, J. C., R. J. Greenspan, and W. A. Harris, *Genetic Neurobiology,* The MIT Press, Cambridge, MA (1982).

Hall, J. L., L. Dudley, P. R. Dobner, S. A. Lewis, and N. J. Cowan, *Mol. Cell Biol.* **3,** 854 (1983).

Hanahan, D. and M. Meselson, *Gene* **10,** 63 (1980).

Harrison, P. R., D. Conkie, N. Affara, and J. Paul, *J. Cell Biol.* **63,** 402 (1974).

Heizmann, C. W., E. M. Arnold, and C. C. Kuenzle, *J. Biol. Chem.* **255,** 11504 (1980).

Heizmann, C. W., E. M. Arnold, and C. C. Kuenzle, *Eur. J. Biochem.* **127,** 57 (1982).

Jacobson, M., in M. Jacobson, Ed., *Developmental Neurobiology,* Plenum, New York, p. 57 (1978).

Jaegar, A. W. and C. C. Kuenzle, *The EMBO J.* **1,** 811 (1982).

Jakowlew, S. B., P. Khandekar, K. Datta, H. H. Arnold, S. K. Narula, and M. A. Q. Siddiqui, *J. Mol. Biol.* **156,** 673 (1982).

Jeffery, W. R., *J. Cell Biol.* **95,** 1 (1982).

Jolly, D. J., H. Okayama, P. Berg, A. C. Esty, D. Filpula, P. Bohlen, G. G. Johnson, J. E. Shively, T. Hunkapillar, and T. Friedmann, *Proc. Natl. Acad. Sci. USA,* **80,** 477 (1983).

Kaplan, B. B. and C. E. Finch, in I. R. Brown, Ed., *Molecular Approaches to Neurobiology,* Academic, New York, p. 71 (1982).

Kaplan, B. B., B. S. Schachter, H. H. Osterburg, J. S. de Vellis, and C. E. Finch, *Biochemistry* **17,** 5516 (1978).

Lajtha, A. and D. Dunlop, *Life Sci.* **29,** 755 (1981).

Lamouroux, A., N. F. Biguet, D. Samolyk, A. Privat, J. C. Salomon, J. F. Pujol, and J. Mallet, *Proc. Natl. Acad. Sci. USA* **79,** 3881 (1982).

Land, H., M. Grez, H. Hauser, W. Lindenmaier, and G. Schütz, *Nucleic Acids Res.* **9,** 2251 (1981).

Land, H., G. Schütz, H. Schmale, and D. Richter, *Nature* **295,** 299 (1982).

Lee, G. T. Y. and D. L. Engelhardt, *J. Mol. Biol.* **129,** 221 (1979).

Lim, L., J. O. White, C. Hall, W. Berthold, and A. N. Davison, *Biochim. Biophys. Acta* **361,** 241 (1974).

Lodish, H. F., *Nature* **251,** 385 (1974).

Lopata, M. A., J. C. Havercroft, L. T. Chow, and D. W. Cleveland, *Cell* **32,** 713 (1983).

Lynn, D. A., L. M. Angerer, A. M. Bruskin, W. H. Klein, and R. C. Angerer, *Proc. Natl. Acad. Sci. USA* **80,** 2656 (1983).

MacDonald, R. J., G. H. Swift, C. Quinto, W. Swain, R. L. Pictet, W. Nikovits, and W. J. Rutter, *Biochemistry* **21,** 1453 (1982).

Mahony, J. B. and I. R. Brown, *J. Neurochem.* **33,** 1019 (1979).

Maniatis, T., E. F. Fritsch, and J. Sambrook, Eds., *Molecular Cloning: A Laboratory Manual,* Cold Spring Harbor Laboratory, Cold Spring Harbor, NY, p. 466 (1982).

Manser, T. and R. F. Gesteland, *J. Mol. Appl. Genet.* **1,** 117 (1981).

Marotta, C. A., P. Strocchi, and J. M. Gilbert, *J. Neurochem.* **33,** 231 (1979).

Matthees, J. and A. T. Campagnoni, *J. Neurochem.* **35,** 867 (1980).

McKay, R., M. Raff, and L. Reichardt, Eds., *Monoclonal Antibodies Against Neural Antigens,* Cold Spring Harbor Laboratory, Cold Spring Harbor, NY (1981).

McMullen, M. D., P. H. Shaw, and T. E. Martin, *J. Mol. Biol.* **132,** 679 (1979).

Michael, A. J., R. B. Gammon, S. Pardue, D. E. Croall, D. R. Sparkman, W. S. T. Griffin, and M. R. Morrison, *Abst. Soc. Neurosci.* **8,** 704 (1982).

Michelson, A. M. and S. H. Orkin, *J. Biol. Chem.* **257,** 14773 (1982).

Milner, R. J. and J. G. Sutcliffe, *Nucleic Acids Res.* **11,** 5497 (1983).

Morrison, M. R., in Annual Clinical Neurochemistry Course, *Genetic Engineering and Neurological Disease,* American Academy of Neurology, p. 90 (1983).

Morrison, M. R., in H. Soreq, Ed., *Methods in the Neurosciences,* Pergamon (in press)

Morrison, M. R. and W. S. T. Griffin, *Anal. Biochem.* **113,** 318 (1981a).

Morrison, M. R. and W. S. T. Griffin, *Abst. Soc. Neurosci.* **7,** 906 (1981b).

Morrison, M. R. and W. S. T. Griffin, in H. Soreq, Ed., *Methods in the Neurosciences,* Pergamon (in press).

Morrison, M. R, and W. S. T. Griffin, in B. J. Meyer, and S. Kramer, Eds., *Neuronal Communications,* Balkema, Cape Town, South Africa, 107 (1984).

Morrison, M. R. and R. N. Rosenberg, *Ann. Neurol.* **14,** 73 (1983).

Morrison, M. R. and D. R. Sparkman, in B. J. Meyer, and S. Kramer, Eds., *Neuronal Communications,* Balkema, Cape Town, South Africa, 264 (1984).

Morrison, M. R., F. Baskin, and R. N. Rosenberg, *Biochim. Biophys. Acta* **476,** 228 (1977).

Morrison, M. R., R. Brodeur, S. Pardue, F. Baskin, C. L. Hall, and R. N. Rosenberg, *J. Biol. Chem.* **254,** 7675 (1979).

Morrison, M. R., C. L. Hall, S. Pardue, R. Brodeur, F. Baskin, and R. N. Rosenberg, *J. Neurochem.* **34,** 50 (1980a).

Morrison, M. R., S. Pardue, N. Prashad, D. E. Croall, and R. Brodeur, *Eur. J. Biochem.* **106,** 463 (1980b).

Morrison, M. R., S. Pardue, and W.S.T. Griffin, *J. Biol. Chem.* **256,** 3550 (1981).

Morrison, M. R., S. Pardue, and W.S.T. Griffin, *J. Neurogenet.* **1,** 105 (1983).

Mullen, R. J., in S. Subtelny and I. M. Sussex, Eds., *The Clonal Basis of Development,* Thirty-Sixth Symposium of the Society for Developmental Biology, Academic, New York, p. 83 (1978).

Murray, J. M., K. E. Davies, P. S. Harper, L. Meredith, C. R. Mueller, and R. Williamson, *Nature* **300,** 69 (1982).

Nakanishi, S., A. Inoue, T. Kita, S. Numa, A. C. Y. Chang, S. N. Cohen, J. Nunberg, and R.T. Schimke, *Proc. Natl. Acad. Sci. USA* **75,** 6021 (1978).

Nakanishi, S., A. Inoue, T. Kita, M. Nakamura, A. C. Y. Chang, S. N. Cohen, and S. Numa, *Nature* **278,** 423 (1979).

Naylor, S. L., A. Y. Sakaguchi, L. P. Shen, G. I. Bell, W. J. Rutter, and T. B. Shows, *Proc. Natl. Acad. Sci. USA* **80,** 2686 (1983).

Noda, M., H. Takahashi, T. Tanabe, M. Toyosato, S. Kikyotani, T. Hirose, M. Asai, H. Takashima, S. Inayama, T. Miyata, and S. Numa, *Nature* **301,** 251 (1983).

Norgard, M. V., M. J. Tocci, and J. J. Monahan, *J. Biol. Chem.* **255,** 7665 (1980).

Nussbaum, R. L., W. E. Crowder, W. L. Nyhan, and C. T. Caskey, *Proc. Natl. Acad. Sci. USA* **80,** 4035 (1983).

Olmstead, J. B. and G.G. Borisy, *Ann. Rev. Biochem.* **42,** 507 (1973).

O'Malley, K. L., A. Mauron, J. Raese, J. D. Barchas, and L. Kedes, *Proc. Natl. Acad. Sci. USA* **80,** 2161 (1983).

Pochet, R., H. Brocas, G. Vassart, G. Toubeau, S. Seo, S. Refetoff, J. E. Dumont, and J. L. Pasteels, *Brain Res.* **211,** 433 (1981).

Pollack, M. S., D. Kirkpatrick, N. Kapoor, B. Dupont, and R. J. O'Reilly, *N. Engl. J. Med.* **307,** 662 (1982).

Prashad, N. K. and S. Kumar, in B. Clarkson and R. Baserga, Eds., *Control of Proliferation of Animal Cells,* Cold Spring Harbor Laboratory, Cold Spring Harbor, NY, p. 581 (1974).

Prashad, N., R. N. Rosenberg, B. Wischmeyer, C. Ulrich, and D. Sparkman, *Biochemistry* **18,** 2717 (1979).

Quinn, W. G. and J. L. Gould, *Nature* **278,** 19 (1979).

Roberts, J. L., P. H. Seeburg, J. Shine, E. Herbert, J. D. Baxter, and H. M. Goodman, *Proc. Natl. Acad. Sci. USA* **76,** 2153 (1979).

Roberts, J. L., C. L. C. Chen, J. H. Eberwine, M. J. Q. Evinger, C. Gee, E. Herbert, and B. S. Schachter, *Rec. Prog. Hormone Res.* **38,** 227 (1982).

Robson, K. J. H., T. Chandra, R. T. A. Macgillivray, and S. L. C. Woo, *Proc. Natl. Acad. Sci. USA* **79,** 4701 (1982).

Rosenberg, R. N., W. L. Nyhan, C. Bay, and P. Shore, *Neurology* **26,** 703 (1976).

Rosenberg, R. N., N. Ivy, J. Kirkpatrick, C. Bay, W. L. Nyhan, and F. Baskin, *Neurology* **31,** 1003 (1981).

Sadjel-Sulkowska, E., J. F. Coughlin, and C. A. Marotta, *J. Neurochem.* **40,** 670 (1983).

Sadler, T., D. Greenberg, P. Coughlin, and J. L. Lessard, *Science* **215,** 172 (1982).

Scheller, R. H., J. F. Jackson, L. B. McAllister, J. H. Schwartz, E. R. Kandel, and R. Axel, *Cell* **28,** 707 (1982).

Scheller, R. H., J. F. Jackson, L. B. McAllister, B. S. Rothman, E. Mayeri, and R. Axel, *Cell* **32,** 7 (1983).

Schmid, D., H. Stadler, and V. P. Whittaker, *Eur. J. Biochem.* **122,** 633 (1982).

Schmitt, H., I. Gozes, and U. Z. Littauer, *Brain Res.* **121,** 327 (1977).

Schrier, B. K., M. Y. Zubairi, C. H. Letendre, and L. D. Grouse, *Differentiation* **12,** 23 (1978).

Shows, T. B., A. Y. Sakaguchi, and S. L. Naylor, *Adv. Hum. Genet.* **12,** 341 (1982).

Sidman, R. L., M. C. Green, and S. H. Appel, *Catalogue of the Neurological Mutants of the Mouse,* Harvard University Press, Cambridge, MA (1965).

Silvers, W. K. and R. E. Billingham, *J. Exp. Med.* **129,** 647 (1969).

Singer, R. H. and D. C. Ward, *Proc. Natl. Acad. Sci. USA* **79,** 7331 (1982).

Soreq, H., A. Safran, and R. Zisling, *Dev. Brain Res.* **3,** 65 (1982).

Streilein, J. W., *Prog. Exp. Tumor Res.* **16,** 396 (1972).

Strocchi, P., J. M. Gilbert, and C. A. Marotta, *J. Neurochem.* **36,** 2044 (1981a).

Strocchi, P., C. A. Marotta, J. Bonventre, and J. M. Gilbert, *Brain Res.* **211,** 206 (1981b).

Sutcliffe, J. G., R. J. Milner, F. E. Bloom, and R. A. Lerner, *Proc. Natl. Acad. Sci. USA* **79,** 4942 (1982).

Sutcliffe, J. G., R. J. Milner, T. M. Shinnick, and F. E. Bloom, *Cell* **33,** 671 (1983).

Szabo, P., R. Elder, D. M. Steffensen, and O. C. Uhlenbeck, *J. Mol. Biol.* **115,** 539 (1977).

Szutowicz, A., M. R. Morrison, and P. A. Srere, *J. Neurochem.* **40,** 1664 (1983).

Thomas, P. S., *Proc. Natl. Acad. Sci. USA* **77,** 5201 (1980).

Van Ness, J. and W. E. Hahn, *Nucleic Acids Res.* **8,** 4259 (1980).

Van Ness, J., I. H. Maxwell, and W. E. Hahn, *Cell* **18,** 1341 (1979).

Venezky, D. L., L. M. Angerer, and R. C. Angerer, *Cell* **24,** 385 (1981).

Villa-Komaroff, L., A. Efstratiadis, S.Broome, P. Lomedico, R. Tizard, S. P. Naber, W. L. Chick, and W. Gilbert, *Proc. Natl. Acad. Sci. USA* **75,** 3727 (1978).

Wilson, M. C., S. G. Sawicki, P. A. White, and J. E. Darnell, Jr., *J. Mol. Biol.* **126,** 23 (1978).

Woods, D., J. Crampton, B. Clarke, and R. Williamson, *Nucleic Acids Res.* **8,** 5157 (1980).

Woodward, D. J., B. J. Hoffer, and L. W. Lapham. in R. Llinas, Ed., *Neurobiology of Cerebellar Evolution and Development,* American Medical Association Press, Chicago, p. 725 (1969).

Zagon, I. S. and P. J. McLaughlin, *Brain Res.* **170,** 443 (1979).

Zeller, N. K., J. A. Sprague, R. A. Lazzarini, Y. T. Yu, and A. T.Campagnoni, *Trans. Amer. Soc. Neurochem.* **14,** 255 (1983).

Zomzely-Neurath, C., C. York, and B. W. Moore, *Proc. Natl. Acad. Sci. USA* **69,** 2326 (1972).

# 4

# *Neuron-Specific Phosphoproteins as Models for Neuronal Gene Expression*

***William C. Wallace, Robert M. Lewis, Selma D. Kanazir, Louis J. DeGennaro, and Paul Greengard***



The nervous system is composed of a great variety of cell types which exhibit a complex array of interconnections and functions. To achieve this diversity of cell structure and function, an exquisite control of gene expression is required during the development of the nervous system. From nucleic acid hybridization experiments, approximately 145,000 different mRNAs have been estimated to be present in the brain (Hahn et al., 1982). This value represents a considerably greater variety than is found in other tissues. The regulation of gene expression during the differentiation of neurons may be investigated at various levels: transcription, processing of transcription product, mRNA stability, translation, or processing of translation product.

In this chapter we summarize our preliminary experimental observations on the expression of genes coding for a unique class of neuronal proteins, neuron-specific phosphoproteins. We have characterized the primary translation products of the mRNAs coding for these proteins. We have also measured the relative levels of translatable polysome-associated mRNA coding for these proteins during brain development. Finally, we have partially purified the mRNAs which code for four neuron-specific proteins. Details of the experimental procedures used in these studies are described in the original publications (DeGennaro et al., 1983, in press; Lewis et al., 1983).

## 1. BACKGROUND

The cellular and molecular complexities within the brain reflect the intricate regulation of gene expression during the development of the nervous system and the differentiation of neurons. The mature adult brain arises from a progression of events that occur during development, which include the induction of the neural crest, the proliferation, migration, aggregation, and differentiation of neurons, synapse formation, selective cell death, elimination of some synapses, and stabilization of others. These events involve alterations of gene expression, in a highly concerted manner, within innumerable neurons. The terminally differentiated neuron also undergoes modulation of gene expression to adapt to changing conditions within the brain. For example, the absolute amounts of neurotransmitter-metabolizing enzymes, synaptic structural proteins, and receptors appear to be controlled by balancing synthesis and degradation of the protein within the adult brain.

A specific group of proteins within the brain, neuron-specific phosphoproteins, provides a convenient experimental model system to study gene expression. Protein phosphorylation appears to regulate the response of neurons to many types of stimulation (Greengard, 1981). We have presented evidence for the existence of over 70 neuron-specific phosphoproteins located throughout the nervous system (Walaas et al., 1983b, 1983c). These cytosolic and membrane proteins exhibit different patterns of regional distribution within the brain. Some phosphoproteins appear to be present in virtually all brain cells, others in only certain classes of brain cells, and still others in only a single type of brain cell. The differences in the regional distributions of these phosphoproteins within the brain indicate the highly specific gene expression that occurs in neurons. Because protein phosphorylation seems to be involved in several aspects of neuronal function, and phosphoproteins are ubiquitous yet exhibit cell-specific distributions within the nervous system, the regulation of gene expression for neuron-specific phosphoproteins represents a convenient experimental system with which to investigate the regulation of gene expression within the brain.

We have characterized four of these neuronal phosphoproteins, synapsin I, protein III, DARPP-32 (a dopamine and cyclic adenosine monophosphate-regulated phosphoprotein of $M_r$ 32,000), and G-substrate (Table 1). These four proteins represent different patterns of phosphoprotein distribution within the nervous system. Synapsin I and protein III are found throughout the entire nervous system. DARPP-32 is concentrated within the basal ganglia. G-substrate is localized exclusively in the Purkinje cells of the cerebellum.

Immunocytochemical studies indicate that synapsin I is located exclusively in neurons and that it occurs in virtually all neurons throughout the peripheral and central nervous systems (DeCamilli et al., 1983). Synapsin I is extrinsically associated with the cytoplasmic surface of synaptic vesicles (Huttner et al., 1983). Protein III is also found in all neurons as an extrinsically associated membrane protein (Huang et al., 1982). Immunocytochemical studies are now underway to determine more precisely the subcellular location of protein III. DARPP-32 exhibits a marked regional variation in concentration and cellular localization. It is highly concentrated in neurons of the caudatoputamen, olfactory tubercle, and nucleus accumbens, as well as in their fibers and terminals in the globus pallidus and substantia nigra. DARPP-32 is especially enriched within dopaminoceptive neurons, specifically in the cytosol of those cells which have dopamine receptors linked to the stimulation of adenylate cyclase (Walaas et al., 1983a). G-substrate is found exclusively within the cytosol of the Purkinje cell of the cerebellum (Nairn et al., 1982; Schlichter et al., 1980).

Synapsin I consists of two nearly identical polypeptides of $M_r$ 86,000 (synapsin Ia) and 80,000 (synapsin Ib). Synapsin Ia and Ib maintain a constant molar proportion of 1:2 throughout purification to homogeneity from bovine and rat brain (Ueda and Greengard, 1977). Protein III consists of two different polypeptides of $M_r$ 74,000 (protein IIIa) and 55,000 (protein IIIb) (Huang et al., 1982). Antibodies raised against synapsin I cross-react with protein III; antibodies raised against protein III cross-react with synapsin I (Browning et al., 1982). DARPP-32 is composed of a single type of peptide of $M_r$ 32,000 (Hemmings et al., 1984). G-substrate is composed of a single type of peptide of $M_r$ 23,000 (Aswad and Greengard, 1981).

These four phosphoproteins provide a convenient model system for the study of neuron-specific gene expression. Each of these proteins is located only in neurons while DARPP-32 and G-substrate are located in specific types of neurons. In addition, the physical and chemical properties of these proteins have been well characterized. Finally, the physiological role of each of these proteins in neuronal function is being investigated.

**TABLE 1. NEURON-SPECIFIC PHOSPHOPROTEINS INVESTIGATED**

| Substrate Protein | Size (Da) | Kinase | Cell Distribution | Subcellular Distribution | Source for Polysomes |
|---|---|---|---|---|---|
| Synapsin I (Ia, Ib) | 86,000; 80,000 | cAMP; $Ca^{2+}$/calmodulin | All neurons | Synaptic vesicles | Rat total brain |
| Protein III (IIIa, IIIb) | 74,000; 55,000 | cAMP | All neurons? | Synaptic vesicles | Rat total brain (minus cerebellum) |
| DARPP-32 | 32,000 | cAMP | Dopaminoceptive cells | Cytosol | Bovine caudate nucleus |
| G-substrate | 23,000 | cGMP | Purkinje cells | Cytosol | Rat or rabbit cerebellum |

## 2. TRANSLATION OF NEURON-SPECIFIC mRNA

In addition to our interest in studying the control of gene expression of neuronal phosphoproteins, there were a number of specific questions about the characteristics of the phosphoproteins that we wanted to answer. What controlling factors at the level of gene expression are responsible for the appearance of the two nearly identical polypeptides that constitute synapsin I? Do different mRNAs code for the independent translation of synapsin Ia and synapsin Ib or are these two polypeptides the result of the post-translational processing of a common precursor? Similar questions were posed for proteins IIIa and IIIb. In addition, we wanted to know if the appearances of DARPP-32 and G-substrate within specific regions of the brain were due to the synthesis of mRNA only in neurons within these regions or whether post-translational modification led to the specific distribution of those proteins.

To answer these questions, we have investigated the translation of neuron-specific mRNA using an *in vitro* translation system to characterize the synthesis of these four neuron-specific proteins by brain polysomes and brain mRNA (DeGennaro et al., 1983, in press; Kanazir et al., 1982; Lewis et al., 1983). Cytosolic (free) polysomes were isolated from the appropriate neuronal tissue (rat whole brain for synapsin I and protein III, bovine caudate nucleus for DARPP-32, and rat or rabbit cerebellum for G-substrate) and incubated with a rabbit reticulocyte lysate *in vitro* translation mixture containing [$^{35}$S]-methionine (Pelham and Jackson, 1976; Walker and Zomzely-Neurath, 1980). In initial experiments we determined that [$^{35}$S]-methionine incorporation into newly synthesized proteins ceased after 30–60 min of incubation at 25°C. The addition of more rabbit reticulocyte lysate to this reaction mixture did not result in the synthesis of additional protein. However, the addition of a second aliquot of polysomes stimulated protein synthesis. Protein synthesis in this system, therefore, probably results from the elongation of existing nascent polypeptide chains without reinitiation.

When polysomes isolated from rat brain were used to direct protein synthesis in this translation assay, numerous polypeptides ranging from $M_r$ 18,000 to $M_r$ 250,000 were detected. Synapsin I and protein III were tentatively identified by immunoprecipitation to be among those proteins newly synthesized by cytosolic (free) polysomes isolated from rat whole brain (Figs. 1 and 2). In experiments in which cytosolic (free) polysomes were isolated from the bovine caudate nucleus, DARPP-32 was tentatively identified by immunoprecipitation to be among the newly synthesized proteins (Fig. 3). G-substrate was immunoprecipitated from proteins newly synthesized by cytosolic (free) polysomes from rat cerebellum (Fig. 4). The protein immunoprecipitated by the appropriate immunoglobulin was identical to authentic

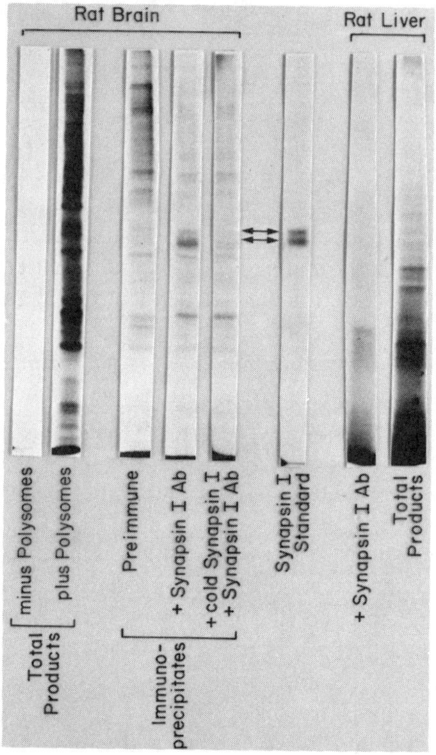


**Figure 1.** *In vitro* translation of polysome-associated synapsin I mRNA. Polysomes were purified from the brains of 14-day-old rats and translated in a reticulocyte lysate translation reaction in the presence of [$^{35}$S]-methionine (DeGennaro et al., in press). The total translation products (Total Products) or translation products precipitated by Synapsin I immunoglobulin (Immunoprecipitates) were separated by SDS–polyacrylamide gel electrophoresis and the [$^{35}$S]-methionine incorporated into protein was visualized by fluorography. The amount of translation reaction mixture used to analyze total products represented 2% of that used for immunoprecipitation. Minus Polysomes: translation products after the reticulocyte lysate mixture had been treated with micrococcal nuclease but without the addition of polysomes. Plus Polysomes: translation products after the reticulocyte lysate mixture had been treated with micrococcal nuclease followed by the addition of cytosolic (free) polysomes. Preimmune: newly synthesized polypeptides that were precipitated from total translation products using preimmune serum. +Synapsin I Ab: newly synthesized polypeptides that were immunoprecipitated from total translation products using affinity-purified rabbit anti-synapsin I immunoglobulin. +Cold synapsin I + synapsin I Ab: newly synthesized polypeptides that were precipitated from total translation products using affinity-purified anti-synapsin I immunoglobulin in the presence of non-radioactive, purified synapsin I. Cytosolic (free) polysomes purified from rat liver were translated in a reticulocyte lysate translation reaction. +Synapsin I Ab: newly synthesized polypeptides that were precipitated from total translation products with affinity-purified anti-synapsin I immunoglobulin. Total Products: total translation products. Synapsin I standard is purified rat synapsin I which had been phosphorylated by the catalytic subunit of cAMP-dependent protein kinase. The arrows indicate the positions of synapsin Ia (upper) and synapsin Ib (lower). (From DeGennaro et al., 1983.) With permission from *Cold Spring Harbor Symposium on Quantitative Biology*. Copyright: Cold Spring Harbor Laboratory.

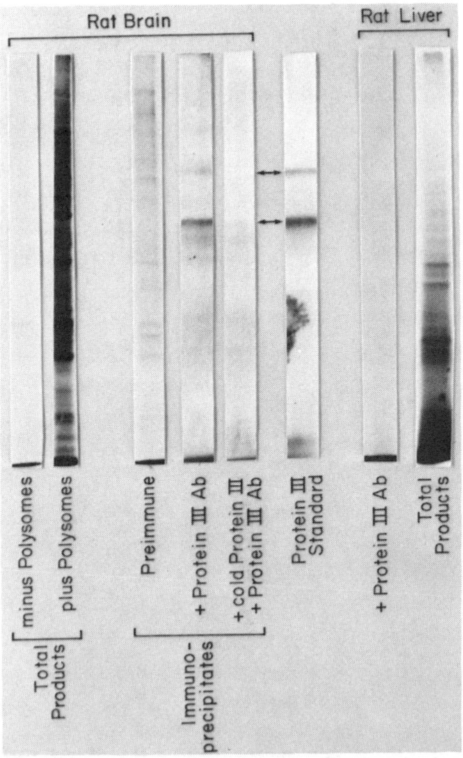


**Figure 2.** *In vitro* translation of polysome-associated protein III mRNA. Polysomes were purified from the brains of 14-day-old rats and translated in a reticulocyte lysate translation reaction in the presence of [$^{35}$S]-methionine (DeGennaro et al., 1983). The total translation products (Total Products) or translation products precipitated by protein III immunoglobulin (Immunoprecipitates) were separated by SDS–polyacrylamide gel electrophoresis and the [$^{35}$S]-methionine incorporated into protein was visualized by fluorography. The amount of translation reaction mixture used to analyze total products represented 2% of that used for immunoprecipitation. Minus Polysomes: translation products after the reticulocyte lysate mixture had been treated with micrococcal nuclease but without the addition of polysomes. Plus Polysomes: translation products after the reticulocyte lysate mixture had been treated with micrococcal nuclease followed by the addition of cytosolic (free) polysomes. Preimmune: newly synthesized polypeptides that were precipitated from total translation products using preimmune serum. +Protein III Ab: newly synthesized polypeptides that were immunoprecipitated from total translation products using affinity-purified rabbit anti-protein III immunoglobulin. +Cold protein III + protein III Ab: newly synthesized polypeptides that were precipitated from total translation products using affinity-purified anti-protein III immunoglobulin in the presence of non-radioactive, purified protein III. Cytosolic (free) polysomes purified from rat liver were translated in a reticulocyte lysate translation reaction. +Protein III Ab: newly synthesized polypeptides that were precipitated from total translation products with affinity-purified anti-protein III immunoglobulin. Total Products: total translation products. Protein III standard is purified rat protein III which had been phosphorylated by the catalytic subunit of cAMP-dependent protein kinase. The arrows indicate the positions of protein IIIa (upper) and protein IIIb (lower). (From DeGennaro et al., 1983.) With permission from *Cold Spring Harbor Symposium on Quantitative Biology*. Copyright: Cold Spring Harbor Laboratory.

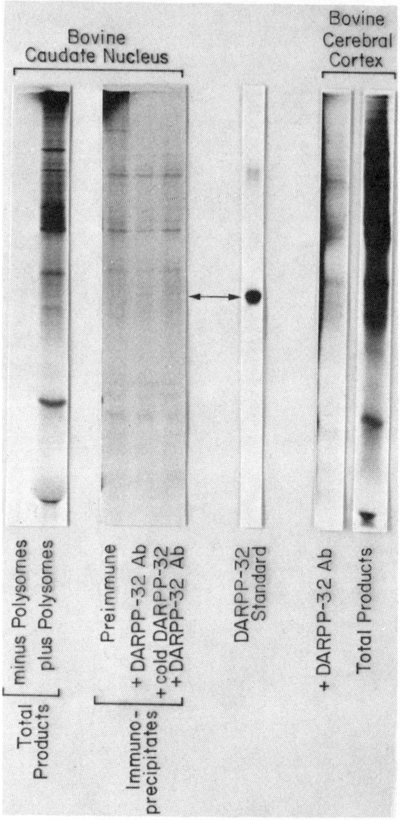


**Figure 3.** *In vitro* translation of polysome-associated DARPP-32 mRNA. Polysomes were purified from the caudate nucleus of bovine calves and translated in a reticulocyte lysate translation reaction in the presence of [$^{35}$S]-methionine (Lewis et al., 1983). The total translation products (Total Products) or translation products precipitated by DARPP-32 immunoglobulin (Immunoprecipitates) were separated by SDS–polyacrylamide gel electrophoresis and the [$^{35}$S]-methionine incorporated into protein was visualized by fluorography. The amount of translation reaction mixture used to analyze total products represented 2% of that used for immunoprecipitations. Minus Polysomes: translation products after the reticulocyte lysate mixture had been treated with micrococcal nuclease but without the addition of polysomes. Plus Polysomes: translation products after the reticulocyte lysate mixture had been treated with micrococcal nuclease followed by the addition of cytosolic (free) polysomes. Preimmune: newly synthesized polypeptides that were precipitated from total translation products using preimmune serum. +DARPP-32 Ab: newly synthesized polypeptides that were immunoprecipitated from total translation products using affinity-purified rabbit anti-DARPP-32 immunoglobulin. +Cold DARPP-32 + DARPP-32 Ab: newly synthesized polypeptides that were precipitated from total translation products with affinity-purified anti-DARPP-32 immunoglobulin in the presence of non-radioactive, purified DARPP-32. Cytosolic (free) polysomes purified from the cerebral cortex of bovine calves were translated in a reticulocyte lysate translation reaction. +DARPP-32 Ab: newly synthesized polypeptides that were precipitated from total translation products with affinity-purified anti-DARPP-32 immunoglobulin. Total Products: total translation products. DARPP-32 standard is purified bovine DARPP-32 which had been phosphorylated by the catalytic subunit of cAMP-dependent protein kinase. The arrow indicates the position of DARPP-32. (From Lewis et al., 1983.) With permission from *Cold Spring Harbor Symposium on Quantitative Biology*. Copyright: Cold Spring Harbor Laboratory.

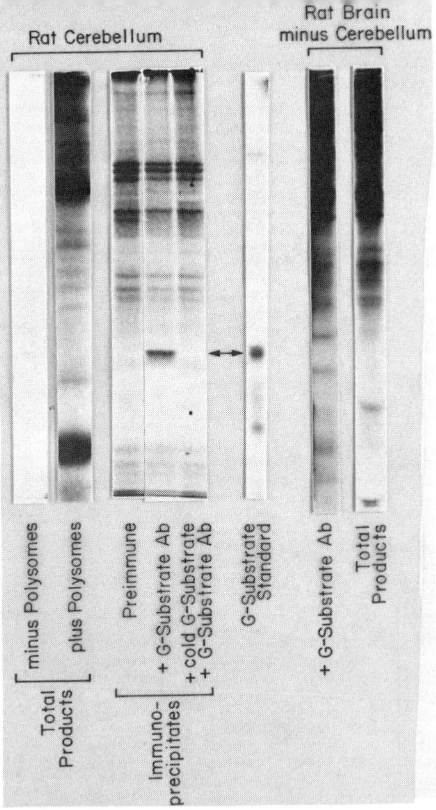


**Figure 4.** *In vitro* translation of polysome-associated G-substrate mRNA. Polysomes were purified from the cerebella of 14-day-old rats and translated in a reticulocyte lysate translation reaction in the presence of [$^{35}$S]-methionine (Lewis et al., 1983). The total translation products (Total Products) or translation products precipitated by G-substrate immunoglobulin (Immunoprecipitates) were separated by SDS–polyacrylamide gel electrophoresis and the [$^{35}$S]-methionine incorporated into protein was visualized by fluorography. The amount of translation reaction mixture used to analyze total products represented 2% of that used for immunoprecipitation. Minus Polysomes: translation products after the reticulocyte lysate mixture had been treated with micrococcal nuclease but without the addition of polysomes. Plus Polysomes: translation products after the reticulocyte lysate mixture had been treated with micrococcal nuclease followed by the addition of cytosolic (free) polysomes. Preimmune: newly synthesized polypeptides that were precipitated from total translation products using preimmune serum. (This preimmune serum precipitates an unidentified protein which is not precipitated by the immune serum.) +G-substrate Ab: newly synthesized polypeptides that were precipitated from total translation products using anti-G-substrate rabbit serum. +Cold G-substrate + G-substrate Ab: newly synthesized polypeptides that were precipitated from total translation products using anti-G-substrate rabbit serum in the presence of non-radioactive, purified G-substrate. Cytosolic (free) polysomes purified from rat brain minus cerebellum were translated in a reticulocyte lysate translation reaction. +G-substrate Ab: newly synthesized polypeptides that were precipitated from total translation products with anti-G-substrate rabbit serum. Total Products: total translation products. G-substrate standard is partially purified rabbit G-substrate which had been phosphorylated by cGMP-dependent protein kinase. The arrow indicates the position of G-substrate. (From Lewis et al., 1983.) With permission from *Cold Spring Harbor Symposium on Quantitative Biology*. Copyright: Cold Spring Harbor Laboratory.

standard phosphoprotein by several other criteria. For all four proteins the newly synthesized protein exhibited the same electrophoretic mobility as the mature protein isolated from adult brains (Figs. 1–4, compare lanes marked "Standard" and "Immunoprecipitates, +Ab"). In the case of synapsin I the two [$^{35}$S]-containing polypeptides appeared to be synthesized in the same 1:2 ratio characteristic of mature synapsin Ia and Ib; synapsin Ia and Ib contain an equal number of methionine residues (Ueda and Greengard, 1977). Newly synthesized protein III also appeared as two polypeptides, protein IIIa and protein IIIb. In addition, the immunoprecipitation of each of these proteins was selectively inhibited by the addition of the corresponding non-radioactive purified protein to the reaction mixture just prior to the additon of immunoglobulin (Figs. 1–4, compare lanes marked "Immunoprecipitates, +Ab" and "Immunoprecipitates, +cold protein + Ab"). Finally, when cytosolic (free) polysomes isolated from rat liver were used to program protein synthesis, neither synapsin I nor protein III, both being neuron-specific proteins, were detected among the numerous proteins synthesized (Figs. 1 and 2, compare lanes marked "Brain, Immunoprecipitates, +Ab" with "Liver, +Ab"). Bovine cerebral cortex polysomes synthesized no detectable DARPP-32 (Fig. 3, compare lanes marked "Caudate Nucleus, Immunoprecipitates, +DARPP-32 Ab" with "Cerebral Cortex, +DARPP-32 Ab"). Polysomes isolated from rat brain minus cerebellum synthesized no detectable G-substrate (Fig. 4, compare lanes marked "Cerebellum, Immunoprecipitates, +G-substrate Ab" with "Rat Brain minus cerebellum, +G-substrate Ab").

We have also investigated the intracellular location of the synthesis of each of these proteins. For the study of synapsin I synthesis, both cytosolic (free) and membrane-bound polysomes were isolated from rat whole brain. When each fraction was incubated with the *in vitro* translation mixture and the resulting newly synthesized proteins were immunoprecipitated with anti-synapsin I immunoglobulin, synapsin I was detected only among the proteins synthesized by the cytosolic (free) polysomes and not among those synthesized by the membrane-bound polysomes (Fig. 5). Using this same procedure, we found that each of these four proteins was synthesized by cytosolic (free) polysomes isolated from the appropriate brain region. This location of synthesis is consistent with the intracellular location of these proteins: Synapsin I and protein III are extrinsically associated with the cytoplasmic surface of synaptic vesicle membrane, whereas DARPP-32 and G-substrate are cytosolic. Newly synthesized synapsin I , protein III, and DARPP-32 were identified as substrates for cAMP-dependent protein kinase, indicating another similarity with the mature phosphoprotein (Fig. 6). Preliminary results indicate that newly synthesized G-substrate can be phosphorylated by cGMP-dependent protein kinase.

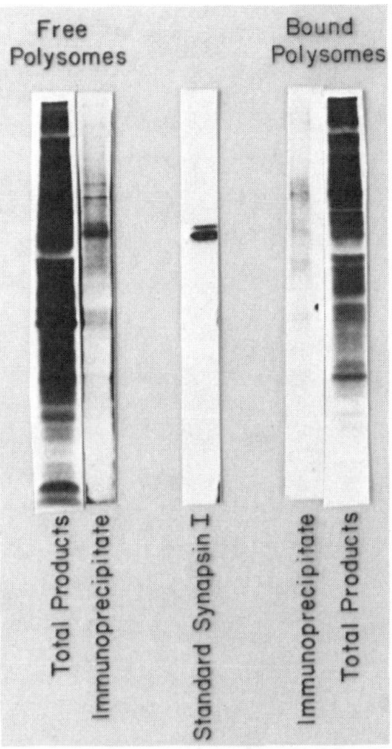


**Figure 5.** *In vitro* translation of synapsin I mRNA: comparison of cytosolic and membrane-bound polysomes. Cytosolic and membrane-bound polysomes were purified from the brains of 14-day-old rats and translated in a reticulocyte lysate translation reaction in the presence of [$^{35}$S]-methionine (DeGennaro et al., in press). The total translation products (Total Products) or translation products precipitated by synapsin I immunoglobulin (Immunoprecipitates) were separated by SDS–polyacrylamide gel electrophoresis and the [$^{35}$S]-methionine incorporated into protein was visualized by fluorography. Free Polysomes: translation products from micrococcal nuclease-treated lysate in the presence of cytosolic (free) polysomes. Bound Polysomes: translation products from micrococcal nuclease-treated lysate in the presence of membrane-bound polysomes. Synapsin I standard is purified rat synapsin I which had been phosphorylated by cAMP-dependent protein kinase. (From DeGennaro et al., 1983.) With permission from *Cold Spring Harbor Symposium on Quantitative Biology*. Copyright: Cold Spring Harbor Laboratory.

These results indicate that these four neuron-specific proteins can be synthesized by brain polysomes in a reticulocyte lysate translation system and that the newly synthesized proteins can be identified by immunoprecipitation. Their identities were based on similarities between the newly synthesized proteins and standard protein in antigenicity, electrophoretic mobility, absence in liver and specific brain regions, intracellular location, and phosphorylatability.

In addition to confirming the similarity between the newly synthesized proteins in this experimental system and the corresponding proteins within the neuron, we wanted to determine if these neuron-specific mRNAs could be

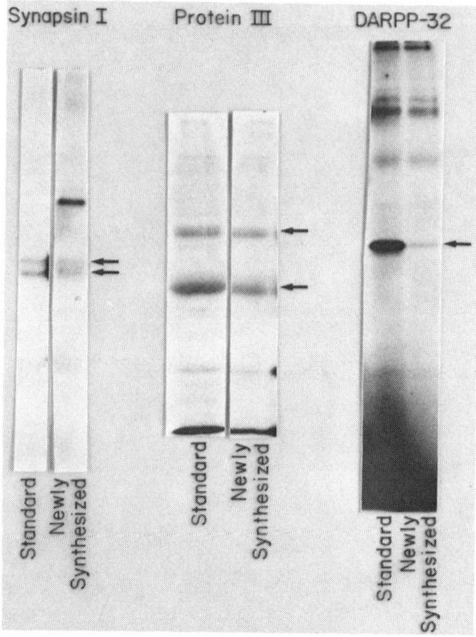


**Figure 6.** Phosphorylation of newly synthesized synapsin I, protein III, and DARPP-32. Polysomes from rat brain (for synapsin I and protein III) or bovine caudate nucleus (for DARPP-32) were translated in the rabbit reticulocyte lysate translation system containing non-radioactive methionine. The newly synthesized synapsin I, protein III, or DARPP-32, was immunoprecipitated from the translation mixture with the corresponding immunoglobulin. The antigen was dissociated from the antibody and phosphorylated using the catalytic subunit of cAMP-dependent protein kinase and [$\gamma$-$^{32}$P]-ATP as described (Nestler and Greengard, 1982). The phosphorylation reaction mixture was separated by SDS–polyacrylamide gel electrophoresis and the proteins that incorporated [$^{32}$P] were visualized by autoradiography. Standard: purified synapsin I, protein III, or DARPP-32 added to the translation reaction mixture, immunoprecipitated, and phosphorylated. Newly synthesized: synapsin I, protein III, or DARPP-32 synthesized by polysomes, immunoprecipitated, and phosphorylated. The arrows indicate the positions of the appropriate phosphoproteins. The higher molecular weight, phosphorylated band which has only appeared in translation mixtures containing polysomes that have been immunoprecipitated with rabbit anti-synapsin I immunoglobulin and goat anti-rabbit immunoglobulin has not been identified. (From DeGennaro et al., 1983 and Lewis et al., 1983.) With permission from *Cold Spring Harbor Symposium on Quantitative Biology*. Copyright: Cold Spring Harbor Laboratory.

expressed in the absence of other neuron-specific factors. Therefore, the mRNAs coding for synapsin I, protein III, DARPP-32, and G-substrate were partially purified and the *in vitro* translation system was programmed with these partially purified mRNAs. Polysomes containing the mRNA for these proteins were isolated by immunoprecipitating cytosolic (free) polysomes from rat whole brain (synapsin I and protein III), bovine caudate nucleus

(DARPP-32), or rabbit cerebella (G-substrate) with the corresponding purified serum antibodies (DeGennaro et al., in press; Lewis et al., 1983). The antibody binds specifically to antigenic sites of nascent polypeptides, thus forming a complex with specific polysomes that are separated from the other polysomes by affinity chromatography with protein A–Sepharose (Shapiro and Young, 1981). The immunoadsorbed polysomes were dissociated and the mRNA was isolated by affinity chromatography with oligo(dT)-cellulose (Aviv and Leder, 1972). Because the mRNAs for all four of these neuron-specific proteins could bind to oligo(dT), these mRNAs most likely contain polyadenylated sequences. The purity of each mRNA fraction was determined by the *in vitro* translation of the partially purified mRNA (Fig. 7). For synapsin I and protein III, the predominant polypeptides synthesized in the presence of partially purified mRNA were either synapsin Ia and synapsin Ib (Fig. 7A) or protein IIIa and protein IIIb (Fig. 7B). Based on the amount of [$^{35}$S]-methionine incorporated into the immunoprecipitated synapsin Ia and Ib (or protein IIIa and IIIb) relative to the amount of [$^{35}$S]-methionine incorporated into the total translation products, the estimates for the purity of these mRNAs varied from 50 to 80%. This calculation is based on the assumption that synapsin I and protein III contain the same number of methionine residues as the average for all other polypeptides synthesized by the partially purified mRNA. This purity represents approximately a 200-fold enrichment of the mRNA. DARPP-32 and G-substrate were also detected by immunoprecipitation of the translation products (Fig. 7C and D). However, the purity of the mRNA for DARPP-32 and G-substrate could not readily be assessed because the background from the protein synthesis due to the endogenous mRNA in the lysate was large compared to the amount of [$^{35}$S] incorporated into new polypeptides synthesized upon addition of partially purified DARPP-32 or G-substrate mRNA.

These results suggest that no neuron-specific factor other than the mRNA is required for the synthesis of these proteins. However, we have not determined whether the efficacy of translation may be affected by neuron-specific factors.

Newly synthesized synapsin I, protein III, and DARPP-32 exhibited electrophoretic mobilities similar to and polypeptide abundance ratios the same as those of their mature counterparts. This was true whether the synthesis was directed by brain polysomes or by partially purified mRNA. Thus, it appears that synapsin I, protein III, and DARPP-32 are synthesized in the absence of any large precursor. However, immunoprecipitation of newly synthesized G-substrate from either cerebellar polysomes or partially purified mRNA yielded an additional larger polypeptide (Fig. 7D, upper arrow). The relationship between G-substrate and its co-immunoprecipitating polypeptide has not yet been determined.

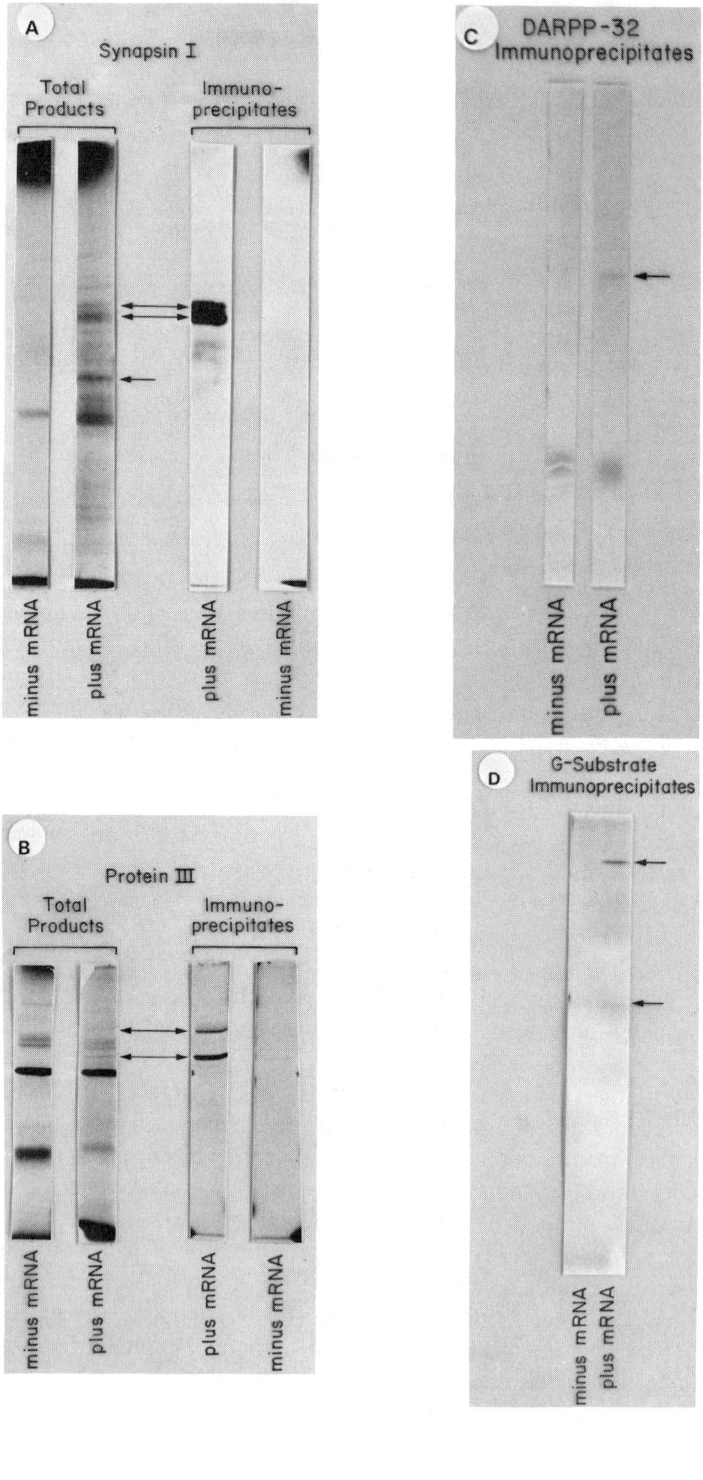
A
Synapsin I
Total Products
Immuno-precipitates
minus mRNA
plus mRNA
plus mRNA
minus mRNA
B
Protein III
Total Products
Immuno-precipitates
minus mRNA
plus mRNA
plus mRNA
minus mRNA
C
DARPP-32 Immunoprecipitates
minus mRNA
plus mRNA
D
G-Substrate Immunoprecipitates
minus mRNA
plus mRNA

Since synapsin I and protein III are antigenically cross-reactive and each consists of two polypeptides, we have investigated the possibility of precursor–product relationships among the four polypeptides (i.e., synapsin Ia, synapsin Ib, protein IIIa, and protein IIIb). Purified synapsin Ia, synapsin Ib, protein IIIa, and protein IIIb were added to a translation reaction mixture containing brain polysomes and the mixture was incubated to facilitate any possible post-translational processing of the exogenously added neuronal proteins. The mature polypeptides exhibited no transformations, which suggested that no post-translational processing had occurred. However, these observations do not exclude the possibility that the mature forms of the polypeptides have been modified to prevent any interconversions. These observations suggest that despite their antigenic similarity, synapsin I and protein III are translation products of different mRNAs.

The similarity between the newly synthesized and the mature forms of the polypeptides of synapsin I (and of protein III) and their appearance in a constant molar ratio may be due to the translation of mRNAs that are unique to each polypeptide. For example, the mRNA coding for synapsin Ib may occur at twice the concentration of that coding for synapsin Ia. Alternatively, either a larger precursor could be processed to yield the polypeptides in the appropriate ratios or the selective post-translational modification of one polypeptide could produce the proper amount of the other. We have not detected the synthesis of any large precursor. Precursor processing and post-translational modification would each require that specific modifying enzymes are either already present in the reticulocyte lysate or that they are synthesized from partially purified mRNA during the translation reaction. In addition, the observation that there was no interconversion of purified mature

---

**Figure 7.** Translation products of partially purified mRNA for (A) Synapsin I, (B) Protein III, (C) DARPP-32, and (D) G-substrate. Messenger RNA for synapsin I, protein III, DARPP-32, and G-substrate was purified by immunoadsorption of polysomes followed by oligo(dT)-cellulose affinity chromatography (DeGennaro et al., 1983; Lewis et al., 1983). [$^{35}$S]-methionine was incorporated into translation products in a translation reaction mixture treated with micrococcal nuclease in the absence (minus mRNA) or presence (plus mRNA) of partially purified mRNA for synapsin I, protein III, DARPP-32, or G-substrate. The translation products were separated by SDS–polyacrylamide gel electrophoresis, either before (Total Products) or after immunoprecipitation with the corresponding immunoglobulin (Immunoprecipitates) and visualized by fluorography. Panel A: the upper two arrows indicate the positions of synapsin Ia and synapsin Ib; the lower arrow indicates the position of protein IIIb, which is antigenically similar to synapsin I. Panel B: the arrows indicate the positions of protein IIIa and protein IIIb. Panel C: the arrow indicates the position of DARPP-32. Panel D: the lower arrow indicates the position of G-substrate; the upper arrow indicates the position of a co-immunoprecipitating polypeptide (see text). (From DeGennaro et al., 1983, and Lewis et al., 1983.) With permission from *Cold Spring Harbor Symposium on Quantitative Biology*. Copyright: Cold Spring Harbor Laboratory.

synapsin Ia, synapsin Ib, protein IIIa, and protein IIIb added to the *in vitro* translation mixture argues against any specific processing or post-translational modification to yield the final, mature form of each of these proteins. Thus, each of these polypeptides appears to be the translation product of separate mRNAs for synapsin Ia and Ib and protein IIa and IIIb.

## 3. DEVELOPMENTAL BIOLOGY OF NEURONAL PHOSPHOPROTEINS

The developing nervous system provides a particularly challenging model for studying cell differentiation because of the large number of different cell types and the complex array of interconnections necessary for the proper functioning of the adult nervous system. This complexity can be exploited for the study of the regulation of gene expression since the diversity of cell types arises from the expression of genes coding for a specialized subset of proteins unique to each cell type. Determining the mechanisms by which individual cells in the nervous system control the expression of specific genes during development would represent a major step toward understanding how cell differentiation occurs. Neuron-specific phosphoproteins are good model proteins for studying control of gene expression during development of the nervous system because they exhibit a high degree of cell-type specificity.

We chose to study neuronal phosphoproteins in the developing postnatal rat brain because of the large amount of anatomical and physiological information available about this system. This can allow a study of possible relationships between the appearance of specific proteins and the development of structures and functions characteristic of specific cell types.

We have examined the synthesis of neuronal phosphoproteins during postnatal development. Polysomes from rat whole brain (for synapsin I), rat whole brain exclusive of cerebellum (for protein III), or rat cerebellum alone (for G-substrate) from successive stages of postnatal development were used to direct the *in vitro* synthesis of these proteins in the presence of [$^{35}$S]-methionine. The specific phosphoproteins were immunoprecipitated, separated on SDS–polyacrylamide gels, and visualized by fluorography. The amount of the specific protein synthesized was quantitated either by densitometric scanning of the bands of the fluorograph or by cutting out the corresponding region of the gel and measuring the incorporated counts in a scintillation counter. The total protein synthesized was determined by counting a trichloroacetic acid-precipitate of an aliquot of the total translation mixture.

The synthesis of synapsin I increased more than eightfold from birth to postnatal day 14 and then decreased to the adult level (Figs. 8 and 9B). The

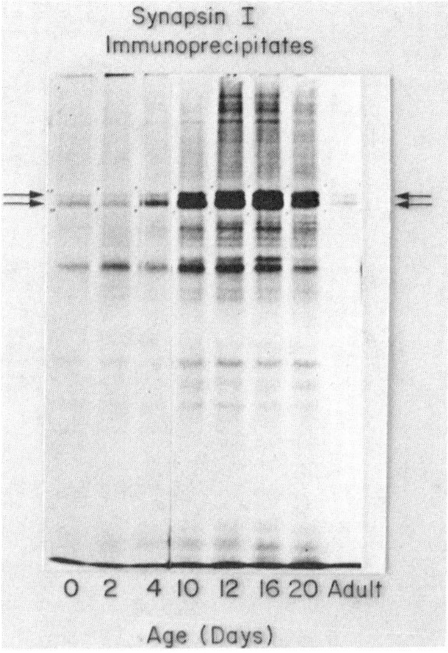


**Figure 8.** Synapsin I synthesized by polysomes isolated from brains of rats of various ages. Equal amounts of cytosolic (free) polysomes were translated in a micrococcal nuclease-treated rabbit reticulocyte lysate translation reaction in the presence of [$^{35}$S]-methionine (DeGennaro et al., in press). Synapsin I was immunoprecipitated from the translation mixture with affinity-purified anti-synapsin I immunoglobulin. The immunoprecipitates were separated by SDS–polyacrylamide gel electrophoresis and visualized by fluorography. A representative experiment showing the translation products present in synapsin I immunoprecipitates is presented above. The arrows indicate the positions of synapsin Ia and synapsin Ib. (From DeGennaro et al., 1983.) With permission from *Cold Spring Harbor Symposium on Quantitative Biology*. Copyright: Cold Spring Harbor Laboratory.

period of rapid synthesis of synapsin I is concurrent with a period of rapid formation of synapses (Aghajanian and Bloom, 1967). The maximal total protein synthesis was observed on postnatal day 2 and steadily decreased with development until it reached a minimum in the adult (Fig. 9A). This difference in the developmental patterns of syntheses of synapsin I and total protein suggested that the change in synapsin I synthesis is a specifically controlled change.

The maximal level of synthesis of synapsin I when expressed per unit of polysomes occurs later during development (postnatal day 14) than the maximal level of synthesis of synapsin I per mass of tissue (as presented in Fig. 9B). In addition, the increase in levels of syntheses of synapsin I and of total protein from postnatal day 0 to 14 occurs more gradually when calculated per unit of polysomes. During development the amount of polysomes isolated per gram wet weight steadily decreases from birth to adult (data not shown).

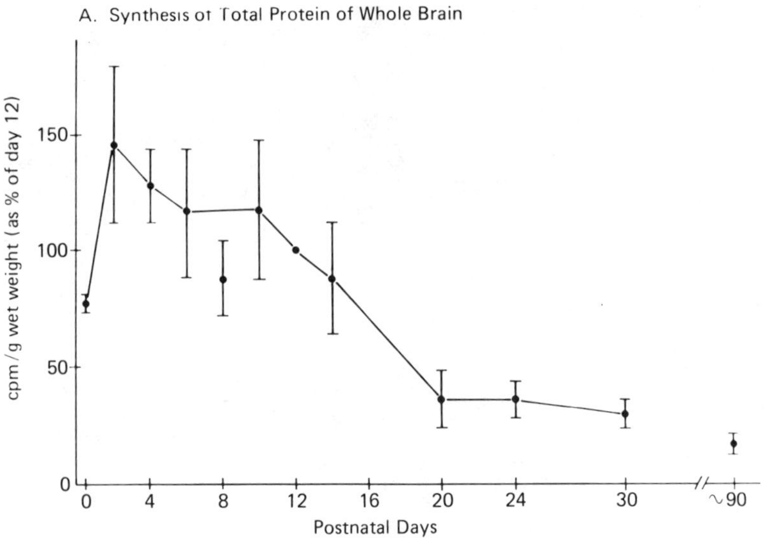
A. Synthesis of Total Protein of Whole Brain
cpm/g wet weight (as % of day 12)
150
100
50
0
0
4
8
12
16
20
24
30
~90
Postnatal Days

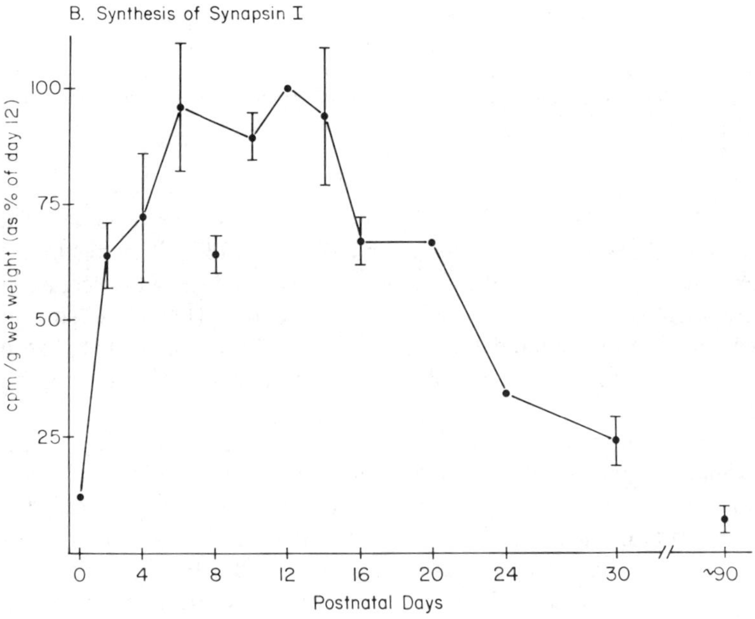
B. Synthesis of Synapsin I
cpm/g wet weight (as % of day 12)
100
75
50
25
0
4
8
12
16
20
24
30
~90
Postnatal Days

Because the rates of synaptogenesis are different in the anterior and posterior regions of the brain (Aghajanian and Bloom, 1967), we have also studied the synthesis of synapsin I in the forebrain (cerebral cortex) and the hindbrain (cerebellum, pons, and medulla) separately. When calculated per unit of polysomes, synthesis of synapsin I in the forebrain was only 50% of that in the hindbrain at birth. At postnatal day 12 synthesis in the two regions was equal. Forebrain activity was 112% of hindbrain activity at postnatal day 16 and 128% at postnatal day 30. In the adult rat the synthesis of synapsin I was the same for both brain regions.

The synthesis of protein IIIa and protein IIIb in the postnatal rat brain increased gradually until it reached a maximum at postnatal day 20. The level of synthesis then decreased, until it reached relatively low levels in the adult brain (Figs. 10, 11). The pattern of change in syntheses during development was only slightly different for protein IIIa and protein IIIb. Increases in the syntheses of protein IIIa and IIIb occurred earlier in development when calculated per unit of polysomes than when calculated per gram wet weight. The maximal level of synthesis of protein III occurred later during development than that of synapsin I (Fig. 11B).

Synthesis of G-substrate could not be detected before postnatal day 6 in the rat cerebellum. Synthesis increased slowly from postnatal day 6 to 12, increased rapidly from postnatal day 12 to 24, and then decreased to the adult level (Fig. 12). It is unlikely that the increased synthesis of G-substrate was due to a general increase in polysome activity since there was no correlation between the relative level of G-substrate synthesis and the relative levels of total protein synthesis during development (Fig. 13). The increase in G-substrate synthesis from postnatal day 12 to 24 was concurrent with an increase in the total activity of cGMP-dependent protein kinase in the cerebellum (Bandle and Guidotti, 1979).

---

**Figure 9.** Incorporation of [$^{35}$S]-methionine into total translation products and into synapsin I as directed by polysomes isolated from brains of rats of various ages. Cytosolic (free) polysomes were purified and translated in a rabbit reticulocyte lysate translation reaction system in the presence of [$^{35}$S]-methionine. Panel A: the total incorporation of [$^{35}$S]-methionine into TCA-insoluble protein was determined by a filter paper assay. Panel B: the reaction mixture was incubated with anti-synapsin I immunoglobulin and the proteins present in the immunoprecipitate were separated by SDS–polyacrylamide gel electrophoresis and visualized by fluorography (see Fig. 8). The [$^{35}$S]-methionine-containing band corresponding to synapsin I was cut from the gel and analyzed by liquid scintillation spectrometry to determine the amount of [$^{35}$S]-synapsin I. The incorporated counts of [$^{35}$S]-methionine for both total translation products and synapsin I were calculated per gram wet weight of brain tissue and are expressed as the percentage of the value determined for postnatal day 12. Data are presented as means ± standard error of the means for 3 separate experiments. (From DeGennaro et al., 1983.) With permission from *Cold Spring Harbor Symposium on Quantitative Biology*. Copyright: Cold Spring Harbor Laboratory.

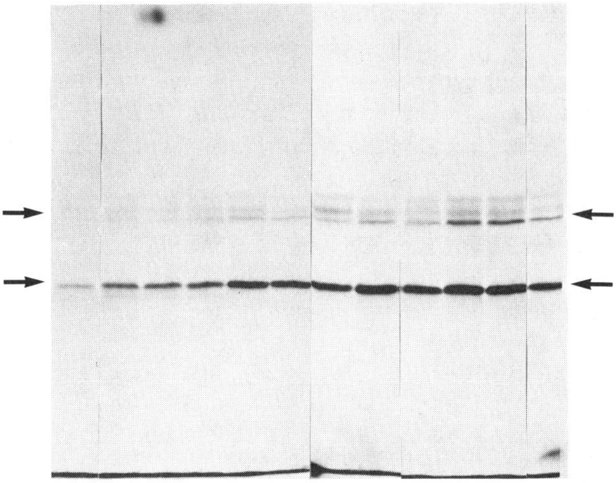


**Figure 10.** Protein III synthesized by polysomes isolated from brains of rats of various ages. Equal amounts of cytosolic (free) polysomes were translated in a micrococcal nuclease-treated rabbit reticulocyte lysate translation reaction in the presence of [$^{35}$S]-methionine (DeGennaro et al., 1983). Protein III was immunoprecipitated from the translation mixture with affinity-purified rabbit anti-protein III immunoglobulin. The immunoprecipitates were separated by SDS–polyacrylamide gel electrophoresis and visualized by fluorography. A representative experiment showing the translation products present in protein III immunoprecipitates is presented. The arrows indicate the positions of protein IIIa and protein IIIb. (From DeGennaro et al., 1983.) With permission from *Cold Spring Harbor Symposium on Quantitative Biology*. Copyright: Cold Spring Harbor Laboratory.

---

**Figure 11.** Incorporation of [$^{35}$S]-methionine into total translation products and into protein III as directed by polysomes isolated from brains (minus cerebella) of rats of various ages. Cytosolic (free) polysomes were purified and translated in a rabbit reticulocyte lysate translation reaction system in the presence of [$^{35}$S]-methionine. Panel A: the total incorporation of [$^{35}$S]-methionine into TCA-insoluble protein was determined by a filter paper assay. Panel B: the reaction mixture was incubated with anti-protein III immunoglobulin and the proteins present in the immunoprecipitate were separated by SDS–polyacrylamide gel electrophoresis and visualized by fluorography (see Fig. 10). Either the [$^{35}$S]-methionine-containing bands corresponding to Protein III were cut from the gel and counted by liquid scintillation spectrometry or the fluorogram was analyzed by densitometry (as described in Fig. 13) to determine the amount of [$^{35}$S]-protein III. The values both for total translation products and for protein III were calculated per gram wet weight of brain tissue and are expressed as the percentage of the value determined for postnatal day 12. Data are presented as means ± standard error of the means for four separate experiments.

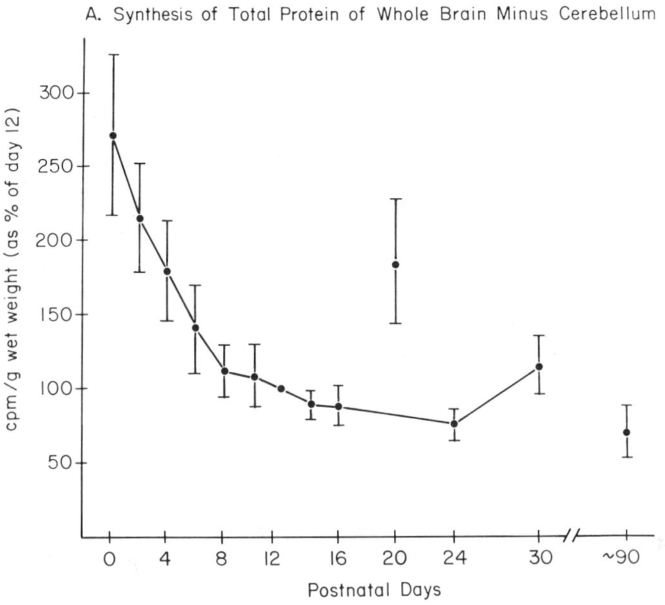
A. Synthesis of Total Protein of Whole Brain Minus Cerebellum
cpm/g wet weight (as % of day 12)
300
250
200
150
100
50
0
4
8
12
16
20
24
30
~90
Postnatal Days

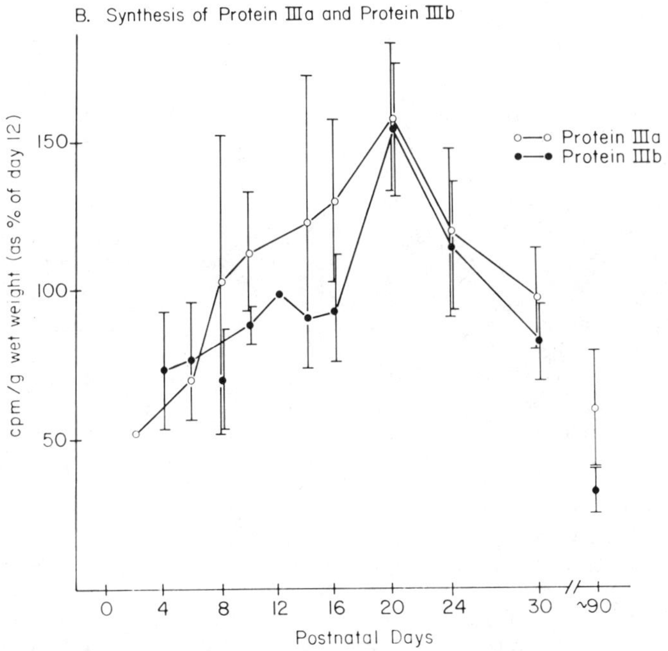
B. Synthesis of Protein IIIa and Protein IIIb
cpm/g wet weight (as % of day 12)
150
100
50
Protein IIIa
Protein IIIb
0
4
8
12
16
20
24
30
~90
Postnatal Days

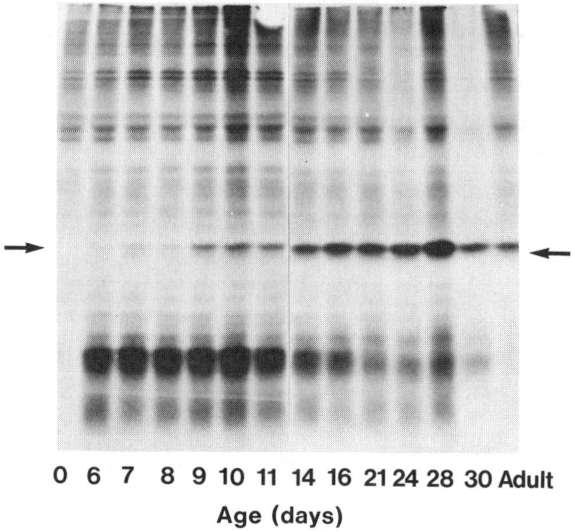


**Figure 12.** G-substrate synthesized by polysomes isolated from cerebella of rats of various ages. Cytosolic (free) polysomes were translated in a micrococcal nuclease-treated rabbit reticulocyte lysate translation reaction in the presence of [$^{35}$S]-methionine (Lewis et al., 1983). G-substrate was immunoprecipitated from the translation mixture with rabbit serum containing immunoglobulin to G-substrate. The immunoprecipitates were separated by SDS–polyacrylamide gel electrophoresis and visualized by fluorography. A representative experiment showing the translation products present in G-substrate immunoprecipitates is presented. The arrows indicate the position of G-substrate. (From Lewis et al., 1983.) With permission from *Cold Spring Harbor Symposium on Quantitative Biology*. Copyright: Cold Spring Harbor Laboratory.

---

**Figure 13.** Incorporation of [$^{35}$S]-methionine into total translation products and into G-substrate as directed by polysomes isolated from cerebella of rats of various ages. Cytosolic free polysomes were purified and translated in a rabbit reticulocyte lysate translation reaction system in the presence of [$^{35}$S]-methionine. Panel A: the total incorporation of [$^{35}$S]-methionine into TCA-insoluble protein was determined by a filter paper assay and is expressed as cpm per $A_{260}$ unit of polysomes, normalized to the value for day 20. Panel B: the reaction mixture was incubated with anti-G-substrate immunoglobulin and the proteins present in the immunoprecipitate were separated by SDS–polyacrylamide gel electrophoresis and visualized by fluorography (see Fig. 12). The band corresponding to G-substrate was scanned by a densitometer. The value for each peak was calculated per unit of total translation products and normalized to the value for day 20. Data are presented as means ± standard error of the means for eight separate experiments. (From Lewis et al., 1983.) With permission from *Cold Spring Harbor Symposium on Quantitative Biology*. Copyright: Cold Spring Harbor Laboratory.

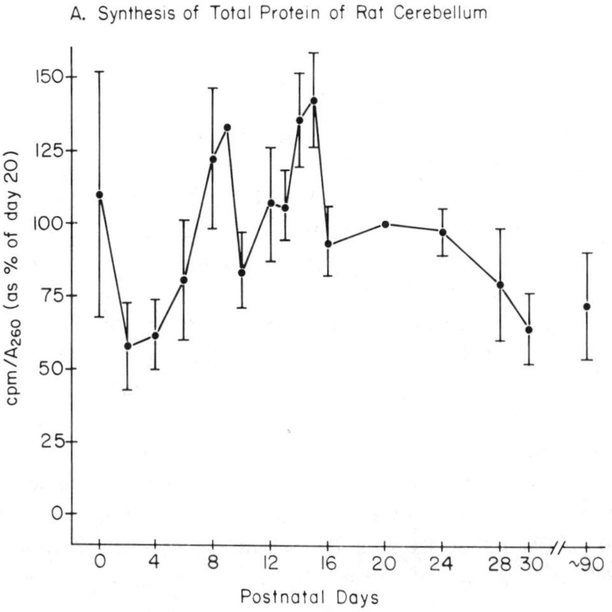
A. Synthesis of Total Protein of Rat Cerebellum
cpm/A260 (as % of day 20)
150
125
100
75
50
25
0
0 4 8 12 16 20 24 28 30 ~90
Postnatal Days

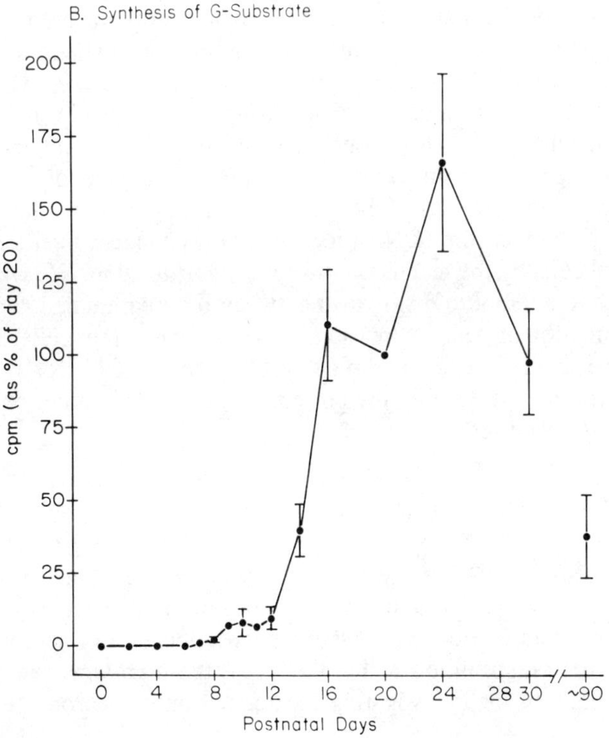
B. Synthesis of G-Substrate
cpm (as % of day 20)
200
175
150
125
100
75
50
25
0
0 4 8 12 16 20 24 28 30 ~90
Postnatal Days

By postnatal day 3 in the rat, Purkinje cells have formed in a monolayer arrangement (Woodward et al., 1969). By postnatal day 12 the Purkinje cells have started to form apical dendritic trees (Berry and Bradley, 1976), contacts with parallel fibers (Woodward et al., 1969), and temporary synapses with the climbing fibers (Shimono et al., 1976). Although the Purkinje cells are well established as a definitive cell type by postnatal day 12, they still do not synthesize G-substrate at a very high rate. This observation that Purkinje cells grow and are maintained in the absence of G-substrate suggests that G-substrate may be involved in a specialized function of the Purkinje cell. It is only after postnatal day 12, when the Golgi, basket, stellate, and granule cells begin to show activity and the Purkinje cells have begun to establish final contact with the climbing fibers (Shimono et al., 1976), that rapid synthesis of G-substrate begins.

These experimental results indicate that the changes in synthesis of the individual phosphoproteins are not directly linked to changes in total protein synthesis. The results imply that the synthesis of the phosphoproteins is controlled either by the amount of mRNA coding for each protein or by some alteration of the mRNA which modifies the rate at which it is translated.

There are clearly limitations of this technique to the study of gene expression during development of the nervous system. The major disadvantage is that no direct information about changes in transcription or processing of mRNA can be obtained. It is planned to search for such changes by cloning cDNA to the purified mRNAs of each of the phosphoproteins. These clones will be used to generate probes for measuring the changes in the number of mRNA molecules for each phosphoprotein during development. Another drawback to the study of the *in vitro* translation of isolated polysomes is that it does not provide information about changes in synthesis in individual cells. We may be able to use the cDNA probes to study changes in gene expression in individual cells by the technique of *in situ* hybridization. We will also use these probes to attempt to distinguish between preprogrammed gene expression and induction of gene expression by contact with other cells. Individual cells will be examined to determine whether the mRNA for the specific protein is present and whether the protein is synthesized before contact with other cells has occurred.

## 4. CONCLUDING REMARKS

Using a specific well-characterized group of neuronal proteins, namely neuron-specific phosphoproteins, we have begun to investigate gene expression within the brain. Because of their specific location in neurons and the variation in their distribution within the brain, these proteins can be used as molecular markers for investigating the regulation of neuron-specific gene expression. We have begun such studies by characterizing the initial translation products using an *in vitro* translation assay to characterize the degree of regulation involved in gene expression at the level of translation product processing. The synthesis of synapsin Ia, synapsin Ib, protein IIIa, protein IIIb, and DARPP-32 apparently involves minimal post-translational processing. The newly synthesized polypeptides exhibit many similarities to the mature polypeptides. The results presented are consistent with the hypothesis that separate mRNAs code for each polypeptide.

We have examined the synthesis of synapsin I, protein III, and G-substrate in the developing postnatal rat brain. The results suggest that the synthesis of each of these three neuron-specific proteins is regulated independently of total protein synthesis within the brain. The increased synthesis of synapsin I and protein III are apparently contemporary with synapse formation; G-substrate appears at approximately the time that Purkinje cells begin to form functional synapses.

Studies of both post-translation processing and synthesis of these neuron-specific proteins suggest that selective expression of these proteins is regulated at the level of transcription or transcription product processing. These forms of regulation may be directly investigated with complementary DNA probes to characterize the mRNAs coding for each polypeptide and the changes in the synthesis of these mRNAs during the development of the rat brain. Numerous specific questions concerning gene expression within the brain may be answered using cDNA probes specific to these neuron-specific phosphoproteins. For example, we may directly determine whether each polypeptide of synapsin I and protein III is synthesized by the translation of an independent mRNA and by the expression of independent genes. We may also determine whether the changes in the level of synthesis of these proteins during development is due to changes in the synthesis of the corresponding mRNAs, a result which would further indicate transcriptional control of gene expression. Once the mechanisms of control of gene expression during development are characterized, subtler questions concerning extracellular versus inherent controls of gene expression in the differentiating neuron may be approached.

## ACKNOWLEDGMENTS

We are indebted to several people for their contributions to this chapter. Annette E. Gwardyak provided editorial assistance. Karen J. Hageman prepared the figures. Philip L. Mobley and Wilson C. S. Wu offered critical comments about the manuscript. Wilson C. S. Wu provided purified synapsin I. Michael D. Browning provided purified protein III and anti-protein III serum. Hugh C. Hemmings, Jr. supplied bovine caudate, purified DARPP-32-Sepharose affinity resin and anti-DARPP-32 serum. Angus C. Nairn supplied purified G-substrate, G-substrate–Sepharose affinity resin and anti-G-substrate serum.

## REFERENCES

Aghajanian, G. K. and F. E. Bloom, *Brain Res.* **6,** 716 (1967).

Aswad, D. W. and P. Greengard, *J. Biol. Chem.* **256,** 3487 (1981).

Aviv, H. and P. Leder, *Proc. Natl. Acad. Sci. USA* **69,** 1408 (1972).

Bandle, E. and A. Guidotti, *J. Neurochem.* **32,** 1343 (1979).

Berry, M. and P. Bradley, *Brain Res.* **112,** 1 (1976).

Browning, M. D., D. K. Huang, and P. Greengard, *Abst. Soc. for Neurosci.* **8,** 794 (1982).

DeCamilli, P., R. Cameron, and P. Greengard, *J. Cell Biol.* **96,** 1337 (1983).

DeGennaro, L. J., W. C. Wallace, and P. Greengard, *J. Biol. Chem.* (in press).

DeGennaro, L. J., S. D. Kanazir, W. C. Wallace, R. M. Lewis, and P. Greengard, *48th Symposium on Quantitative Biology,* Cold Spring Harbor Laboratory, Cold Spring Harbor, NY, Chap. 25 (1983).

Greengard, P., *The Harvey Lectures*, **75,** 277 (1981).

Hahn, W. E., J. Van Ness, and N. Chaudhari, in F. O. Schmitt, S. J. Bird, and F. E. Bloom, Eds., *Molecular Genetic Neuroscience,* Raven Press, New York, p. 323 (1982).

Hemmings, H. C., Jr., A. C. Nairn, D. W. Aswad, and P. Greengard, *J. Neurosci.* **4,** 99 (1984).

Huang, C. K., M. D. Browning, and P. Greengard, *J. Biol. Chem.* **257,** 6524 (1982).

Huttner, W. B., W. Schiebler, P. Greengard, and P. DeCamilli, *J. Cell Biol.* **96,** 1374 (1983).

Kanazir, S. D., R. M. Lewis, W. C. Wallace, L. J. DeGennaro, and P. Greengard, *Abst. Soc. for Neurosci.* **8,** 174 (1982).

Lewis, R. M., W. C. Wallace, S. D. Kanazir, and P. Greengard, *48th Symposium on Quantitative Biology,* Cold Spring Harbor Laboratory, Cold Spring Harbor, NY, Chap. 26 (1983).

Nairn, A. C., J. A. Detre, J. E. Casnellie, and P. Greengard, *Nature* **299,** 734 (1982).

Nestler, E. J. and P. Greengard, *J. Neurosci.* **2,** 1011 (1982).

Pelham, H. R. B. and R. J. Jackson, *Eur. J. Biochem.* **67,** 247 (1976).

Schlichter, D. J., J. A. Detre, D. W. Aswad, B. Chehrazi, and P. Greengard, *Proc. Natl. Acad. Sci. USA* **77,** 5537 (1980).

Shapiro, S. Z. and J. R. Young, *J. Biol. Chem.* **256,** 2495 (1981).

Shimono, T., S. Nosaka, and K. Sasaki, *Brain Res.* **108,** 279 (1976).

Ueda, T. and P. Greengard, *J. Biol. Chem.* **252,** 5155 (1977).

Walaas, S. I., D. W. Aswad, and P. Greengard, *Nature* **301,** 69 (1983a).

Walaas, S. I., A. C. Nairn, and P. Greengard, *J. Neurosci.* **3,** 291 (1983b).

Walaas, S. I., A. C. Nairn, and P. Greengard, *J. Neurosci.* **3,** 302 (1983c).

Walker, W. A. and C. Zomzely-Neurath, *Neurochem. Res.* **5,** 361 (1980).

Woodward, D. J., B. J. Hoffer, and L. W. Lapham, in R. Llinas, Ed., *Neurobiology of Cerebellar Evolution and Development,* AMA, Chicago, p. 725 (1969).

# 5

# *Expression of Microtubule Proteins in Brain*

***Uriel Z. Littauer and Irith Ginzburg***

Eukaryotic cells contain an intricate filamentous network that is involved in a variety of cellular functions. The various filaments have been classified according to their diameter and include microfilaments (5–7 nm), intermediate filaments (8–11 nm), and microtubules (24–25 nm). The distribution and subunit composition of these filaments and their associated molecules may vary between different cell types at various stages of differentiation. They may also vary in their attachment to cell organelles. Electron microscopic visualization, two-dimensional gel electrophoresis, and immunofluorescence studies have been used to distinguish between different types of filaments. Protein sequences and immunological cross-reactivities have shown that actins and tubulins, the major constituents of microfilaments and microtubules, are conserved throughout evolution. Five different types of intermediate filaments have been defined: (1) vimentin filaments, (2) keratin filaments, (3) desmin filaments, (4) glial filaments, and (5) neurofilaments (Cleveland, 1982; Davison, 1980–1981; Dustin, 1978; Firtel, 1981; Goldman et al., 1976; Gozes, 1982; Gozes and Littauer, 1982; Kirschner, 1978, Lazarides, 1980, 1981; Olmsted and Borisy, 1973; Roberts and Hyams, 1979; Schiff and Horwitz, 1981; Timasheff and Grisham, 1980; Wilson and Bryan, 1974).

Microtubules are ubiquitous cytoskeletal filaments believed to be involved in the formation of a large variety of structures such as cytoplasmic networks, mitotic spindles, and highly ordered organelles such as centrioles and flagella axonema. Functionally, microtubules are involved in a number of dynamic processes such as chromosomal movement, cell migration, intracellular transport, formation of cell shape, and secretion (Bray, 1973; Bray and Gilbert, 1981; Dustin, 1978; Roberts and Hyams, 1979). In the nervous system microtubules are highly abundant in nerve processes. Their number per axon and per length increases with development, and axon outgrowth as well as axoplasmic transport are also thought to depend on microtubule integrity (Daniels, 1972; Marchisio et al., 1980; Seeds et al., 1970; Shelanski and Feit, 1972; Yamada et al., 1970).

## 1. MICROTUBULE STRUCTURE AND PROPERTIES

### 1.1. Structure of Microtubules

Cytoplasmic microtubules are long cylinders with an outer diameter of 24–25 nm and a hollow core of about 14–15 nm across. The wall of the tubule usually consists of thirteen longitudinal protofilaments in a helical arrangement lying parallel to the long axis of the tubule (Amos et al., 1976). In other organelles such as flagella, centrioles, and basal bodies a more complex arrangement of the protofilaments has been observed. The major component of protofilaments is tubulin, a heterodimer (Luduena et al., 1977) composed of $\alpha$ and $\beta$ subunits, each having a molecular mass of approximately 50 kDa. In addition to tubulin, microtubules are composed of several other proteins (approximately 10–20% depending on the conditions of the *in vitro* assembly) designated microtubule-associated proteins (MAPs). Two major groups of MAPs have been identified in neural tissues. These include high-molecular-mass components (350 and 270 kDa) termed MAP1 and MAP2, respectively, (Borisy et al., 1975; Murphy and Borisy, 1975; Murphy et al., 1977; Sloboda et al., 1976) and a number of polypeptides in the molecular mass range of 55–62 kDa known collectively as TAU proteins (Cleveland et al., 1977; Weingarten et al., 1975). The most prominent MAP components in neuronal tissue are the high-molecular-mass MAP1 and MAP2. The latter was found by electron microscopy to associate with microtubules at a fixed periodicity forming laterally projecting arms (Amos, 1977; Kim et al., 1979; Vallee and Borisy, 1977).

An increasing body of evidence suggests that the microtubules interact with other cytoskeletal structures such as intermediate filaments or with cellular organelles including mitochondria or pigment granules (Ball and Singer, 1982; Berkwitz et al., 1977; Burton and Fernandez, 1973; Geiger and Singer, 1980; Leterrier et al., 1981; Masurovsky et al., 1981; Rice et al., 1980; Schliwa, 1979; Soiefer and Czosnek, 1980; Wuerker and Palay, 1969). Although the molecular nature of these interactions is not clear, it has been suggested that at least some of them might be mediated by MAP2 molecules (Griffith and Pollard, 1978; Schliwa and van Blerkom, 1981).

Microtubules from different sources show different degrees of stability in response to physical and chemical treatments. For example, in living cells most cytoplasmic microtubules depolymerize rapidly after exposure to low temperature, high pressure, or cytostatic drugs such as colchicine or vinblastine. In contrast, microtubules found in cilia and flagella are more stable to the above treatments (Olmsted and Borisy, 1973) although little is known about the factors that promote such stabilization.

### 1.2. *In Vitro* Polymerization of Microtubules

Reconstitution of microtubules from tubulin has been achieved *in vitro* (Weisenberg, 1972) and requires GTP, the removal of calcium ions, and the presence of oligomeric ring structures of tubulin that serve as nucleation centers (Borisy and Olmstead, 1972). Efficient *in vitro* polymerization occurs at 37°C in organic buffers (pH 6.6–6.9) in the presence of GTP, magnesium ions, and EGTA and is stimulated by glycerol and the presence of MAPs. The polymerization reaction requires a minimal tubulin concentration (designated the critical concentration) which varies according to the conditions used. The assembly of tubulin appears to involve the hydrolysis of one GTP molecule per tubulin monomer added to a growing microtubule. The microtubules assembled *in vitro* can be depolymerized at low temperatures or by the addition of calcium as well as by the antimitotic drugs colchicine and vinblastine (Kirschner, 1978; Timasheff and Grisham, 1980). On the other hand, the drug taxol promotes the rate, extent, and number of nucleations in the polymerization of tubulin. In addition, it stabilizes microtubules and renders them resistant to depolymerization by cold and $CaCl_2$ treatments (Schiff and Horwitz, 1981; Shiff et al., 1979).

Two distinct phases can be distinguished during the *in vitro* polymerization of tubulin: a slow nucleation step and a faster elongation process (Gaskin et al., 1974; Johnson and Borisy, 1977). The intermediate steps of the *in vitro* polymerization reaction are not yet clear. It has been suggested that during the initiation of tubulin polymerization, oligomeric ring strutures (that are present along with free tubulin dimers) open to form short protofilaments and that this is followed by their lateral aggregation to yield helical ribbons. Additional protofilaments are added until there are thirteen in all, and the ribbons then fold to assume a hollow cylindrical form. Elongation takes place presumably by the addition of tubulin dimers to the ends of these structures to form extended microtubules (for a discussion of the various polymerization models see Kirschner, 1978).

### 1.3. Organization and Dynamic Properties

In living cells microtubules seem to grow from distinct nucleating regions found at or near centrioles, kinetochores, amorphous electron-dense bodies or basal bodies. Such regions have been collectively termed microtubule-organizing centers (MTOCs) (Brinkley et al., 1981; Kirschner, 1978; Raff, 1979; Sharp et al., 1981). The growth from the organizing center is polar; asymmetric tubulin molecules are all oriented in the same way along the microtubule axis (Hyams, 1980, 1981). In differentiating neuroblastoma cells microtubules have been visualized by immunofluorescent techniques using antibodies directed against tubulin (Isenberg et al., 1977; Marchisio et al., 1978; Scherson et al., 1982) or antibodies directed against MAP2 (Scherson et al., 1982) or observed after microinjection of fluorescently labeled MAPs (Scherson et al., 1984). In these cells the microtubules form an elaborate network arising near the cytocenter and radiating over long distances to the cell periphery and into the elongating processes where they are organized mainly in parallel bundles. The microtubules diverge from the bundles at the growth cone and form a fan-like pattern which does not reach the extreme edge of the growth cone (Marchisio et al., 1980; Geiger et al., unpublished results) as shown previously by electron microscopy in nerve cells (Yamada et al., 1971).

The origin of cytoplasmic microtubules could not be readily traced in neuroblastoma cells because of their rounded morphology and the complexity of the microtubule network. However, this technical difficulty was overcome by observing microtubules in cells recovering from colchicine poisoning. Following depolymerization of microtubules and retraction of the cell neurites by colchicine, the cultures were exposed to colchicine-free medium; the regrowth of microtubules and processes was observed by indirect immunofluorescence. Such studies revealed several MTOCs in differentiating neuroblastoma cells (Gozes, 1982; Marchisio et al., 1979; Spiegelman et al., 1979), probably related to multiple centrioles clustered together in the perinuclear areas of these cells (Sharp et al., 1981). Unlike the neuroblastoma cells many non-neuronal cells display only one or two MTOCs per cell (Brinkley et al., 1979; Watt and Harris, 1980).

The dynamic nature of microtubules and the broad spectrum of functions in which they are involved suggest that the growth of microtubules is controlled not only temporally but also spatially in all cells. The detailed mechanisms by which microtubules regulate this process are not yet well understood (Scherson et al., 1984). Since individual microtubules are not contractile in nature, it is conceivable that the generation of movement by microtubules is based on polymerization-depolymerization events (Inoue and Sato, 1967), or sliding movements between individual microtubules (McIntosh et al., 1969; Warner, 1976), or on interactions with other cell organelles (Murphy and Tilney, 1974). Polymerization-dependent movement could be involved in the generation of forces by at least two mechanisms: (1) elongation or shortening at one or both ends of the microtubule (Allen and Borisy, 1974) and (2) treadmilling [i.e., the net addition of tubulin heterodimers at one end of the microtubule ("plus" end) and a similar net dissociation of at the "minus" end (Bergen and Borisy, 1980; Cote and Borisy, 1981; Kirschner, 1980; Margolis and Wilson, 1978, 1981)]. The treadmilling model implies that at steady state there is a flux of tubulin in subunits along the filaments from the plus end to the minus end. Moreover, it has been speculated that organelle translocation along microtubules could be related to a treadmilling process (Margolis and Wilson, 1981). However, the treadmilling model is based predominantly on *in vitro* reconstitution studies. In living cells the situation appears to be more complex. Microinjection of fluorescently tagged MAP2 has recently been combined with fluorescence photobleaching recovery experiments. It was shown that bleached spots along fluorescently labeled microtubules in neurites of living neuroblastoma-liver hybrid cells did not show "movement" along the filaments. This suggests that treadmilling of MAP2-containing microtubules, at least as a process which is supposed to translocate the entire microtubule structure and possibly different cellular organelles associated with it, either does not occur in the living cultured cells tested or else is limited to a fairly slow maximal rate of about 4 $\mu$m/hr (Scherson et al., 1984).

## 2. COMPOSITION OF TUBULIN

### 2.1. Tubulin Structure

The major building block of the microtubules is the protein tubulin, a heterodimer composed of two non-identical polypeptides designated $\alpha$ and $\beta$, each having a molecular mass of about 50 kDa (Bryan and Wilson, 1971; Krauhs et al., 1981; Luduena et al., 1977; Ponstingl et al., 1981; Valenzuela et al., 1981). However, the possibility of a dynamic equilibrium existing between heterodimer and homodimer tubulin structures cannot be excluded (Gozes and Barnstable, 1982; Luduena et al., 1977). The tubulin subunits which are of similar molecular mass and amino acid composition exhibit functional differences as indicated for example by the binding of cAMP to $\alpha$-tubulin from blood platelets (Steiner, 1978) and the binding of exchangeable GTP to $\beta$-tubulin (Geahlen and Haley, 1977). The tubulin subunits also differ with respect to their N-terminal amino acid sequence (Luduena and Woodward, 1973), immunological specificity (Fulton et al., 1971; Piperno and Luck, 1977), and peptide mapping (Cleveland et al., 1977; Gozes and Littauer, 1978; Marotta et al., 1979). In addition, tubulin mRNA from brain was found to direct the synthesis of both $\alpha$- and $\beta$-tubulin subunits indicating the presence of at least two distinct tubulin mRNAs (Bryan et al., 1978; Cleveland et al., 1978; Gozes et al., 1975; Marotta et al., 1979; Saborio et al., 1978). These studies as well as the participation of tissue-specific $\beta$-tubulin in the formation of the meiotic spindle and other stages of spermatogenesis in *Drosophila* (Kemphues et al., 1979, 1980) support the view that $\alpha$- and $\beta$-tubulin subunits are products of separate genes. More recently amino acid sequence analysis of porcine brain tubulin (Krauhs et al., 1981; Ponstingl et al., 1981) as well as nucleotide sequence analysis of chick brain cDNA tubulin clones (Valenzuela et al., 1981) has shown that the $\alpha$- and $\beta$-tubulin subunits are indeed different.

Complete amino acid sequence analysis of $\alpha$-tubulin and $\beta$-tubulin from porcine brain (Krauhs et al., 1981; Ponstingl et al., 1981) shows that they consist of 451 and 445 residues, respectively. Despite the close similarity of their molecular masses the $\alpha$ and $\beta$ subunits can be separated by electrophoresis in SDS-polyacrylamide gels (or urea-SDS-polyacrylamide gels) displaying abnormally high apparent molecular masses of approximately 55,000 and 53,000, respectively (Amir-Zaltsman et al., 1982; Feit et al., 1971). The overall shape of the $\alpha$- and $\beta$-tubulin subunits is also similar (Amos, 1977) with about 41–42% amino acid sequence identity (Krauhs et al., 1981; Valenzuela et al., 1981). The high degree of homology between the two subunits suggests that they arise from a common ancestral gene. This homology, although distributed throughout the entire length of the two polypeptide chains, appears in more than 15 clusters of conserved amino acid sequences (Krauhs et al., 1981; Valenzuela et al., 1981). These conserved regions may be important in giving the two subunits their overall similarity in shape while the variable regions may be important for their more specialized functions such as binding of ligands and MAPs (Krauhs et al., 1981).

The tubulin subunits are rich in glutamic acid and aspartic acid (Bryan, 1974; Bryan and Wilson, 1971) with a pI of about 5.3 for the $\alpha$ subunit and 5.2 for the $\beta$ subunit as measured by isoelectric focusing in the presence of urea (Gozes and Littauer, 1978, 1982). The C-terminus of both subunits is a highly acidic region; the last 40 amino acid residues of $\alpha$-tubulin have 47% acidic side chains, while in the last 45 positions of $\beta$-tubulin, glutamate and aspartate comprise 42% of the residues and basic amino acid chains are absent. The highly acidic C-terminal region of $\alpha$- and $\beta$-tubulin resembles the $NH_2$-terminus of troponin T. It may be involved in the binding of divalent cations such as $Ca^{2+}$ or may be part of the GTP binding site; alternatively, it may be involved in interactions with basic MAPs which separately or together may control the assembly of tubulin (Ginzburg et al., 1981; Krauhs et al., 1981; Ponstingl et al., 1979). It is interesting to note that part of the acidic C-terminus region of $\alpha$-tubulin defined by Glu-Glu-Glu-Gly-Glu-Glu-Tyr (Ginzburg et al., 1981) is very similar to the amino acid sequence Glu-Glu-Glu-Glu-Glu-Glu-Tyr found close to the C-terminus of polyoma virus middle T-antigen (Soeda et al., 1980). The latter antigen has been found to be associated with plasma membranes (Ito et al., 1977). Perhaps the glutamic acid-rich region of polyoma virus middle T-antigen interacts with some basic membrane components, and this may be analogous to the preferential association of $\alpha$-tubulin with brain presynaptic membranes (Gozes and Littauer, 1979). Four regions of $\alpha$-tubulin are similar to actin sequences (Collins and Elzinga, 1975) and may denote a similar function for these two proteins that otherwise appear to have no evolutionary relationship (Ponstingl et al., 1981).

Tubulin is highly labile in solution and rapidly loses its capacity to polymerize and to bind colchicine which may be due to the low levels of secondary structure. The $\alpha$-tubulin appears to be rather irregularly folded. It has been predicted that only 26% of the chain is helical and that 33% has a $\beta$-sheet conformation (Ponstingl et al., 1981). Correspondingly low values have been calculated (Krauhs et al., 1981) for the $\beta$ subunit (24 and 29%) and are consistent with the results of circular dichroism studies on native tubulin (Lee and Timasheff, 1977; Ventilla et al., 1972).

**TABLE 1. DIVERGENCE OF THE AMINO ACID SEQUENCE OF α-TUBULIN**

| | Amino Acid Position | | | | | | | | | |
|---|---|---|---|---|---|---|---|---|---|---|
| Source | 136 | 175 | 265 | 266 | 271 | 272 | 273 | 340 | 358 | References |
| Pig | Ser | Pro | Ile | His | Thr | Tyr | Ala | Thr | Glu | Ponstingl et al. (1981) |
| | | | Ala | Ile | Arg | Phe | Asx | Ser | | |
| | | | Gly | | | | | | | |
| Rat | Leu | Pro | Ile | His | Thr | Tyr | Ala | Thr | Gln | Lemischka et al. (1981) |
| | | | | | | | | Ser | Gln | Ginzburg et al. (1981) |
| Chick | Ser | Arg | Ile | His | Thr | Tyr | Ala | Thr | Gln | Valenzuela et al. (1981) |

The amino acid sequence of α-tubulin from porcine brain shows that positions 265, 266, 271–273, and 340 are heterogenous (Ponstingl et al., 1981). The amino acid sequence of rat (Lemischka et al., 1981) and chick α-tubulin (Valenzuela et al., 1981) were deduced from the nucleotide sequence of cDNA clones. The amino acid sequence of rat α-tubulin between positions 307–451 was deduced from the nucleotide sequence of an independent DNA clone (Ginzburg et a., 1981). The two rat α-tubulin amino acid sequences differ at position 340 which is also heterogenous in pig α-tubulin. Rat and chick α-tubulin differ at amino acid positions 136 and 175, while pig and chick α-tubulin differ at positions 175 and 358. The first 39 amino terminal residues of chick α-tubulin have not yet been determined. Although tyrosine was reported at position 295 of chick α-tubulin, it has now been corrected to cysteine as is the case for α-tubulin from pig and rat brain (Lemischka et al., 1981). Amino acid positions are numbered as published for pig α-tubulin.

**TABLE 2. DIVERGENCE OF THE AMINO ACID SEQUENCE OF β-TUBULIN**

| | Amino Acid Position | | | | | | | | | |
|---|---|---|---|---|---|---|---|---|---|---|
| Source | 37 | 48 | 55 | 56 | 57 | 170 | 275 | 296 | 316 | References |
| Pig | His | Asn | Ala | Gly | Asn | Val | Ser | Ala | Val | Krauhs et al. (1981) |
| | Val | Ser | Ser | Ser | His | | Ala | | | |
| Chick | His | Asn | Thr | Gly | Asn | Met | Ser | Ser | Ile | Valenzuela et al. (1981) |

The amino acid sequence of β-tubulin from porcine brain shows that positions 37, 48, 55, 56, 57, and 275 are heterogenous (Krauhs et al., 1981). The amino acid sequence of chick β-tubulin was deduced from the nucleotide sequence of a cDNA clone (Valenzuela et al., 1981). Pig and chick β-tubulin differ in their amino acid sequence at positions 55, 170, 296, 316.

Tubulin is a highly conserved protein and when isolated from different sources appears to have the same overall structure. Thus, tubulin from yeast (Water and Kleinsmith, 1976), *Aspergillus* (Sheir-Neiss et al., 1978), or algae (McKeithan and Rosenbaum, 1981) can copolymerize *in vitro* with brain tubulin. Comparison of the amino acid sequence of pig tubulin (Krauhs et al., 1981; Ponstingl et al., 1981) with that deduced from the nucleotide sequence of chick (Valenzuela et al., 1981) and rat tubulin cDNA (Ginzburg et al., 1981; Lemischka et al., 1981) shows very few differences. Pig and chick α-tubulin differ at two positions (Table 1) while the difference in the β-tubulin from these sources involves four amino acids (Table 2). It should be noted, however, that due to the heterogeneity of the various isotubulins the differences between chick and pig tubulin will increase by a few more amino acid residues (see Section 2.2). The cleavage peptide patterns obtained by limited proteolytic digestion of α- and β-tubulin subunits from various sources have been compared. It was shown that the β subunits yield similar peptide patterns. However, differences were found between the cleavage products of α-tubulin from low and high eukaryotes suggesting that the β subunit may be subject to more rigorous functonal constraints than the α subunit (Little et al., 1981).

### 2.2. Tubulin Microheterogeneity

Considerable evidence points to the existence of multiple isoforms of α- and β-tubulin. Brain α- and β-tubulin subunits display extensive microheterogeneity as compared to tubulin from other tissues (Gozes and Littauer, 1978). The microheterogeneity of brain tubulin may arise from the presence of multiple isoforms in a single neuron (Gozes and Sweadner, 1981) as well as from differences in the distribution of tubulin forms present in the various brain cell types (Gozes et al., 1979, 1980). Microheterogeneity has also been observed in $\alpha$- and $\beta$-tubulin isolated from various sources (Bibring et al., 1976; Feit and Shelanski, 1975; Feit et al., 1971; Kobayashi and Mohri, 1977; Morris et al., 1979; Sheir-Neiss et al., 1978; Witman et al., 1972). In several cases separate isotubulins have been shown to be associated with specific organelles. Thus, differences have been noted in the composition and electrophoretic mobility of tubulins from mitotic apparatus, cilia, and flagella in sea urchin (Bibring et al., 1976; Stephens, 1978) and differences between cytoplasmic and flagellar isotubulins from the alga *Polytomella* have been reported (McKeithan and Rosenbaum, 1981). These results suggest that within a single cell there may be different control mechanisms for microtubule polymerization that involve specific isotubulins. Using isoelectric focusing gels, between 9 and 11 isoforms could be resolved from rat brain tubulin (Dahl and Weibel, 1979; Gozes and Littauer, 1978). More refined techniques have recently enabled the identification of up to 15–21 isoforms (Table 3) in mouse and calf brain tubulin (George et al., 1981; Stromska, 1982; von Hungen et al., 1981; Wolff et al., 1982). By contrast there are only two actin isoforms in mammalian brain (Buckingham and Minty, in press; Elzinga et al., 1976; Vandekerckhove and Weber, 1978).

Detailed studies show that within each group the various $\alpha$-isotubulins as well as the $\beta$-forms differ from each other by only a few amino acid replacements. Amino acid analysis of porcine $\alpha$- and $\beta$-tubulin shows that at least six positions carry amino acid exchanges (Tables 1 and 2) for each of the $\alpha$- or $\beta$-tubulins examined (Krauhs et al., 1981; Ponstingl et al., 1981). Most of the amino acid exchanges can be explained in terms of substitution of a single base and appear to concentrate in two variable regions of the $\alpha$-chain (positions 265–273 and 340). Analysis of the tryptic peptides show that position 265 of $\alpha$-tubulin can occupy one of three amino acid residues in four linkage groups between positions 265 and 273 (Ponstingl et al., 1981). The heterogeneity of the $\beta$ subunit is also confined to two variable regions whose positions differ from those observed in the $\alpha$ subunit. The tryptic peptides comprising residues 47–58 of $\beta$-tubulin have been found in two linkage groups (Krauhs et al., 1981). Thus, amino acid analysis of pig tubulin indicates that the $\alpha$ and $\beta$ subunits each contain about four isoforms. The considerably higher number of isotubulins resolved by isoelectric focusing (Table 3) may indicate that some of the tubulin forms are present in minute quantitites or are the result of post-translational modifications. It is predicted that some of the heterogeneous positions in both $\alpha$- and $\beta$-tubulin are located in turns of the polypeptide chain at the surface of the protein where they are available for interaction and binding with ligands or other proteins (Krauhs et al., 1981; Ponstingl et al., 1981).

TABLE 3. TUBULIN AND ACTIN ISOFORMS IN MAMMALIAN TISSUE

| Protein | Tissue | Isoform | | References[a] |
|---|---|---|---|---|
| | | Number | Type | |
| Tubulin | Adult brain | 11–21 | $\alpha$(6–7 forms), $\beta$(5–14 forms) | 1–11 |
| | Newborn brain | 9–13 | $\alpha$(6–7 forms), $\beta$(3–6 forms) | 1, 3, 6, 8–10 |
| | Prenatal brain | 6–9 | $\alpha$(4 forms), $\beta$(2–5 forms) | 5, 6, 9, 10 |
| | Cultured single neuron | 7 | $\alpha$(4 forms), $\beta$(3 forms) | 12 |
| | Mouse neuroblastoma | 4–6 | $\alpha$(1–2 forms), $\beta$(2–5 forms) | 13–15 |
| | Rat glioma | 4 | $\alpha$(2–3 forms), $\beta$(1 form) | 13 |
| Actin | Skeletal muscle | 1 | $\alpha_{SK}$ | 16, 17 |
| | Cardiac muscle | 1 | $\alpha_{C}$ | 17 |
| | Smooth muscle | 2 | $\alpha_{SM}$, $\gamma_{SM}$ | 17, 18 |
| | Non-muscle (including brain) | 2 | $\beta_{NM}$, $\gamma_{NM}$ | 19, 20, 21 |

[a]1. Gozes and Littauer (1978). 2. Marotta et al. (1978). 3. Dahl and Weibel (1979). 4. Nelles and Bamburg (1979). 5. Gozes et al. (1980). 6. Littauer et al. (1980). 7. George et al. (1981). 8. Von Hungen et al. (1981). 9. Denoulet et al. (1982). 10. Wolff et al. (1982). 11. Gozes and Littauer (1982). 12. Gozes and Sweadner (1981). 13. Gozes et al. (1979). 14. Edde et al. (1981). 15. Edde et al. (1982). 16. Collins and Elzinga (1975). 17. Vandekerckhove and Weber (1979). 18. Vandekerckhove and Weber (1981). 19. Elzinga et al. (1976). 20. Vandekerckhove and Weber (1978). 21. Buckingham and Minty (in press).

### 2.3. Post-Translational Modifications of Tubulin

One possible way of regulating the multiple functions of tubulin is through post-translational modifications. Tubulin is reportedly a substrate for several modifying enzymes. Glucosamine has been observed to be incorporated into mouse brain tubulin (Feit and Shelanski, 1975) and into tubulin secreted from mouse neuroblastoma cells in culture (Truding et al., 1975). One or both of the tubulin subunits can be phosphorylated (Eipper, 1972; Forgue and Dahl, 1979; Reddington et al., 1976). A phosphorylated tryptic peptide of rat brain $\beta$-tubulin was isolated after *in vivo* [$^{32}$P] labeling (Eipper, 1972) and was apparently located at the C-terminus (Krauhs et al., 1981). Phosphorylation of $\alpha$-tubulin from *Chlamydomonas reinhardii* has also been reported (Piperno and Luck, 1976). On the other hand, phosphorylation was not detected with either brain or thyroid tubulin (Rappaport et al., 1972); likewise, amino acid sequence analysis of $\alpha$- and $\beta$-tubulin from pig brain also failed to detect any carbohydrate moiety or phosphorylated residues (Krauhs et al., 1981; Ponstingl et al., 1981), nor was neuroblastoma tubulin found to be phosphorylated *in vivo* (Edde et al., 1982; Littauer et al., 1976; Schmitt, 1976; Solomon et al., 1976).

Nucleotide sequence analysis of cDNA clones shows that a C-terminal tyrosine residue is coded in the neucleotide sequence of $\alpha$-tubulin mRNA from chick (Valenzuela et al., 1981) and rat brain (Ginzburg et al., 1981; Lemischka et al., 1981). However, analysis of the amino acid sequence of $\alpha$-tubulin reveals variable amounts of tyrosine at its C-terminus. Thus, pig brain $\alpha$-tubulin contains 15% of tyrosine at is C-terminal end (Ponstingl et al., 1979), while in membrane-bound tubulin the terminal tyrosine residue is completely lacking (Nath and Flavin, 1978). This variability in the level of C-terminal tyrosine probably results from the action of tyrosine ligase which catalyzes the reversible detyrosylation and tyrosylation of a tyrosine residue at the C-terminal end of $\alpha$-tubulin in rat or bovine extracts (Arce et al., 1975; Argarana et al., 1977; Raybin and Flavin, 1975). The reaction is viewed as a post-translational modification specific to the $\alpha$-isotubulin forms (Gozes and Littauer, 1978; Murofushi, 1980) and may play a role in modulating the activity of tubulin in the cell.

ADP-ribosylation has emerged as a potentially important mechanism for controlling the activity of several proteins. Endogenous ADP-ribosyltransferases and several bacterial toxins have been shown to catalyze the NAD-dependent ADP-ribosylation of key regulatory proteins. For example, cholera toxin will catalyze the ADP-ribosylation of the GTP-binding regulatory protein of the hormone-sensitive adenylate cyclase system, thereby leading to more persistent cyclase stimulation (Moss and Vaughan, 1979). Our own work (Amir-Zaltsman et al., 1982) and that of others (Hawkins and

Browning, 1982; Kaslow et al., 1981) has shown that tubulin serves as a substrate for cholera-toxin-mediated NAD-dependent ADP-ribosylation. MAP1, MAP2, and non-muscle actin (but not TAU proteins) are similarly modified (Amir-Zaltsman etal., 1982). The cholera-toxin-mediated ADP-ribosylation of tubulin has also been demonstrated in permeated rat neuroblastoma cells (Amir-Zaltsman et al., 1982). However, experiments aimed at demonstrating cholera-toxin-dependent ADP-ribosylation in rat glioma cells labeled with either [$^{32}$P]-Pi or [$^{3}$H]-adenine were not successful (Hawkins and Browning, 1982). It remains to be determined whether tubulin or the MAPs are modified by endogenous ADP-ribosyltransferases present in animal cells.

## 3. MODULATION OF TUBULIN DURING BRAIN DEVELOPMENT

### 3.1. Changes in Level and in Isoform Distribution

Brain development is characterized in its early stages by intensive cell proliferation followed by later stages of cell differentiation and neurite outgrowth. During the early stages of development microtubules are necessary for cell division; later on they are involved in the formation of the neuronal networks. The concentration of tubulin is much higher in the central nervous system (CNS) than in other tissues (Borisy and Taylor, 1967); it comprises about 30–34% of the total soluble proteins of newborn rat brain (Lennon et al., 1980; Schmitt et al., 1977a). The relative amount of soluble tubulin increases during prenatal brain development, thereafter decreasing in the postnatal stage. The time period during which this decrease occurs varies in different animals and corresponds to the maturation process of that animal's brain (Bamburg et al., 1973; Dutton and Barondes, 1969; Fellous et al., 1975; Lennon et al., 1980; Littauer et al., 1980; Morrison et al., 1981; Schmitt et al., 1977b). Thus, during the first ten days after birth the percentage of cytoplasmic rat brain tubulin increases somewhat (Lennon et al., 1980) or remains almost constant (Schmitt et al., 1977a). From about that time on until the rat reaches adulthood the percentage of cytoplasmic brain tubulin decreases progressively to a value of 20%. The changes in tubulin content during brain development appear in parallel for both neurons and glia (Burgoyne et al., 1981; Gozes et al., 1977) and result from changes in the proportion of the corresponding mRNA (Gozes et al., 1975, 1977; Morrison et al., 1981; Schmitt, 1976).

Tubulin is not only confined to the cytoplasm but is also found to be associated with various membranes (Bhattacharyya and Wolff, 1975, 1976;

Estridge, 1977; Gozes and Richter-Landsberg, 1978; Zisapel and Littauer, 1979). During brain development there is a marked increase in the particulate brain proteins, paralleled by an increase in the particulate tubulin in the mature rat brain (Schmitt et al., 1977a). Some of the particulate tubulin is associated with the myelin membrane (Gozes and Richter-Landsberg, 1978), synaptosomes, postsynaptic densities, presynaptic membranes and synaptic vesicles (Chan and Bunt, 1978; Feit et al., 1977; Gozes and Littauer, 1979; Kelly and Cotman, 1978; Kornguth and Sunderland, 1975; Mahler et al., 1977; Walters and Matus, 1975; Wang and Mahler, 1976; Zisapel et al., 1980). Tubulin also appears to be associated with the axolemma of the squid giant axon (Baumgold et al., 1981). A striking feature is the observation that the $\alpha$- and $\beta$-tubulin subunits differ in their association with brain presynaptic membranes (Gozes and Littauer, 1979) as well as with synaptic vesicle membranes (Zisapel et al., 1980). In the mammalian neuron it appears that the $\alpha$ subunit can be easily dissociated from these membranes. Similarly, in rat sympathetic neurons there is an enrichment of the soluble $\beta$-tubulin in the axoplasm relative to that in the cytoplasm (Gozes, 1982), and it is interesting to note that $\beta$-tubulin is also the main subunit identified in axonal transport (Brady et al., 1980). In agreement with the above results, some differences have been noted in the distribution of isotubulin species between the particulate and cytoplasmic fractions of differentiating neuroblastoma cells (Chan and Baxter, 1979).

Although the overall synthesis of tubulin decreases during rat brain development, there is an increase in the number of isotubulin forms and a change in their relative proportions during the maturation process (Dahl and Weibel, 1979; Gozes and Littauer, 1978; Gozes et al., 1980; see also Table 3). These changes in brain tubulin microheterogeneity are regulated, at lease in part, at the mRNA level (Gozes et al., 1980). They appear to be completed by the time the rat is 10–12 days old (Dahl and Weibel, 1979) which coincides with intense elaboration of neuronal processes. However, the modulation of the relative porportions of the $\alpha$- and $\beta$-isotubulins follows a different pattern during brain development. Most of the changes in the proportion of the mouse brain $\alpha$-isotubulins occur just before birth, whereas the evolution of the more acidic $\beta$-tubulin isoforms takes place throughout brain development. It thus appears that the $\alpha$- and $\beta$-isotubulins are regulated independently (Denoulet et al., 1982; Wolff et al., 1982). In addition, changes in the rate of synthesis of isotubulins have been observed in the developing chick brain (Bryan et al., 1981; Sullivan et al., 1979). There is some dissimilarity in the distribution of isotubulins isolated from various regions of the rat brain which may to some extent reflect differences in tubulin microheterogeneity within the different brain cell types (Gozes et al., 1980). Indeed, the presence of more than one $\beta$-tubulin form specific for different cells has been established for the ovaries and testes of *Drosophila* (Kemphues et al., 1979, 1980). Most of the tubulin heterogeneity appears to be associated with neurons because analysis of cultured glia cells from mouse embryos indicates that they are able to synthesize only a small number of isotubulins (Moura Neto et al., 1982). However, the results of the analysis of mature glia cells are not yet available. It is interesting to note that a single neuron was found to contain as many isotubulins as are found in the whole brain (Gozes and Sweadner, 1981). These results suggest an alternative hypothesis, namely, that the various isoforms may be differentially utilized in different subcellular structures. This hypothesis is supported by analyses of tubulin heterogeneity in sea urchin and *Chlamydomonas* in which different isotubulins are present in the cilia and flagella, respectively, and in the cytoplasm (Lefebvre et al., 1980; Stephens, 1978).

### 3.2. Organization and Expression of Tubulin Genes

In order to study the organization of the tubulin genes and the regulation of their expression, homogenous DNA probes are necessary. cDNA clones were constructed from mRNA of rat brain (Ginzburg et al., 1980, 1981; Lemischka et al., 1981) and from embryonic chick brain (Valenzuela et al., 1981). Nucleotide sequence analysis of the tubulin cDNA clones from both rat and chick revealed some striking features. The chick $\alpha$-tubulin cDNA clone contains 1450 base pairs, which include 85% of the coding sequence and the 3′-nontranslated region. The chick $\beta$-tubulin cDNA clone is longer and contains 1730 base pairs including the 5′- and 3′-noncoding regions and the presumed polyadenylation site (ACATAAA) located 22 nucleotides 5′- to the beginning of the poly (A) tract. The rat $\alpha$-tubulin cDNA clone isolated in our laboratory contains 569 base pairs of which 125 comprise the 3′-noncoding region (Ginzburg et al., 1981). The rat cDNA clone isolated by Lemischka et al. (1981) includes the complete coding region, the 5′- and the 3′-noncoding regions as well as the AAATAAA presumed polyadenylation site. When the coding regions of all these $\alpha$-tubulin cDNA clones are compared, a strong homology is observed between the chick and rat nucleotide sequence. There is only 13% divergence in the coding part of the nucleotide sequence and, except for two amino acid substitutions, all the other changes are silent. On the other hand, when the 5′- and the 3′-noncoding regions of the above chick and rat cDNA clones were compared, little homology was observed and up to 64% diveregence was detected. Comparison of the two rat $\alpha$-tubulin cDNA clones shows that even within the same species there is a high divergence in the nucleotide sequence of the 3′-noncoding regions (Ginzburg et al., 1981; Lemischka et al., 1981). The strict evolutionary conservation of the coding region may point to the need for the preservation of many functional roles played by tubulin within the cell such as the multiplicity of binding sites, contacts between the $\alpha$ and $\beta$ subunits, and interaction with MAPs and with other filaments in the cell.

As described above, the tubulin coding sequence is highly conserved, thus enabling estimation of the number of tubulin genes in various species using rat or chick tubulin cDNA clones (Table 4). By employing a labeled rat cDNA probe we were able to observe between 10 and 20 hybridizable bands in DNA isolated from *Tetrahymena*, plant, mouse, rat, and human cells. The DNA was cleaved by means of several restriction endonucleases which do not cut within the nucleotide sequence of the cDNA clone (Ginzburg and Littauer, unpublished results). With chick tubulin cDNA probes eight specific sequences were detected in cleaved chicken DNA, 15–20 fragments in human DNA, 18–40 sequences in sea urchin DNA, and eight sequences in *Drosophila* DNA (Table 4). However, under the strict hybridization conditions employed no hybridization was detected in DNA isolated from yeast (Sanchez et al., 1980). The presence of multiple bands containing tubulin-specific sequences is indicative of a tubulin multiple gene family. It should be noted, however, that this method does not discriminate between functional, expressed genes, and silent pseudogenes. In many gene families genomic DNA sequences have been described which show homology to functional genes but do not give rise to functional RNA transcripts. Two types of pseudogenes

**TABLE 4. Tubulin Gene Number**

| | Gene Number | | | |
|---|---|---|---|---|
| Organism | Total | $\alpha$-Subunit | $\beta$-Subunit | Reference[a] |
| *Aspergillus* | >3 | >1 | >2 | 1, 2 |
| *Chlamydomonas* | 4 | 2 | 2 | 3–5 |
| *Trypanosomes* | 28–33 | 15–17 | 13–16 | 20 |
| *Drosophila* | 8 | 4 | 4 | 6–10 |
| Sea urchin | 18–40 | 9–20 | 9–20 | 11, 12 |
| Chicken | 8 | 4 | 4 | 11, 13 |
| Rat | 17–20 | 10–11 | 11 | 11, 14–16 |
| Human | 15–20 | 7–10 | 7–10 | 11, 17–19 |

[a]1. Sheir-Neiss et al. (1978). 2. Morris et al. (1979). 3. Silflow and Rosenbaum (1981). 4. McKeithan and Rosenbaum (1981). 5. Brunke et al. (1982). 6. Sanchez et al. (1980). 7. Kalfayan and Wensink (1981). 8. Kalfayan and Wensink (1982). 9. Raff et al. (1982). 10. Mischke and Pardue (1982). 11. Cleveland et al. (1980). 12. Alexandraki and Ruderman (1981). 13. Cleveland et al. (1981b). 14. Ginzburg et al. (1983b). 15. Lemischka et al. (1981). 16. Gozes et al. (1980). 17. Cowan et al. (1981). 18. Wilde et al. (1982a). 19. Wilde et al. (1982b). 20. Thomashow et al. (1983).

have been described: "processed" pseudogenes which presumably originate via the reverse transcription of active mRNA followed by an integration event, possibly involving a retrovirus-like sequence (Hollis et al., 1982), and "duplicated" pseudogenes derived by duplication and subsequent mutation of a functional gene (Little, 1982). A rat $\alpha$-tubulin (Lemischka and Sharp, 1982) and two human "processed" $\beta$-tubulin pseudogenes (Wilde et al., 1982b, 1982c) have been described, each of them terminating in a poly(A) tract. The rat $\alpha$-tubulin pseudogene is highly homologous to the expressed $\alpha$-tubulin gene described by this group (Lemischka and Sharp, 1982). We have sequenced another $\alpha$-tubulin pseudogene (Ginzburg and Littauer, unpublished results) which contains several small deletions, does not contain intervening sequences, and has three termination codons. Of the two human $\beta$-pseudogenes isolated, one contains an intevening sequence; the other does not but contains the polyadenylation signal and an oligoadenylate tract at its 3′ end. In addition, it is characterized by the presence of short flanking repeat sequences (Wilde et al., 1982a).

An important question in a multiple gene family in which the various members are coordinately expressed is whether this property requires the clustering of the genes. Using cloned cDNA sequences specific for chick $\alpha$- or $\beta$-tubulin it was shown that the genes coding for these proteins are unlinked and their members are dispersed throughout the chicken genome; $\alpha$- and $\beta$-tubulin sequences were located on four and two different chromosomes, respectively (Cleveland et al., 1981a). Upon more detailed analysis using a cell sorter to fractionate metaphase chromosomes one $\alpha$-tubulin gene was localized on chromosome 1 and another on chromosome 8, whereas two $\beta$-tubulin genes were localized on chromosome 2 (Cleveland et al., 1980). *In situ* hybridization of tubulin cDNA to *Drosophila melanogaster* chromosomes shows that both the $\alpha$- and the $\beta$-tubulin are dispersed as well (Sanchez et al., 1980). Thus, the controlled expression of tubulin genes is not necessarily based on the activation of a single chromatin domain. In contrast with the dispersed organization of tubulin genes described above, trypanosome $\alpha$ and $\beta$-tubulin genes are physically linked and clustered in tandem repeats of approximately 13–17 copies per haploid genome of alternating $\alpha$ and $\beta$-tubulin sequences (Thomashow et al., 1983).

The significance of the tubulin gene family is still obscure. It is possible that each functional gene will code for the synthesis of a tubulin isoform that may be more efficient in performing certain functions. Another possibility involves the control of the expression of tubulin synthesis so that if low amounts are required, only one gene is activated provided that the regulatory signals in the different genes are different. The differences in the sequences of the various isotubulins would then represent evolutionary changes during the periods that these genes originated from a common ancestral gene.

Partial resolution of tubulin mRNAs into fractions containing $\alpha$- and $\beta$-tubulin has also been obtained (Bryan et al., 1978; Cleveland et al., 1978; Gozes et al., 1980; Portier et al., 1980a; Saborio et al., 1978). Although the molecular masses of the $\alpha$- and $\beta$-tubulin subunits are very similar, it is possible to separate the corresponding mRNA species in non-denaturing gels, methylmercury-containing gels, or sucrose gradients (Bryan et al., 1978; Ginzburg et al., 1982; Gozes et al., 1980). The results obtained indicate a high degree of secondary structure for the tubulin mRNA molecules. The $\alpha$-tubulin mRNA is longer than the $\beta$-tubulin mRNA sequence. Thus, the mRNA molecules coding for $\alpha$- and $\beta$-tubulin from *Drosophila* chick or rat brain were estimated to be between 1800 and 2000 nucleotides long. This would imply that the noncoding regions contain about 400–600 nucleotides (Ginzburg et al., 1980; Lemischka et al., 1981; Valenzuela et al., 1981). However, in the case of $\beta$-tubulin mRNA an additional species of about 2900 bases has also been detected (Cleveland et al., 1980). Using a rat brain tubulin cDNA probe it was shown that the synthesis of tubulin during brain development is controlled at the mRNA level (Ginzburg et al., 1983a). Thus, it was found that the level of tubulin mRNA is higher (by about 165%) in 12-day-old rat brain than in newborn brain. At 30 days of age, however, the level of tubulin mRNA decreases to about 75% of the newborn level (Ginzburg et al., 1983a). These results are in agreement with previous studies which showed a correlation between the changes in tubulin levels and the amount of translationally active mRNA in developing rat brain as well as in the cerebellum (Morrison et al., 1981; Schmitt et al., 1977a). It appears that the changes in the level of tubulin mRNA sequences are even more pronounced than the modulation of the steady-state tubulin amounts observed in the the developing brain (Bryan et al., 1981; Schmitt et al., 1977a). Transcriptionally active genes were observed to be located in chromatin structures that are more sensitive to DNase I digestion (Weintraub and Groudine, 1976). We have found that some of the tubulin genes are more sensitive to nuclease digestion of nuclei isolated from rat brain than from muscle cells. An age-dependent decrease in sensititivity to digestion was also observed which correlates with the changes described above for the tubulin mRNA synthesis (Ginzburg et al., 1983b).

Despite the overall decrease in total tubulin synthesis more isotubulin mRNA species are expressed in the adult than in the newborn rat brain; this may offer a hint as to their specific functions (Gozes et al., 1980). Relative to newborn brain, enhancement in the level of at least one isotubulin mRNA species was observed in preparations isolated from mature brain. The mRNA species coding for several isotubulin species could be partially separated by centrifugation through sucrose gradients. From these studies it appears that brain mRNA may consist of at least five different mRNA species coding for distinct isotubulins and that these are developmentally controlled (Gozes et al., 1980).

Recently a novel mechanism for the regulation of microtubule levels was proposed. It was shown that treatment of cells with colchicine or nocodazole which depolymerize microtubules and raise the level of unpolymerized tubulin cause directly or indirectly a decrease in the level of tubulin mRNA in many types of cells (Ben-Zeev et al., 1979; Cleveland et al., 1981b). Using cDNA probes for $\alpha$- and $\beta$-tubulin it was shown in several rodent cell lines (Cleveland et al., 1981b) as well as in human lymphoblastoid cells (Fellous et al., 1982) that there is a rapid decrease in tubulin mRNA levels in response to treatment with depolymerizing alkaloids. The presence of a subpopulation of tubulin mRNA molecules which exhibits a short half-line of 1–2 hr may allow the cell to respond rapidly to these environmental changes. It is suggested that this mechanism of autoregulation exists to keep tight control on the concentration of unpolymerized tubulin levels and hence prevent spontaneous (non-nucleated) assembly (Cleveland et al., 1981b). A similar rapid response was detected following flagella excision in *C. reinhardii*. Increased tubulin mRNA synthesis was detected within 5–10 min following deflagellation and flagellar resorption, reaching 15- to 35-fold over the control level within 45–50 min. After 3–4 hr, by which time flagellar regeneration was complete, the level of tubulin mRNA decreased and reached the control level (Minami et al., 1981; Silflow and Rosenbaum, 1981).

### 3.3. Expression of Tubulin in Cultured Cell Lines

The multiplicity of tubulin isoforms and the changes which occur in their relative proportions during brain maturation could be related to the different roles in isotubulins within the nerve cells. Another source for this heterogeneity was thought to be the large number of cell types in the brain and the changing ratio between these populations during development. In the rat brain the glial cells continue to divide postnatally while most neuronal cells cease to divide at birth. Thus, the ratio between these two cell types, which at birth is higher for neurons, changes in favor of glial cells postnatally. Cloned cells originating from nervous tissue provide a suitable model system for studying specific changes that take place during cell differentiation (Kimhi, 1981). The different cell lines exhibiting properties of neurons or glial cells can be induced to differentiate morphologically and functionally following the addition of various inducers (Hamprecht, 1977; Littauer et al., 1978; Nelson et al., 1976; Prasad, 1975; Soreq et al., 1983). Under these conditions mouse neuroblastoma cells extend neurites (Schubert et al., 1969; Seeds et al., 1970), enzymes involved in neurotransmitter synthesis are induced (Waymire et al., 1972), and their membranes exhibit electrical properties (Kimhi et al.,

1976). On the molecular level it was shown using cDNA probes that the sequence complexity of RNA isolated from both neuroblastoma and glioma cells is greater than in other somatic cell lines. This observation is in accord with the greater degree of complexity found in total brain mRNA than in mRNA from other organs (Grouse et al., 1979, 1980).

Tubulin microheterogeneity was also observed in mouse neuroblastoma cells although to a lesser extent than in brain (Edde et al., 1981, 1982; Gozes et al., 1979). Various chemicals which induce differentiation have differing effects on tubulin heterogeneity, thus suggesting various patterns of differentiation in these cells. Dimethylsulfoxide, which induces morphological differentiation, does not affect either the level or the rate of synthesis of tubulin, implying that in this case the assembly of microtubules necessary for neurite outgrowth proceeds from a pre-existing pool of tubulin. On the other hand, the use of other agents such as L-methyl cyclohexane carboxylic acid (CCA) or serum withdrawal and dibutyryl cAMP ($Bt_2cAMP$) did affect tubulin synthesis and the distribution of the various isotubulins (Chan and Baxter, 1979; Edde et al., 1981; Portier et al., 1980b; Schmitt, 1976). Pulse-chase experiments with [$^{35}S$]-methionine indicate that one isotubulin form ($\beta_2$) is regulated at the post-translational level during serum deprivation of neuroblastoma cells (Edde et al., 1981). It is interesting to note that various inducers also differ in their effects on vimentin levels in neuroblastoma cells. Thus serum deprivation will result in a decrease (Edde et al., 1982) and treatment with CCA will induce an increase in vimentin levels. However, vimentin accumulation in CCA-treated cells is unrelated to neurite extension and appears as the result of interaction between the cell membrane and the culture substrate (Portier et al., 1980b). There is a reduction in the level of tubulin and actin mRNAs during differentiation of neuroblastoma cells in culture. The amount of synthesis of these two proteins analyzed by the products of an *in vitro* cell-free system directed by polysomal mRNAs is lower in $Bt_2cAMP$-treated cells than in untreated neuroblastoma cells (Morrison et al., 1980).

Treatment of Ramos cells, a human cell line of lymphoblastoid origin, with either $\alpha$ or $\beta$-interferon induced a marked increase in the amount of tubulin mRNA sequences (Fellous et al., 1982). The level of tubulin mRNA increased rapidly after cells were exposed for as little as 1–4 hr and was between four- and sevenfold higher than the control level. Expression of $\alpha$- and $\beta$-tubulin does not seem to be coordinately controlled since the treatment with $\beta$-interferon induced a more pronounced increase in the level of the $\alpha$- than the $\beta$-tubulin mRNA. The specific induction of tubulin mRNA in interferon-treated cells was abolished when the cells were pretreated with colchicine suggesting the existence of two independent regulatory sites for tubulin mRNA expression. It should be noted that cytoplasmic actin mRNA levels were not affected following interferon treatment of these cells indicating that the two cytoskeletal elements, tubulin and actin, are differently regulated (Fellous et al., 1982). A reduction in expression of both tubulin and actin mRNA during the morphological differentiation of 3T3 adipocytes has also been reported (Spiegelman and Farmer, 1982).

Recently, using cDNA probes a biphasic regulation by $Bt_2cAMP$ of tubulin and actin mRNA levels was observed in neuroblastoma cells (Ginzburg et al., 1983a). $Bt_2cAMP$ initially increases and later decreases the level of tubulin mRNA in a neuroblastoma-glioma hybrid cell line as well as in the parent cells. A significant increase in tubulin mRNA sequences is already evident 1 hr after adding the inducer. It then levels off and is followed by a down-regulation of the tubulin mRNA level to half of the value found in untreated cells. The expression of actin mRNA during the differentiation of these cells differed from that of tubulin mRNA, suggesting again that the expression of these two cytoskeletal proteins are independently regulated.

## 4. MICROTUBULE-ASSOCIATED PROTEINS

Microtubules isolated from cell homogenates by successive cycles of warm assembly and cold disassembly steps are composed of tubulin and several additional proteins known as MAPs. These include high-molecular-weight proteins (MAP1 and MAP2), TAU factors, and several other proteins. Both MAP2 and TAU factors are heat stable and can be purified and separated from tubulin and other MAPs by thermic treatment of microtubules assembled *in vitro* (Fellous et al., 1977; Weingarten et al., 1975). MAP2 and TAU factors have been shown to promote nucleation and elongation of microtubules *in vitro*, but their role in the control of microtubule assembly and their interaction with other cell structures in the living cell is still uncertain.

SDS–polyacrylamide gel electrophoresis resolves MAP2 into two bands with apparent molecular masses of 288 and 271 kDa. The MAP2 molecule is highly asymmetric, and electron microscopic studies with reconstituted microtubules indicate that these molecules are bound to the microtubular core with fixed periodicity and project radially from it (Kim et al., 1979; Murphy and Borisy, 1975; Vallee, 1980; Vallee and Borisy, 1977). Proteolytic digestion of MAP2-containing microtubules results in the specific removal of the projecting arms leaving behind a microtubule-binding peptide of about 32 kDa (Vallee, 1980). Besides the tubulin binding domain of MAP2, other functions have been assigned to the projecting "arm" of the molecule. These include a binding site for cAMP-dependent protein kinase which is associated with that region, the capacity to undergo phosphorylation (Coughlin et al., 1980; Theurkauf and Vallee, 1982; Vallee, 1980; Vallee et al., 1981), and perhaps the capacity to mediate interactions between the microtubules and other cytoplasmic elements (Ball and Singer, 1982; Geiger and Singer, 1980; Griffith and Pollard, 1978; Pytela and Wiche, 1980; Schliwa and van Blerkom, 1981). Indirect immunofluorescence experiments with monoclonal antibodies directed against terminal domains distal to the microtubule binding site of MAP2 indicate that this portion of the "arm" is exposed when MAP2 associates with microtubules in the cell body and neurites of differentiated rat brain neuroblastoma cells (Scherson et al., 1982). It is interesting to note that out of a number of different MAP2 monoclonal antibodies none reacted against the microtubule binding regions. It was inferred that such regions may be ubiquitous and conserved during evolution and may thus have a tendency to be immunologically silent. On the other hand, it has been suggested that the structure of the arm region of MAP2 may vary in different species (Scherson et al., 1982).

The tissue and species distribution of MAP2 is still controversial. Immunocytochemical data suggest that this protein is associated with microtubules from a wide variety of cells and tissues (Cleveland et al., 1979; Connolly et al., 1978; Sherline, 1978; Sherline and Schiavone, 1977; Sloboda and Dickersin, 1980). Other groups, on the other hand, have used multispecific and monoclonal antibodies to show that MAP2 is present only in cells of neuronal origin (Izant and McIntosh, 1980). Various neuroblastoma clones differ in their MAP2 content. It appears to be absent from cell lines derived from mouse neuroblastoma C1300 (Pallas and Solomon, 1982; Schmitt et al., 1977b) while significant amounts are found in rat neuroblastoma B104 cells (Izant and McIntosh, 1980; Scherson et al., 1982). Other cells such as HeLa mainly contain distinct types of MAP components with molecular masses of 125, 210, and 255 kDa (Bulinski and Borisy, 1980, 1980b; De Brabander et al., 1981; Vallee, 1982; Weatherbee et al., 1980). It should be noted that MAP2 molecules in cultured cells are easily lost during the detergent extraction used for its visualization by indirect immunofluorescence staining and may therefore be missed altogether (Scherson et al., 1982; Schliwa and van Blerkom, 1981). Another cause for ambiguity could arise from the purity and specificity of the antibodies used.

The distribution of tubulin and high-molecular-weight MAPs (HMWs consisting of a mixture of MAP1 and MAP2) in brain were determined by immunoperoxidase histochemistry with specific antisera. Tubulin was found in microtubules of both neurons and glia and in axons as well as in dendrites. HMWs, on the other hand, were detected only in neurons where they associated with dendritic but not with axonal microtubules (Matus et al., 1981). These results are consistent with the observation that there is a fivefold lower level of MAP2 relative to tubulin in white matter (from corpus callosum) than in dendrite-rich gray matter (from cerebral cortex). On the other hand, MAP1 and TAU proteins were present in both white and gray matter at roughly equal ratios to tubulin (Vallee, 1982). The finding that HMWs are not transported with tubulin in peripheral axons (Lasek, 1970) is consistent with their low content in these processes. It was also shown that Purkinje cell dendrites contain neither tubulin nor microtubules at early stages of their growth. However, the developing dendrites are rich in HMW suggesting that these proteins act as a framework upon which the microtubules are subsequently built (Bernhardt and Matus, 1982). The above results suggest that the distribution of MAPs and their mode of association with microtubules in neurons may vary in different subcellular structures.

The TAU factors appear to be composed of four to six closely spaced bands (55–52 kDa) when subjected to electrophoresis in SDS–polyacrylamide gels and a multiplicity of forms when analyzed by two-dimensional gel electrophoresis (Cleveland et al., 1977). In contrast to MAP1 and MAP2, which are already found in young rat brain (about 16%) (Schmitt et al., 1977b), there is a marked increase in the diversity of TAU factors during brain development and an enhancement of TAU ability to induce *in vitro* microtubule assembly (Mareck et al., 1980). Newborn rat brain contains two TAU forms whose molecular masses are somewhat different from those of the four to six TAU components associated with microtubules from developing brain (Ginzburg et al., 1982; Mareck et al., 1980). However, the two groups of proteins are immunologically related (Ginzburg et al., 1982). The TAU factors from adult brain have a better ability to promote *in vitro* tubulin polymerization (Fellous et al., 1976; Francon et al., 1978), and their appearance coincides with an increase in tubulin microheterogeneity in the adult brain (Gozes and Littauer, 1978). It appears that the various TAU forms are different gene products since the major differences between the TAU isoforms during brain development are controlled at the mRNA level (Ginzburg et al., 1982). It is possible that the distinct isotubulins and TAU subspecies are nonrandomly distributed within a given microtubular system and may generate microtubule subclasses with different stabilities and functions. Indeed, enhanced synthesis of $\beta_2$-isotubulin and two TAU forms were found in goldfish retinae during optic nerve regeneration suggesting that these isoforms play a role in the formation of the new axonal matrix (Neuman et al., unpublished results).

## 5. EXPRESSION OF MICROTUBULE PROTEINS DURING NERVE REGENERATION

Cytoskeletal elements play a fundamental role in nerve regeneration. The proteins of the axon are synthesized in the nerve cell body and then delivered by axonal transport to the axonal elongation site (Estridge and Bunge, 1978).

The axon itself does not contain protein-synthesizing machinery and depends on the synthesis and transport of these macromolecules from the neuronal cell body into the growing tip (Lasek, 1970).

Among the initial phases of axonal regeneration, sprouting and elongation depend on the correct organization of microfilaments and microtubules. Application of colchicine, which causes depolymerization of microtubules, was shown to inhibit axonal outgrowth (Daniels, 1972; Estridge and Bunge, 1978; Yamada et al., 1970). Motility of the growth cone, which depends on interaction between the growing end of microfilaments and the cell membranes, is inhibited by cytochalasin B which blocks actin polymerization. As a result the movement of the growth cone stops and the microfilaments either depolymerize or aggregate (Bray and Gilbert, 1981; Cribbs et al., 1982). In both cases the cytoskeletal architecture is disrupted, thus preventing functional axonal recovery. Injection of radioactive amino acid precursors into the nerve cell bodies enables newly synthesized proteins to be labeled and the axonal flow of the labeled proteins to be followed (Black and Lasek, 1980). Such analysis reveals five classes of proteins having different rates of axonal flow. In the group containing the most slowly moving proteins two subclasses were identified. The slowest group, designated slow component a (Sca), consists of tubulin and neurofilament proteins as well as the MAPs of the TAU family. The high-molecular-weight MAPs do not enter the axons and are found only in cell bodies and dendrites (Lasek and Brady, 1981; Matus et al., 1981). The other group, designated slow component b (Scb), contains many polypeptides, including actin, and thus represents the transport of the microfilament network together with several other proteins complexed to it (Black and Lasek, 1980). It is interesting to note that although the rates of Sca and Scb may differ in different species, the same proteins will remain in each defined class of axonal flow. The Scb proteins are particularly interesting because their rate of movement corresponds to the rate of axonal elongation during nerve regeneration (Lasek et al., 1981). Both $\alpha$-and $\beta$-tubulin were found in the slowest phase of axoplasmic transport in mouse retinal ganglion cells. However, a regional heterogeneity along the axon was observed for several $\alpha$-tubulin forms suggesting post-translational processing of these subunits during axoplasmic transport (Brown et al., 1982).

The regulation of tubulin expression during optic nerve regeneration has been investigated in the goldfish retina. Neurons of the CNS of lower vertebrates, unlike most of the neurons in mammalian CNS, regenerate following optic nerve axotomy and re-establish appropriate connections in the optic tectum. This model system has been studied by electrophysiological, anatomical, and more recently by biochemical methods (Heacock and Agranoff, 1976; Jacobson, 1978). During the recovery process changes in the composition of several proteins and the rate of their axonal transport were reported. Analysis of labeled proteins synthesized in the retinal ganglia cells (i.e., the cell bodies of the injured axons) as well as of the proteins undergoing axonal transport has revealed enhanced labeling of tubulin and of several other acidic proteins having a molecular mass of 68–70 kDa (Heacock and Agranoff, 1982). In addition, an increase was noted in both the level and transport of tubulin under these conditions (Giulian et al., 1980). An increase in translatable tubulin mRNA isolated from regenerating retinae was detected 10 days following the lesion when maximal labeling of tubulin protein was observed (Burrell et al., 1979). Hybridization experiments with tubulin cDNA probes have shown that the increase of tubulin labeling during regeneration of the goldfish nerve is due to an enhanced level of tubulin mRNA molecules and does not arise from increased translatability of pre-existing mRNA (Neuman et al., unpublished results). Further analysis of the enhancement in [$^{35}$S]-methionine labeling of tubulin during nerve regeneration indicates that this increase is more pronounced in one of the $\beta$-isoforms of tubulin (Neuman et al., unpublished results). It is interesting to note that in differentiating neuroblastoma cells extension of neurites was associated with preferential synthesis of a particular $\beta$-isoform of tubulin (Croall and Morrison, 1980; Edde et al., 1981; Portier et al., 1982). The involvement of the TAU MAPs was also investigated during the regeneration process in goldfish retinae. Enhanced synthesis of two specific TAU factors was observed in cultured regenerating retinae (Neuman et al., unpublished results). These results are in accord with the observation that both tubulin and TAU factors are both constituents of the same slow component of axonal transport (Black and Lasek, 1980). The enhanced synthesis of two TAU factors could be associated with the increased level of the $\beta_2$-tubulin isoform; together they may be involved in the organization of specific microtubules within the regenerating axon.

## ACKNOWLEDGMENTS

Research that originated in the authors' laboratory was supported by grants from the Muscular Dystrophy Association and the U.S.–Israel Binational Science Foundation (BSF). U. Z. Littauer was the recipient of an EMBO fellowship and would like to thank Drs. P. Marks and D. Rifkind for their kind hospitality during his stay at the Memorial Sloan–Kettering Cancer Center and to Dr. I. Gozes, T. Scherson for her helpful suggestions, and D. Giveon for his help in preparing the manuscript.

## REFERENCES

Alexandraki, D. and J. V. Ruderman, *Mol. Cell Biol.* **1,** 1125 (1981).

Allen, C. and G. G. Borisy, *J. Mol. Biol.* **90,** 381 (1974).

Amir-Zaltsman, Y., E. Ezra, T. Scherson, A. Zutra, U. Z. Littauer, and Y. Salomon, *EMBO J.* **1,** 181 (1982).

Amos, L. A., *J. Cell Biol.* **72,** 642 (1977).

Amos, L. A., R. W. Linck, and A. Klug, in R. Goldman, T. Pollard, and J. Rosenbaum, Eds., *Cell Motility,* Vol. 3, Cold Spring Harbor Laboratory, Cold Spring Harbor, NY, p.847 (1976).

Arce, C. A ., J. A. Rodriguez, H. S. Barra, and R. Caputto, *Eur. J. Biochem.* **59,** 145 (1975).

Argarana, C. E., C. A. Arce, H. S. Barra, and R. Caputto, *Arch. Biochem. Biophys.* **180,** 264 (1977).

Ball, E. H. and S. J. Singer, *Proc. Natl. Acad. Sci. USA* **79,** 123 (1982).

Bamburg, J. R., E. M. Shooter, and L. Wilson, *Biochemistry* **12,** 1476 (1973).

Baumgold, J., S. Terakawa, K. Iwasa, and H. Gainer, *J. Neurochem.* **36,** 759 (1981).

Ben-Zeev, A., S. R. Farmer, and S. Penman, *Cell* **17,** 319 (1979).

Bergen, L. G. and G. G. Borisy, *J. Cell Biol.* **84,** 141 (1980).

Berkwitz, S. A., J. Katagiri, H. H. Binder, and R. C. Williams, Jr., *Biochemistry* **16,** 5610 (1977).

Bernhardt, R. and A. Matus, *J. Cell Biol.* **92,** 589 (1982).

Bhattacharyya, B. and J. Wolff, *J. Biol. Chem.* **250,** 7639 (1975).

Bhattacharyya, B. and J. Wolff, *Nature* **264,** 576 (1976).

Bibring, T., J. Baxandall, S. Denslow, and B. Walker, *J. Cell Biol.* **69,** 301 (1976).

Black, M. M. and R. J. Lasek, *J. Cell Biol.* **86,** 616 (1980).

Borisy, G. G. and J. B. Olmsted, *Science* **177,** 1196 (1972).

Borisy, G. G. and E. W. Taylor, *J. Cell Biol.* **34,** 525 (1967).

Borisy, G.G., J. M. Marcum, J. B. Olmsted, D. B. Murphy, and R. A. Johnson, *Ann. NY Acad. Sci.* **253,** 107 (1975).

Brady, S. T., M. Tytell, I. G. McQuarrie, and R. J. Lasek, *J. Cell Biol.* **87,** 250 (1980).

Bray, D., *Nature (London)* **224,** 93 (1973).

Bray, D. and D. Gilbert, *Ann. Rev. Neurosci.* **4,** 505 (1981).

Brinkley, B. R., S. M. Cox, and D. A. Pepper, *J. Cell Biol.* **83,** 349a (1979).

Brinkley, B. R., S. M. Cox, and S. H. Fistel, *Neurosci. Res. Prog. Bull.* **19,** 108 (1981).

Brown, B. A., R. A. Nixon, and C. A. Marotta, *J. Cell Biol.* **94,** 159 (1982).

Brunke, K. J., E. E. Young, B. U. Buchbinder, and D. P. Weeks, *Nucleic Acids Res.* **10,** 1295 (1982).

Bryan, J., *Fed. Proc.* **33,** 152 (1974).

Bryan J. and L. Wilson, *Proc. Natl. Acad. Sci. USA* **68,** 1762 (1971).

Bryan, R. N., G. A. Cutter, and M. Hayashi, *Nature* **272,** 81 (1978).

Bryan, R. N., J. Bossinger, and M. Hayashi, *Dev. Biol.* **81,** 349 (1981).

Buckingham, M. E. and A. J. Minty, in N. Maclean, S. George, and R. Flavell, Eds., *Eukaryotic Genes; Their Structure, Activity and Regulation,* Butterworths, London, (in press).

Bulinski, J. C. and G. G. Borisy, *J. Cell Biol.* **87,** 792 (1980a).

Bulinski, J. C. and G. G. Borisy, *J. Cell Biol.* **87,** 802 (1980b).

Burgoyne, R. D., J. S. Rudge, and S. Murphy, *J. Neurochem.* **36,** 661 (1981).

Burrell, H. R., A. M. Heacock, R. D. Water, and B. W. Agranoff, *Brain Res.* **168,** 628 (1979).

Burton, P. R. and H. L. Fernandez, *J. Cell Sci.* **12,** 567 (1973).

Chan, K. Y. and C. F. Baxter, *Brain Res.* **174,** 135 (1979).

Chan, K. Y. and A. H. Bunt, *J. Neurocytol.* **7,** 137 (1978).

Cleveland, D. W., *Cell* **28,** 689 (1982).

Cleveland, D. W., S. Y. Hwo, and M. W. Kirschner, *J. Mol. Biol.* **116,** 227 (1977).

Cleveland, D.W., M. W. Kirschner, and N. J. Cowan, *Cell* **15,** 1021 (1978).

Cleveland, D. W., B. M. Spiegelman, and M. W. Kirschner, *J. Biol. Chem.* **254,** 12670 (1979).

Cleveland, D. W., M. A. Lopata, R. J. MacDonald, N. J. Cowan, W. J. Rutter, and M. W. Kirschner, *Cell* **20,** 95 (1980).

Cleveland, D. W., S. H. Hughes, E. Stubblefield, M. W. Kirschner, and H. E. Varmus, *J. Biol. Chem.* **256,** 3130 (1981a).

Cleveland, D. W., M. A. Lopata, P. Sherline, and M. W. Kirschner, *Cell* **25,** 537 (1981b).

Collins, J. H. and M. Elzinga, *J. Biol. Chem.* **250,** 5915 (1975).

Connolly, J. A., V. I. Kalnins, D. W. Cleveland, and M. W. Kirschner, *J. Cell Biol.* **78,** 781 (1978).

Cote, R. H. and G. G. Borisy, *J. Mol. Biol.* **150,** 577 (1981).

Coughlin, B. A., H. D. White, and D. L. Purich, *Biochem. Biophys. Res. Commun.* **92,** 89 (1980).

Cowan, N. J., C. D. Wilde, L. T. Chow, and F. C. Wefald, *Proc. Natl. Acad. Sci. USA* **78,** 4877 (1981).

Cribbs, D. H., J. R. Glenney, Jr., P. Kaulfus, K. Weber, and S. Lin, *J. Biol. Chem.* **257,** 395 (1982).

Croall, D. E. and M. R. Morrison, *J. Mol. Biol.* **140,** 549 (1980).

Dahl, J. L. and V. J. Weibel, *Biochem. Biophys. Res. Commun.* **86,** 822 (1979).

Daniels, M. P., *J. Cell Biol.* **53,** 164 (1972).

Davison, P. F., in H. Schweiger, Ed., *International Cell Biology,* Springer-Verlag, New York, p. 286 (1980–1981).

De Brabander, M., J. C. Bulinski, G. Geuens, J. De Mey, and G. G. Borisy, *J. Cell Biol.* **91,** 438 (1981).

Denoulet, P., B. Edde, C. Jeantet, and F. Gros, *Biochimie* **64,** 165 (1982).

Dustin, D., in P. Dustin, Ed., *Microtubules,* Springer-Verlag, New York (1978).

Dutton, G. R. and S. Barondes, *Science* **166,** 1637 (1969).

Edde, B., C. Jeantet, and F. Gros, *Biochem. Biophys. Res. Commun.* **103,** 1035 (1981).

Edde, B., M. M. Portier, C. Sanuquillo, C. Jeantet, and F. Gros, *Biochimie* **64,** 141 (1982).

Eipper, B. A., *Proc. Natl. Acad. Sci. USA* **69,** 2283 (1972).

Elzinga, M., B. J. Maron, and R. S. Adelstein, *Science* **191,** 94 (1976).

Estridge, M., *Nature* **268,** 60 (1977).

Estridge, M. and R. Bunge, *J. Cell Biol.* **79,** 138 (1978).

Feit, H. and M. L. Shelanski, *Biochem. Biophys. Res. Commun.* **66,** 920 (1975).

Feit, H., L. Slusarek, and M. L. Shelanski, *Proc. Natl. Acad. Sci. USA* **68,** 2028 (1971).

Feit, H., P. Kelly, and C. W. Cotman, *Proc. Natl. Acad. Sci. USA* **74,** 1047 (1977).

Fellous, A., J. Francon, A. Virion, and J. Nunez, *FEBS Lett.* **57,** 5 (1975).

Fellous, A., J. Francon, A. M. Lennon, and J. Nunez, *FEBS Lett.* **64,** 400 (1976).

Fellous, A., J. Francon, A. M. Lennon, and J. Nunez, *Eur. J. Biochem.* **78,** 167 (1977).

Fellous, A., I. Ginzburg, and U. Z. Littauer, *EMBO J.* **1,** 835 (1982).

Firtel, R. A., *Cell* **24,** 6 (1981).

Forgue, S. T. and J. L. Dahl, *J. Neurochem.* **32,** 1015 (1979).

Francon, J., A. Fellous, A.M. Lennon, and J. Nunez, *Eur. J. Biochem.* **85,** 43 (1978).

Fulton, C., R. E. Kane, and R. E. Stephens, *J. Cell Biol.* **50,** 762 (1971).

Gaskin, F., C. R. Cantor, and M. L. Shelanski, *J. Mol. Biol.* **89,** 737 (1974).

Geahlen, R. L. and B. E. Haley, *Proc. Natl. Acad. Sci. USA* **74,** 4375 (1977).

Geiger, B. and S. J. Singer, *Proc. Natl. Acad. Sci. USA* **77,** 4769 (1980).

George, H. J., L. Misra, D. J. Field, and J. C. Lee, *Biochemistry* **20,** 2402 (1981).

Ginzburg, I., A. de Baetselier, M. Walker, L.Behar, H. Lehrach, A. M. Frischauf, and U. Z. Littauer, *Nucleic Acids Res.* **8,** 3553 (1980).

Ginzburg, I., L. Behar, D. Givol, and U.Z. Littauer, *Nucleic Acids Res.* **9,** 2691 (1981).

Ginzburg, I., T. Scherson, D. Giveon, L. Behar, and U. Z. Littauer, *Proc. Natl. Acad. Sci. USA* **79,** 4892 (1982).

Ginzburg, I., S. Rybak, Y. Kimhi, and U. Z. Littauer, *Proc. Natl. Acad. Sci. USA* **80,** 4243 (1983a).

Ginzburg, I., T. Scherson, S. Rybak, Y. Kimhi, D. Neuman, M. Schwartz, and U. Z. Littauer, Cold Spring Harbor Symposia on Quantitative Biology, Vol. 48: *Molecular Neurobiology,* Cold Spring Harbor Laboratory, Cold Spring Harbor, NY, (1983b).

Giulian, D., H. Des Ruisseaux, and D. Cowburn, *J. Biol. Chem.* **255,** 6494 (1980).

Goldman, R., T. Pollard, and J. Rosenbaum, Eds., *Cell Motility* Vol. 3, Cold Spring Harbor Laboratory, Cold Spring Harbor, (1976).

Gozes, I. *Neurochem. Intl.* **4,** 101 (1982).

Gozes, I. and C. J. Barnstable, *Proc. Natl. Acad. Sci. USA* **79,** 2579 (1982).

Gozes, I., and U. Z. Littauer, *Nature* **276,** 411 (1978).

Gozes, I. and U. Z. Littauer, *FEBS Lett.* **99,** 86 (1979).

Gozes, I. and U. Z. Littauer, *Scand. J. Immunol.* **15,** Suppl. 9, 299 (1982).

Gozes, I. and C. Richter-Landsberg, *FEBS Lett.* **95,** 169 (1978).

Gozes, I. and K. J. Sweadner, *Nature* **294,** 477 (1981).

Gozes, I., H. Schmitt, and U. Z. Littauer, *Proc. Natl. Acad. Sci. USA* **72,** 701 (1975).

Gozes, I., M. D. Walker, A. M. Kaye, and U. Z. Littauer, *J. Biol. Chem.* **252,** 1819 (1977).

Gozes, I., D. Saya, and U. Z. Littauer, *Brain Res.* **171,** 171 (1979).

Gozes, I., A. de Baetselier, and U. Z. Littauer, *Eur. J. Biochem.* **103,** 13 (1980).

Griffith, L. M. and T. D. Pollard, *J. Cell Biol.* **78,** 958 (1978).

Grouse, L. D., C. H. Letendre, and B. K. Schrier, *J. Neurochem.* **33,** 583 (1979).

Grouse, L. D., B. K. Schrier, C. H. Letendre, M. Y. Zubairi, and P. G. Nelson, *J. Biol. Chem.* **255,** 3871 (1980).

Hamprecht, B., *Int. Rev. of Cytology* **49,** 99 (1977).

Hawkins, D. J. and E. T. Browning, *Biochemistry* **21,** 4474 (1982).

Heacock, A. M. and B. W. Agranoff, *Proc. Natl. Acad. Sci. USA* **73,** 828 (1976).

Heacock, A. M. and B. W. Agranoff, *Neurochem. Res.* **7,** 771 (1982).

Hollis G. F., P. A. Hieter, O. W. McBride, D. Swan, and P. Leder, *Nature* **296,** 321 (1982).

Hyams, J., *Nature (London)* **284,** 402 (1980).

Hyams, J., *Nature (London)* **291,** 107 (1981).

Inoue, S. and H. Sato, *J. Gen. Physiol. Suppl.* **50,** 259 (1967).

Isenberg, G., E. Rieske, and G. W. Kreutzberg, *Cytobiologie* **15,** 382 (1977).

Ito, Y., J. R. Brocklehurst, and R. Dulbecco, *Proc. Natl. Acad. Sci. USA* **74,** 4666 (1977).

Izant, J. G. and J. R. McIntosh, *Proc. Natl. Acad. Sci. USA* **77,** 4741 (1980).

Jacobson, M., in M. Jacobson, Ed., *Developmental Neurobiology,* Plenum, New York (1978).

Johnson, K. A. and G. G. Borisy, *J. Mol. Biol.* **117,** 1 (1977).

Kalfayan, L. and P. C. Wensink, *Cell* **24,** 97 (1981).

Kalfayan, L. and P. C. Wensink, *Cell* **29,** 91 (1982).

Kaslow, H. R., V. E. Groppi, M. E. Abood, and H. R. Bourne, *J. Cell Biol.* **91,** 410 (1981).

Kelly, P. T. and C. W. Cotman, *J. Cell Biol.* **79,** 173 (1978).

Kemphues, K. J., R. A. Raff, T. C. Kaufman, and E. C. Raff, *Proc. Natl. Acad. Sci. USA* **76,** 3991 (1979).

Kemphues, K. J., E. C. Raff, R. A. Raff, and T. C. Kaufman, *Cell* **21,** 445 (1980).

Kim, H., L. I. Binder, and J. L. Rosenbaum, *J. Cell Biol.* **80,** 266 (1979).

Kimhi, Y., in P. G. Nelson and M. Lieberman, Eds., *Excitable Cells in Tissue Culture,* Plenum, New York, p.173 (1981).

Kimhi, Y., C. Palfrey, I. Spector, Y. Barak, and U. Z. Littauer, *Proc. Natl. Acad. Sci. USA* **73,** 462 (1976).

Kirschner, M. W., *Intl. Rev. of Cytology* **54,** 1 (1978).

Kirschner, M. W., *J. Cell Biol.* **86,** 330 (1980).

Kobayashi, Y. and H. Mohri, *J. Mol. Biol.* **116,** 613 (1977).

Kornguth, S. E. and E. Sunderland, *Biochem. Biophys. Acta* **393,** 100 (1975).

Krauhs, E., M. Little, T. Kempf, R. Hofer-Warbinek, W. Ade, and H. Ponstingl, *Proc. Natl. Acad. Sci. USA* **78,** 4156 (1981).

Lasek, R. J., *Intl. Rev. Neurobiol.* **13,** 289 (1970).

Lasek, R. J. and S. T. Brady, *Cold Spring Harbor Symp. Quant. Biol.* **46,** 113 (1981).

Lasek, R. J., I. G. McQuarrie, and J. R. Wujek, in A. Gorio, Ed., *Posttraumatic Peripheral Nerve Regeneration,* Raven Press, New York, p. 59 (1981).

Lazarides, E., *Nature* **283,** 249 (1980).

Lazarides, E., *Cell* **23,** 649 (1981).

Lee, J. C. and S. D. Timasheff, *Biochemistry* **16,** 1754 (1977).

Lefebvre, P. A., C. D. Silflow, E. D. Wieben, and J. L. Rosenbaum, *Cell* **20,** 469 (1980).

Lemischka, I. and P. A. Sharp, *Nature* **300,** 330 (1982).

Lemischka, I. R., S. Farmer, V. R. Racaniello, and P. A. Sharp, *J. Mol. Biol.* **151,** 101 (1981).

Lennon, A. M., J.Francon, A. Fellous, and J. Nunez, *J. Neurochem.* **35,** 804 (1980).

Leterrier, J. F., R. K. H. Liem, and M. L. Shelanski, *J. Cell Biol.* **90,** 755 (1981).

Littauer, U. Z., H. Schmitt, and I. Gozes, *J. Natl. Cancer Inst.* **57,** 647 (1976).

Littauer, U. Z., C. Palfrey, Y. Kimhi, and I. Spector, *Natl. Cancer Inst. Monograph* **48,** 333 (1978).

Littauer, U. Z., A. de Baetselier, I. Ginzburg, and I. Gozes, in U. Z. Littauer, Ed., *Neurotransmitters and Their Receptors,* Wiley, Chichester, U. K. p. 547 (1980).

Little, M., R. F. Luduena, G. M. Langford, C. F. Asnes, and K. Farrell, *J. Mol. Biol.* **149,** 95 (1981).

Little, P. F. R., *Cell* **28,** 683 (1982).

Luduena, R. F. and D. O. Woodward, *Proc. Natl. Acad. Sci. USA* **70,** 3594 (1973).

Luduena, R. F., E. M. Shooter, and L. Wilson, *J. Biol. Chem.* **252,** 7006 (1977).

Mahler, H. R., U. J. Wang, A. De Blas, and G. Crawford, in S. Roberts, A. Lajtha, and W. H. Gispen, Eds., *Mechanisms, Regulation and Special Function of Protein Synthesis in the Brain,* Elsevier/North Holland, New York, p. 205 (1977).

Marchisio, P. C., M. Osborn, and K. Weber, *Brain Res.* **155,** 229 (1978).

Marchisio, P. C., K. Weber, and M. Osborn, *Eur. J. Cell Biol.* **20,** 45 (1979).

Marchisio, P. C., K. Weber, and M. Osborn, in E. Giacobini, A. Vernadakis, and A. Shahar, Eds., *Tissue Culture in Neurobiology,* Raven Press, New York, p. 99 (1980).

Mareck, A., A. Fellous, J. Francon, and J. Nunez, *Nature* **284,** 353 (1980).

Margolis, R. L. and L. Wilson, *Cell* **13,** 1 (1978).

Margolis, R. L. and L. Wilson, *Nature* **293,** 705 (1981).

Marotta, C. A., J. L. Harris, and J. M. Gilbert, *J. Neurochem.* **30,** 1431 (1978).

Marotta, C. A., P. Strocchi, and J. M. Gilbert, *J. Neurochem.* **33,** 231 (1979).

Masurovsky, E. B., E. R. Peterson, S. M. Crain, and S. B. Horwitz, *Brain Res.* **217,** 392 (1981).

Matus, A., R. Bernhardt, and T. Hugh-Jones, *Proc. Natl. Acad. Sci. USA* **78,** 3010 (1981).

McIntosh, J. R., P. K. Hepler, and D. G. van Wie, *Nature* **224,** 659 (1969).

McKeithan, T. W. and J. L. Rosenbaum, *J. Cell Biol.* **91,** 352 (1981).

Minami, S. A., P. S. Collis, E. E. Young, and D. P. Weeks, *Cell* **24,** 89 (1981).

Mischke, D. and M. L. Pardue, *J. Mol. Biol.* **156,** 449 (1982).

Morris, N. R., M. H. Lai, and C. E. Oakley, *Cell* **16,** 437 (1979).

Morrison, M. R., S. Pardue, N. Prashad, D. E. Croall, and R. Brodeur, *Eur. J. Biochem.* **106,** 463 (1980).

Morrison, M. R., S. Pardue, and W. S. T. Griffin, *J. Biol. Chem.* **256,** 3550 (1981).

Moss, J. and M. Vaughan, *Ann. Rev. Biochem.* **48,** 581 (1979).

Moura Neto, V., C. Jeantet, J. Glowinski, and A. Prochiantz, *Proceedings of the First European Congress on Cell Biology,* Reprinted in, *Biology of the Cell* **145,** 145 (1982).

Murofushi, H., *J. Biochem. (Tokyo)* **87,** 979 (1980).

Murphy, D. B. and G. G. Borisy, *Proc. Natl. Acad. Sci. USA* **72,** 2696 (1975).

Murphy, D. B. and L. G. Tilney, *J. Cell Biol.* **61,** 757 (1974).

Murphy, D. B., K. A. Johnson, and G. G. Borisy, *J. Mol. Biol.* **117,** 33 (1977).

Nath, J. and M. Flavin, *FEBS Lett.* **95,** 335 (1978).

Nelles, L. P. and J. R. Bamburg, *J. Neurochem.* **32,** 477 (1979).

Nelson, P., C. Christian, and M. Nirenberg, *Proc. Natl. Acad. Sci. USA* **73,** 123 (1976).

Olmsted, J. B. and G. G. Borisy, *Ann. Rev. Biochem.* **42,** 507 (1973).

Pallas, D. and F. Solomon, *Cell* **30,** 407 (1982).

Piperno, G. and D. S. L. Luck, *J. Biol. Chem.* **251,** 2161 (1976).

Piperno, G. and D. J. L. Luck, *J. Biol. Chem.* **252,** 383 (1977).

Ponstingl, H., M. Little, E. Krauhs, and T. Kempf, *Nature* **282,** 423 (1979).

Ponstingl, H., E. Krauhs, M. Little, and T. Kempf, *Proc. Natl. Acad. Sci. USA* **78,** 2757 (1981).

Portier, M. M., C. Jeantet, and F. Gros, *Eur. J. Biochem.* **112,** 610 (1980a).

Portier, M. M., B. Edde, F. Berthelot, B. Croizat, and F. Gros, *Biochem. Biophys. Res. Commun.* **96,** 1610 (1980b).

Portier, M. M., B. Croizat, and F. Gros, *FEBS Lett.* **146,** 283 (1982).

Prasad, M. K., *Biol Rev.* **50,** 129 (1975).

Pytela, R. and G. Wiche, *Proc. Natl. Acad. Sci. USA* **77,** 4808 (1980).

Raff, E. C., *Intl. Rev. Cytol.* **59,** 1 (1979).

Raff, E. C., M. T. Fuller, T. C. Kaufman, K. J. Kemphues, J. E. Rudolph, and R. A. Raff, *Cell* **28,** 33 (1982).

Rappaport, L., J. F. Leterrier, and J. Nunez, FEBS Lett. **26,** 349 (1972).

Raybin, D. and M. Flavin, *Biochem. Biophys. Res. Commun.* **65,** 1088 (1975).

Reddington, M., L. P. Tan, and J. R. Lagnado, *J. Neurochem.* **27,** 1229 (1976).

Rice, R. V., P. F. Roslansky, N. Pascoe, and S. M. Houghton, *J. Ultrastruct. Res.* **71,** 303 (1980).

Roberts, K. and J. Hyams, Eds., *Microtubules,* Academic, New York (1979).

Saborio, J. L., E. Palmer, and I. Meza, *Exptl. Cell Res.* **114,** 365 (1978).

Sanchez, F., J. E. Natzle, D. W. Cleveland, M. W. Kirschner, and B. J. McCarthy, *Cell* **22,** 845 (1980).

Scherson, T., B. Geiger, Z. Eshhar, and U. Z. Littauer, *Eur. J. Biochem.* **129,** 295 (1982).

Scherson, T., T. E. Kreis, J. Schleisinger, U. Z. Littauer, G. G. Borisy, and B. Geiger, *J. Cell Biol.*, 425 (1984).

Schiff, P. B. and S. B. Horwitz, in A. C. Sartorelli, J. S. Lazo, and J. R. Bertino, Eds., *Molecular Actions and Targets for Cancer Chemotherapeutic Agents,* Academic, New York, p. 483 (1981).

Schiff, P. B., J. Fant, and S. B. Horwitz, *Nature* **277,** 665 (1979).

Schliwa, M., *Exptl. Cell Res.* **118,** 323 (1979).

Schliwa, M. and J. van Blerkom, *J. Cell Biol.* **90,** 222 (1981).

Schmitt, H., *Brain Res.* 115, 165 (1976).

Schmitt, H., I. Gozes, and U. Z. Littauer, *Brain Res.* **121,** 327 (1977a).

Schmitt, H., R. Josephs, and E. Reisler, *Nature* **265,** 653 (1977b).

Schubert, D., S. Humphreys, C. Baroni, and M. Cohn, *Proc. Natl. Acad. Sci. USA* **64,** 316 (1969).

Seeds, N. W., A. G. Gilman, T. Amano, and M. W. Nirenberg, *Proc. Natl. Acad. Sci. USA* **66,** 160 (1970).

Sharp, G. A., M. Osborn, and K. Weber, *J. Cell Sci.* **47,** 1 (1981).

Sheir-Neiss, G., M. H. Lai, and N. R. Morris, *Cell* **15,** 639 (1978).

Shelanski, M. L. and H. Feit, in G. H. Bourne, Ed., *The Structure and Function of the Nervous Tissue,* Vol. 6, Academic, New York, p. 47 (1972).

Sherline, P., *Exptl. Cell Res.* **115,** 460 (1978).

Sherline, P. and K. Schiavone, *Science* **198,** 1038 (1977).

Silflow, C. D. and J. L. Rosenbaum, *Cell* **24,** 81 (1981).

Sloboda, R. D. and K. Dickersin, *J. Cell Biol.* **87,** 170 (1980).

Sloboda, R. D., W. L. Dentler, R. A. Bloodgood, B. R. Telzer, S. Granett, and J. L. Rosenbaum, in R. Goldman, T. Pollard, and J. Rosenbaum, Eds., *Cell Motility,* Vol. 3, Cold Spring Harbor Laboratory, Cold Spring Harbor, NY, p. 1171 (1976).

Soeda, E., J. R. Arrand, N. Smolar, J. E. Walsh, and B. E. Griffin, *Nature* **283,** 445 (1980).

Soifer, D. and H. Czosnek, *J. Neurochem.* **35,** 1128 (1980).

Solomon, F., R. Gysin, M. Rentsch, and D. Monard, *FEBS Lett.* **63,** 316 (1976).

Soreq, H., R. Miskin, A. Zutra, and U. Z. Littauer, *Dev. Brain Res.* **7,** 257 (1983).

Spiegelman, B. M. and S. R. Farmer, *Cell* **29,** 53 (1982).

Spiegelman, B. M., M. A. Lopata, and M. W. Kirschner, *Cell* **16,** 253 (1979).

Steiner, M., *Nature* **272,** 834 (1978).

Stephens, R. E., *Biochemistry* **17,** 2882 (1978).

Stromska, D. P., *Anal. Biochem.* **119,** 387 (1982).

Sullivan, K. F., K. Farrel, and L. Wilson, *J. Cell Biol.* **83,** 351a (1979).

Theurkauf, W. E. and R. B. Vallee, *J. Biol. Chem.* **257,** 3284 (1982).

Thomashow, L. S., M. Milhausen, W. J. Rutter, and N. Agabian, *Cell* **32,** 35 (1983).

Timasheff, S. N. and L. M. Grisham, *Ann. Rev. Biochem.* **49,** 565 (1980).

Truding, R., M. L. Shelanski, and P. Morell, *J. Biol. Chem.* **250,** 9348 (1975).

Valenzuela, P., M. Quiroga, J. Zaldivar, W. J. Rutter, M. W. Kirschner, and D. W. Cleveland, *Nature* **289,** 650 (1981).

Vallee, R. B., *Proc. Natl. Acad. Sci. USA* **77,** 3206 (1980).

Vallee, R. B., *J. Cell Biol.* **92,** 435 (1982).

Vallee, R. B. and G. G. Borisy, *J. Biol. Chem.* **252,** 377 (1977).

Vallee, R. B., M. J. DiBartolomeis, and W. E. Theurkauf, *J. Cell Biol.* **90,** 568 (1981).

Vandekerckhove, J. and K. Weber, *Eur. J. Biochem.* **90,** 451 (1978).

Vandekerckhove, J. and K. Weber, *Differentiation* **14,** 123 (1979).

Vandekerckhove, J. and K. Weber, *Eur. J. Biochem.* **113,** 595 (1981).

Ventilla, M., C. R. Cantor, and M. Shelanski, *Biochemistry* **11,** 1554 (1972).

von Hungen, K., R. C. Chin, and C. F. Baxter, *J. Neurochem.* **37,** 511 (1981).

Walters, B. B. and A. I. Matus, *Nature* **275,** 496 (1975)

Wang, Y. J. and H. R. Mahler, *J. Cell Biol.* **71,** 639 (1976).

Warner, F. D., in R. Goldman, T. Pollard, and J. Rosenbaum, Eds., *Cell Motility,* Vol. 3, Cold Spring Harbor Laboratory, Cold Spring Harbor, NY, p. 891 (1976).

Water, R. D. and L. J. Kleinsmith, *Biochem. Biophys. Res. Commun.* **70,** 704 (1976).

Watt, F. M. and H. Harris, *J. Cell Sci.* **44,** 103 (1980).

Waymire, J. C., N. Weiner, and K. N. Prasad, *Proc. Natl. Acad. Sci. USA* **69,** 2241 (1972).

Weatherbee, J. A., R. B. Luftig, and R. R. Weihing, *Biochemistry* **19,** 4116 (1980).

Weingarten, M. D., A. H. Lockwood, S. Y. Hwo, and M. W. Kirschner, *Proc. Natl. Acad. Sci. USA* **72,** 1858 (1975).

Weintraub, H. and M. Groudine, *Science* **193,** 848 (1976).

Weisenberg, R. C., *Science* **177** 1104 (1972).

Wilde, C. D., C. E. Crowther, and N. J. Cowan, *J. Mol. Biol.* **155,** 533 (1982a).

Wilde, C. D., C. E. Crowther, T. P. Cripe, M. Gwo-Shu Lee, and N. J. Cowan, *Nature* **297,** 83 (1982b).

Wilde, C. D., C. E. Crowther, and N. J. Cowan, *Science* **217,** 549 (1982c).

Wilson, L. and J. Bryan, *Adv. Cell. Molec. Biol.* **3,** 21 (1974).

Witman, G. B., K. Carlson, and J. L. Rosenbaum, *J. Cell Biol.* **54,** 540 (1972).

Wolff, A., P. Denoulet, and C. Jeantet, *Neurosci. Lett.* **31,** 323 (1982).

Wuerker, R. B. and S. L. Palay, *Tissue and Cell* **1,** 337 (1969).

Yamada, K. M., B. S. Spooner, and N. K. Wessells, *Proc. Natl. Acad. Sci. USA* **66,** 1206 (1970).

Yamada, K. M., B. S. Spooner, and N. K. Wessells, *J. Cell Biol.* **49,** 614 (1971).

Zisapel, N. and U. Z. Littauer, *Eur. J. Biochem.* **95,** 51 (1979).

Zisapel, N., M. Levi, and I. Gozes, *J. Neurochem.* **34,** 26 (1980).

# 6

# *Modification of Gene Expression in the Mammalian Brain after Hyperthermia*

***Ian R. Brown***



Physiologically relevant increases in body temperature (i.e., hyperthermia) induce immediate and very specific changes in gene expression in the mammalian brain. Elevation of body temperature to levels similar to that attained in fever reactions induces the activation of a translational control mechanism which results in a rapid global inhibition of protein synthesis in the mammalian brain. Hyperthermia also induces the selective expression of a gene coding for a brain protein which is similar in molecular weight to one of the major heat shock proteins previously reported in unicellular organisms and tissue culture systems following elevation of ambient temperature (Ashburner and Bonner, 1979; Schlesinger et al., 1982). Heat shock proteins are thought to be involved in homeostatic control mechanisms which enable cells to survive stressful conditions. It has been suggested that the synthesis of heat shock proteins is correlated to the phenomenon of acquired thermotolerance (Li et al., 1982).

During brain development or following perturbation of brain protein synthesis by various chemical and physical stimuli, relatively few changes in the synthesis of abundant brain proteins have been noted (Brown and Cosgrove, 1983). The few apparent changes tend to be quantitative rather than qualitative. Two of the more prominent changes are the decline in the relative synthesis of tubulin and actin during postnatal development of the mammalian brain (Hall and Lim, 1981; Morrison et al., 1981) and the induction of synthesis of a heat shock protein in the mammalian brain following elevation of body temperature (Brown et al., 1982a, 1982b; Cosgrove and Brown, 1983).

The synthesis of heat shock protein in the mammalian brain provides a convenient inducible system for analyzing mechanisms of control of gene expression in the brain at both the transcriptional and the translational levels. A physiologically relevant stimulus (i.e., hyperthermia) induces the rapid and selective synthesis of a highly abundant brain protein which is readily detected by two-dimensional gel fluorography within 1 hr of the elevation of body temperature. During this selective induction of the heat shock protein, the synthesis of other brain proteins is inhibited by a translational control mechanism. The induction of the brain heat shock protein also involves selective changes at the transcriptional level since the total cellular level of mRNA coding for the heat shock protein is greatly increased within 1 hr of the elevation of body temperature to levels similar to that attained during fever reactions.

## 1. EFFECT OF HYPERTHERMIA ON BRAIN PROTEIN SYNTHESIS

### 1.1. Drug-Induced Hyperthermia

In a series of studies the psychotropic drug LSD has been utilized as an experimental probe to perturb protein synthesis in the mammalian brain (Brown et al., 1982b). Following intravenous administration to young adult rabbits the drug binds to neurotransmitter receptors, altering patterns of nerve firing and inducing profound changes in sensory perception. At a biochemical level the binding of the drug to neurotransmitter receptors triggers a transient inhibition of brain protein synthesis which is rapid in onset (i.e., detectable within 10–15 min with maximal inhibition by about 30–60 min and a gradual recovery to normal protein synthesis levels by 4–6 hr.

The inhibitory effect of LSD on the protein synthesis apparatus of the mammalian brain has been confirmed by various experimental approaches, for example, the disaggregation of brain polysomes to monosomes (Heikkila and Brown, 1979a, 1981; Holbrook and Brown, 1976), the decreased incorporation of labeled amino acids *in vivo* (Freedman et al., 1981; Holbrook and Brown, 1976), and the transient inhibition of brain protein synthesis assayed in a cell-free system which is prepared at various time points following intravenous administration of the drug (Cosgrove et al., 1981). These studies have also revealed that LSD induces a dose-dependent increase in body temperature in rabbits and that the degree of drug-induced hyperthermia correlates to the degree of inhibition of brain protein synthesis (Heikkila and Brown, 1979a).

LSD-induced hyperthermia results from an inhibitory action on heat loss mechanisms plus an excitatory action on heat production (Bligh, 1979). A disruption in heat loss occurs because the drug interacts with central nervous system receptors involved in vasoconstriction. This leads to a reduction of blood flow through the ears which decreases heat loss.

As shown in Figure 1, the time course of the rise and fall in body temperature induced by the intravenous injection of LSD parallels the kinetics of brain polysome disaggregation. Both rectal temperature and brain polysome disaggregation show maximal levels 1 hr after drug administration and return to near normal levels by 4 hr. Procedures which accentuate drug-induced disaggregation of brain polysomes, such as application of mild stress, are associated with greater increases in body temperatures while procedures which prevent LSD-induced disaggregation of polysomes, such as prior injection of neurotransmitter-blocking agents, produce no increase in body temperature (Heikkila and Brown, 1979a).

### 1.2. Hyperthermia Induced by Fever or Elevation of Ambient Temperature

An obvious question is the determination of whether hyperthermia induced by means other than LSD can influence protein synthesis in the mammalian brain. Fever can be readily induced in rabbits by the intravenous injection of a bacterial lipopolysaccharide pyrogen (Heikkila and Brown, 1979b). As shown in Figure 2, injection of a bacterial pyrogen, which elevates body temperature by 2°C, induces a massive disaggregation of brain polysomes to monosomes (compare panels A and D).

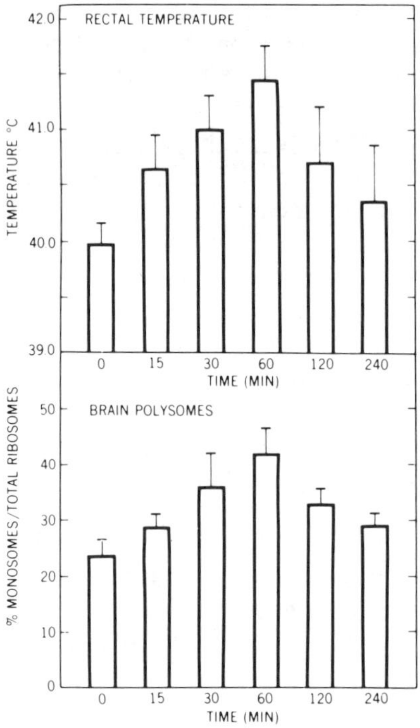


**Figure 1.** Time course of LSD-induced hyperthermia and brain polysome disaggregation. LSD was injected intravenously to young adult rabbits at 50 $\mu$g/kg body weight, and at the time points indicated rectal temperature was recorded. Cerebral hemisphere polysomes were also isolated and analyzed on sucrose gradients. The area under polysome and monosome regions of the gradients was determined by planimetry. Reprinted from Heikkila, J. J., and Brown, I. R., *Neurochem. Res.* **4,** 763–776 (1979). Copyright: Plenum Publishing Corp.

Body temperature in rabbits can also be increased by placement of animals at an elevated ambient temperature. Rabbits control their body temperature via heat loss through their ears and when placed in an incubator, this mechanism is hampered and body temperature rises. As shown in Figure 2, when body temperature is increased by 2°C following elevation of ambient temperature, a massive disaggregaton of brain polysomes to monosomes is observed (compare panels A and G).

These data together with the previously mentioned results on LSD-induced hyperthermia suggest that polysomes in the mammalian brain are sensitive to physiologically relevant increases in body temperature. Hyperthermia induced by three quite different procedures causes a rapid inhibitory effect on the translational apparatus of the mammalian brain.

## 2. MECHANISM OF TRANSLATIONAL INHIBITION

The hyperthermia-induced inhibition of protein synthesis in the mammalian brain could be due to a temperature-induced lesion at a specific stage in protein synthesis, premature termination of translation, or degradation of

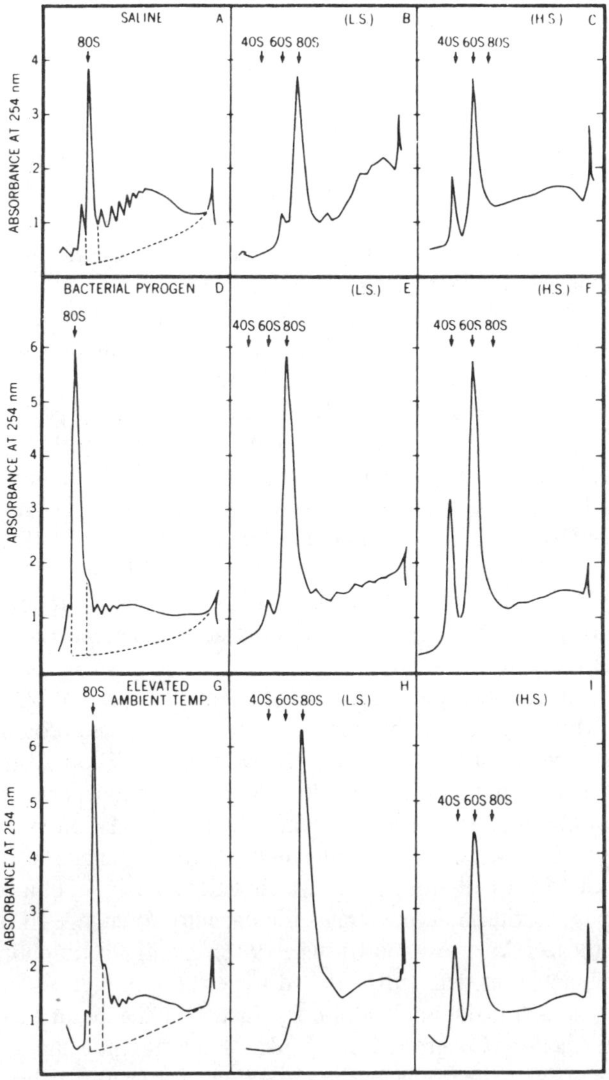


**Figure 2.** Effect of hyperthermia on brain polysomes. Body temperature of young adult rabbits was elevated from 39.5 to 41.5°C either by intravenous injection of the bacterial pyrogen *Salmonella abortus equi* lipopolysaccharide (5 $\mu$g/kg body weight) or by placement of animals at an elevated ambient temperature of 37°C. Two hours after treatment cerebral hemisphere polysomes were isolated and fractionated on 15–45% sucrose gradients under either low-salt (LS = 25 m*M* KCl) or high-salt (HS = 500 m*M* KCl) conditions. Centrifugation in a Beckman SW-41 rotor at 40,000 rpm at 4°C was either for 75 min to show the entire polysome profile (panels a, d, and g) or for 3.5 hr to display ribosomal subunits (panels b, c, e, f, h, and i). The vertical dashed lines indicate the separation between monosomes and polysomes while the horizontal line is a blank gradient. Reprinted from Heikkila, J. J., and Brown, I. R., *Life Sci.* **25,** 347–352 (1979). Copyright: Pergamon Press.

polysomal mRNA. Our results suggest that the hyperthermia-induced disaggregation of brain polysomes to monosomes occurs by a non-RNase mechanism (Heikkila and Brown, 1979b; Holbrook and Brown, 1976). The increased pool of monosomes generated following polysome disaggregation is dissociable into 40S and 60S ribosomal subunits by high salt treatment [Fig. 2—compare panels C, F, and I (high salt treatment) to panels B, E, and H, (controls)]. These results suggest that the monosomes are not stabilized against dissociating conditions by associated nascent peptides produced as a result of ribonuclease degradation of polysomes.

Analysis of the peptidyl tRNA content of the monosome pool indicates that normal run-off of ribosomes rather than premature termination occurs during polysome disaggregation (Holbrook and Brown, 1977). Quantitation of mRNA levels has revealed that mRNA is preserved during polysome disaggregation (Heikkila and Brown, 1979b; Mahony and Brown, 1979). Fractionation of brain polysomes on sucrose gradients indicates that there is a shift of polyadenylated mRNA from the polysome to the monosome region of the gradients following the disaggregation event (Cosgrove et al., 1982; Mahony and Brown, 1979). These results suggest that inhibition of reinitiation of protein synthesis rather than cleavage of mRNA or premature termination is the mechanism whereby brain polysomes dissociate to monosomes following elevation of body temperature.

To facilitate a more detailed analysis of the steps in brain protein synthesis which are affected by hyperthermia, an initiating cell-free protein synthesis system was developed (Cosgrove and Brown, 1981). The advantage of the cell-free system is that it can be fractionated into components to characterize translational control factors. Most brain cell-free translation systems which have been developed are defective in their ability to engage in initiation of protein synthesis. A cell-free system was developed using rabbit brain which by four separate criteria demonstrates the capacity to engage in initiation of protein synthesis (Cosgrove and Brown, 1981). Using this initiating cell-free system, it is apparent that drug-induced hyperthermia induces the transient appearance of a translational inhibitory factor in the brain postribosomal supernatant fraction (Cosgrove et al., 1981). Hyperthermia appears to inhibit protein synthesis in the brain by the induction of a lesion in the reinitiation of translation which results in a decreased rate of formation of 40S and 80S initiation complexes (Cosgrove and Brown, 1984).

## 3. INDUCTION OF HEAT SHOCK PROTEIN IN THE BRAIN

While hyperthermia generated by LSD, a bacterial pyrogen, or elevation of ambient temperature induces an overall inhibition of protein synthesis in the

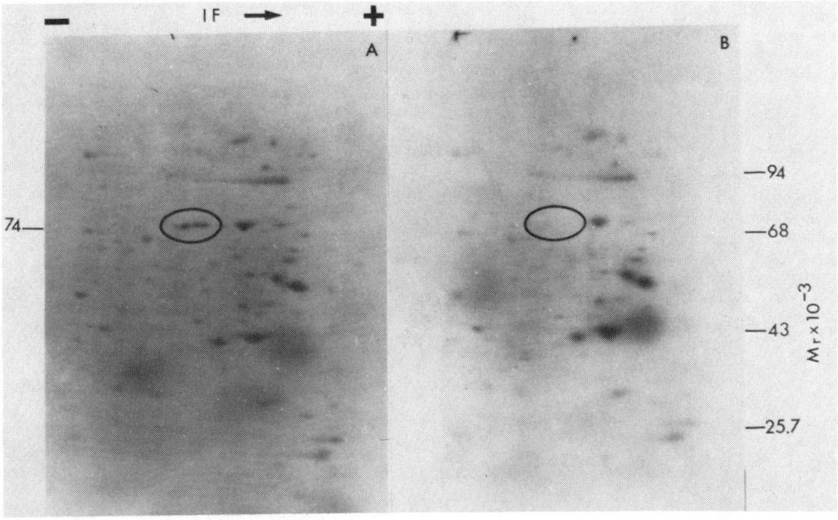


**Figure 3.** Induction of a heat shock protein in the brain after hyperthermia. Drug-induced hyperthermia (A) and saline control (B). The 74K heat shock protein is encircled. Elevation of body temperature in rabbits from 39.7 to 42.5°C was induced by the intravenous injection of LSD at 100 μg/kg body weight. *In vivo* labeled cerebral hemisphere proteins were obtained by pulse labeling for 1 hr with [$^{35}$S]-methionine 20 min after drug injection. Equal amounts of acid-precipitable radioactivity from drug-treated and saline-treated control animals were analyzed by two-dimensional gel electrophoresis and fluorography. IF, isoelectric focusing. (From Cosgrove and Brown, 1983.)

rabbit brain, the question might be asked "are all brain proteins similarly affected?" To address this point, *in vivo* labeled brain proteins were resolved by two-dimensional gel fluorography. As shown in Figure 3, an increase of body temperature from 39.7 to 42.5°C after the intravenous injection of LSD produced selective effects on brain protein synthesis. The induction of synthesis of a brain protein of molecular weight 74,000 (74K, encircled in Fig. 3A) was apparent relative to controls whose body temperature remained at 39.7°C (Fig 3B). This protein is similar in molecular weight to one of the major heat shock proteins previously reported to be induced in a wide range of tissue culture systems and unicellular organisms after increase of ambient temperature (Ashburner and Bonner, 1979; Schlesinger et al., 1982). The 74K brain protein appears to be simliar to the heat shock protein induced in other systems since it is precipitated by antibodies prepared against the chicken 70K heat shock protein (Cosgrove and Brown, 1983).

Induction of the 74K brain protein is correlated to drug-induced hyperthermia since prevention of hyperthermia by injection of the drug into rabbits maintained in a cold room to 4°C eliminated the phenomenon. However,

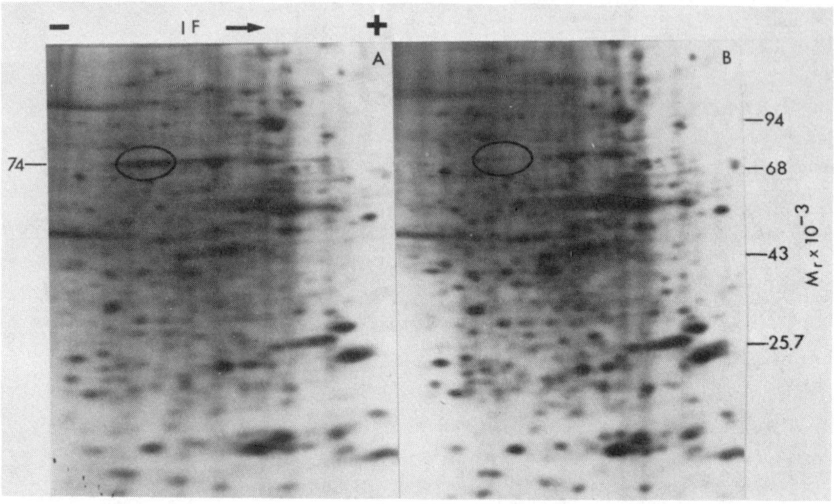


**Figure 4.** Selective increase in polysomal heat shock mRNA after hyperthermia. Drug-induced hyperthermia (A) and saline control (B). Cerebral hemisphere polysomes were isolated 1 hr after the intravenous injection of LSD which increased body temperature from 39.7 to 42.5°C. Polysomes were then translated in an mRNA-dependent reticulocyte lysate with [$^{35}$S]-methionine and nascent polypeptide analyzed by two-dimensional gel electrophoresis and fluorography. The 74K heat shock protein is encircled. (From Cosgrove and Brown, 1983.)

behavioral effects of the drug and overall inhibition of brain protein synthesis were still apparent (Cosgrove et al., 1981; Cosgrove and Brown, 1983). Increase of body temperature to 42.5°C by means other than LSD (such as placement of rabbits in a 37°C incubator) also induced the synthesis of the 74K brain protein (data identical to Fig. 3A).

To determine whether the increased synthesis of the 74K heat shock protein was due to a selective increase in polysomal mRNA coding for this protein, purified brain polysomes were translated in an mRNA-dependent reticulocyte lysate. The resultant [$^{35}$S]-methionine-labeled nascent proteins were then resolved by two-dimensional gel fluorography. As shown in Figure 4, a selective increase in labeling of a 74K protein (encircled in panel A) is observed in the cell-free translation products of polysomes isolated 1 hr following elevation of body temperature by 3°C. The increased synthesis of the 74K protein which is observed in these cell-free polysome translation experiments suggests that the change is due to an increase in the relative abundance in polysomes of mRNA coding for this protein and not to a change in post-translational modification.

Analysis of the cell-free translation products of cerebellar polysomes isolated at various time points following drug injection indicates that the time

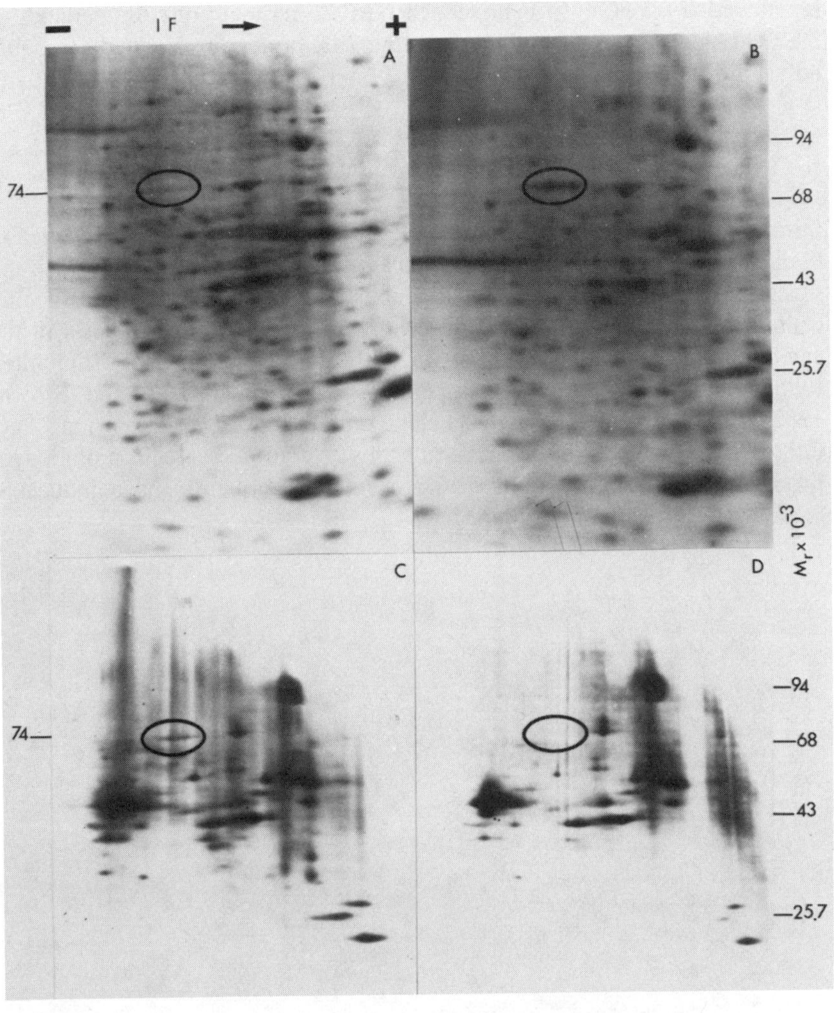


**Figure 5.** Temporal appearance of heat shock mRNA in cerebellar polysomes. Polysomes were isolated from the cerebellum of the rabbit brain 1, 4, or 8 hr after drug administration (panels B, C, and D). Control animals received a saline injection (A). Polysomes were translated in a reticulocyte cell-free system and the labeled products analyzed as in Figure 4. The position of the 74K heat shock protein is encircled. (From Cosgrove and Brown, 1983.)

course of appearance of heat shock mRNA in brain polysomes parallels the rise and fall of body temperature induced by the drug (Fig. 5). Synthesis of the 74K protein is maximal in cerebellar polysomes isolated 1 hr after injection of LSD when body temperature is at its peak of 42.5°C (Fig. 5B). By 4 hr body temperature has decreased to 41°C and synthesis of the 74K protein is reduced (Fig. 5C). At 8 hr recovery to the normal body temperature of 39.7°C has occurred and synthesis of the 74K protein by isolated cerebellar polysomes is not detectable (Fig 5D).

## 4. SYNTHESIS OF HEAT SHOCK PROTEIN IN THE RETINA AND CEREBRAL MICROVASCULAR SYSTEM

Initially studies on the effect of hyperthermia on protein synthesis in the rabbit brain were carried out on total cerebral hemispheres, a tissue in which the functional and topographical heterogeneity is great. Recently more specific cellular systems in the brain have been analyzed. Protein synthesis in the cerebral microvascular system, the site of the blood–brain barrier, is inhibited following elevation of body temperature by 2–2.5°C (Inasi and Brown, 1982). Analysis of *in vivo* labeled proteins by two-dimensional gel fluorography demonstrates that during this overall decrease in protein synthesis in the cerebral microvascular system hyperthermia also induces a

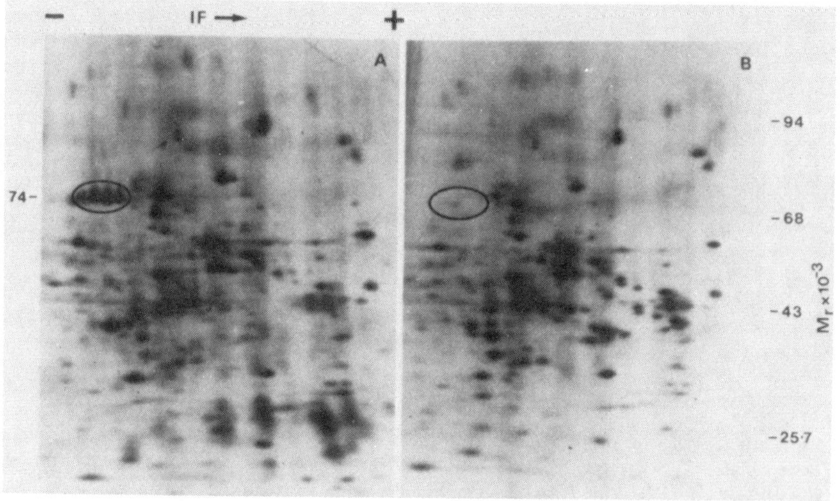


**Figure 6.** Hyperthermia-induced synthesis of heat shock protein in the cerebral microvascular system. Following elevation of body temperature from 39.6 to 42°C by intravenous injection of LSD, *in vivo* labeled proteins from cerebral microvessels were obtained by pulse labeling with [$^{35}$S]-methionine. Nascent proteins were then analyzed by two-dimensional gel electrophoresis and fluorography. Drug-induced hyperthermia (A) and saline control (B). The position of the 74K heat shock protein is encircled. Reprinted from Inasi, B. S., and Brown, I. R., *Biochem. Biophys. Res. Commun.* **106,** 88–887 (1982). Copyright: Academic Press, Inc.

marked increase in synthesis of a 74K heat shock protein (encircled in Fig. 6A).

In the retina of the rabbit eye the kinetics of polysome disaggregation parallels the rise and fall of body temperature induced by the intravenous injection of LSD (Clark and Brown, 1982). *In vivo* labeling studies and analysis of the cell-free translation products of isolated retinal polysomes demonstrate that during the phase of overall reduction in retinal protein synthesis, synthesis of a 74K retinal protein is selectively increased (Clark and Brown, 1982a). We have recently analyzed the fate of the induced heat shock protein in the retina and found that it is transported down the optic nerve to synaptic termini in the visual projection areas by slow axonal transport (Clark and Brown, 1984.)

The retinal system is particularly useful since it facilitates *in vitro* experiments designed to investigate the mechanism of induction of the 74K protein. The retina can be dissected from the eye and subjected to heat shock while incubated in a tissue culture medium in the presence and absence of RNA synthesis inhibitors (Clark and Brown, 1982b). Such experiments suggest that the induction of the 74K retinal protein is dependent on the synthesis of new RNA.

## 5. DEVELOPMENTAL CHANGES IN THE INDUCTION OF BRAIN HEAT SHOCK PROTEIN

The newborn mammalian brain appears to be particularly sensitive to the induction of the 74K heat shock protein. In the 2-day-old rat, elevation of body temperature from 32 to 38°C induces synthesis of the 74K protein as determined by the cell-free translation of isolated brain polysomes (Brown, 1983). At the 7-day postnatal stage, however, body temperature must be elevated to 43°C in order to induce synthesis of the 74K heat shock protein. This suggests that as early development proceeds, higher body temperatures are required to induce synthesis of this protein.

An interesting parallel is apparent between the sensitivity of the 2-day-old rat pup to induction of the heat shock protein and the sensitivity of the neonatal rat to hyperthermia-induced seizures. In the 2-day-old rat, elevation of body temperature from 32 to 37°C has been reported to induce seizure activity, whereas in 7-day-old pups seizures do not occur until body temperature is increased to 43°C (Holtzman et al., 1981).

Elevation of maternal body temperature has a dramatic effect on protein synthesis in the fetal brain of the rabbit. Hyperthermia induces a disaggregation of polysomes in the fetal brain (Heikkila et al., 1979) and a marked induction of synthesis of a 74K heat shock protein (Brown, 1983).

## 6. ANALYSIS OF mRNA CODING FOR BRAIN HEAT SHOCK PROTEIN

Elevation of body temperature of 2–3°C induces a rapid overall inhibition of protein synthesis in the rabbit brain and a selective increase in synthesis of an abundant protein of molecular weight 74,000 (referred to as the 74K heat shock protein). Heat shock proteins have been highly conserved in evolution. Antibody to the chicken 70K heat shock protein cross-reacts with proteins in systems ranging from yeast to mammalian tissue culture cells (Kelley and Schlesinger, 1982). Cloned DNA sequences of the coding region of the *Drosophila* 70K heat shock gene hybridize to DNA sequences in the mammalian genome (Lowe et al., 1983). Since the coding sequences of the heat shock genes have been highly conserved in evolution, we have been able to use existing recombinant DNA probes to assay for hyperthermia-induced changes in the level of brain mRNA coding for the heat shock protein. These studies have been facilitated by the fact that the brain heat shock protein is synthesized by an abundant class mRNA.

In collaboration with David Lowe and Larry Moran of the Department of Biochemistry at the University of Toronto, we have assayed for changes in the level of rabbit brain heat shock mRNA following hyperthermia (Brown et al., 1983). This involves the hybridization of nick-translated cloned DNA sequences of the coding region of the major *Drosophila* heat shock gene to purified brain RNA by Northern blot techniques. A prominent increase in the total cellular content of brain heat shock mRNA was apparent 1 hr following the elevation of body temperature by 2°C. This suggests that the induction of synthesis of the 74K brain heat shock protein is controlled at the level of transcription.

## 7. CONCLUDING REMARKS

Elevation of ambient temperature has been reported to induce the synthesis of heat shock proteins in a wide range of tissue culture systems and unicellular organisms (Ashburner and Bonner, 1979; Kelley and Schlesinger, 1978; Schlesinger et al., 1982). Few studies, however, have addressed the question of whether heat shock proteins are induced within organs of the intact mammal following elevation of body temperatures such as is attained during fever reactions. Our studies suggest that the heat shock phenomenon, which in mammalian cells is frequently studied in somewhat artificial laboratory conditions in tissue culture systems, is a natural cellular response to physiological stress in the intact animal.

Fever in susceptible children can provoke convulsions, hallucinations, and a variety of neurological disorders. However, little is known of the consequences of elevated body temperature on neurochemical events. Seizures induced by increases in body temperature (i.e., febrile convulsions) are an important clinical problem in about 3–5% of children between the ages of 6 months and 5 years. Seizures induced by fever occur in normal children with a genetic predisposition and in children with brain damage or epilepsy. Our results provide new data on the effects of physiologically relevant increases in body temperature on molecular events in both the developing brain and in specific neural systems such as the retina.

In tissue culture systems, other factors in addition to temperature shock have been reported to induce the synthesis of proteins similar in molecular weight to heat shock proteins, for example, amino acid analogs, sulfhydryl reagents, and chelating agents (Ashburner and Bonner, 1979; Burdon, 1982; Hightower and White, 1981; Kelley and Schlesinger, 1978). A trauma-induced protein of molecular weight similar to our 74K heat shock protein has been reported to be induced in incubated brain slices and in tissues of the hyperthermic rat (Currie and White, 1981; Hightower and White, 1981; White, 1980, 1981). It is possible that a common group of proteins is induced by diverse treatments as a general cellular response to stress.

Comparatively few examples of gene induction in neural systems have been noted which result in selective changes in the synthesis of abundant class proteins (Brown and Cosgrove, 1983). The induction of the 74K heat shock protein in the mammalian brain by a physiologically relevant increase in body temperature provides a clear example of this type of phenomenon. A marked increase in the synthesis of a highly abundant protein is apparent within 1 hr while the synthesis of other brain proteins is inhibited. Elements of both transcriptional and translational control of gene expression appear to be operative in this induction system.

Since the coding region of heat shock genes has been highly conserved in evolution, existing recombinant DNA probes can be utilized to assay for hyperthermia-induced changes in levels of brain heat shock mRNA. Questions that await investigation are: (1) which brain cell types transcribe the heat shock gene and (2) do differences exist in the threshold temperature required to induce the heat shock gene in different brain cell types? We plan to investigate these questions using *in situ* hybridization to visualize those cell types in a brain tissue slice which hybridize to a nick-translated heat shock gene probe. *In situ* hybridization usually employs autoradiographic detection of hybrid molecules. However, recent refinements of the technique in which a biotinated analog of dUTP is incorporated into a recombinant DNA molecule by nick-translation permit the detection of hybridization sites in a tissue slice by immunological histochemical procedures (Singer and Ward, 1982; Szabo and Ward, 1982). *In situ* hybridization should prove to be very useful in the field of neuroscience since it provides an approach to determine which cell types in complex neural tissue transcribe specific genes.

## ACKNOWLEDGMENTS

I thank postdoctoral fellow James Cosgrove, research technician Sheila Rush, and graduate students Bruce Clark and John Heikkila for their contributions to experiments cited in this review article. These studies were supported by grants from the Medical Research Council of Canada.

## REFERENCES

Ashburner, M. and J. J. Bonner, *Cell* **17,** 241 (1979).

Bligh, J., *Neuroscience* **4**, 1212 (1979).

Brown, I.R., *J. Neurochem.* **40**, 1490 (1983).

Brown, I. R. and J. W. Cosgrove, in A. Lajtha, Ed., *The Handbook of Neurochemistry* Vol. 5, Plenum, New York, (1983).

Brown, I. R., J. W. Cosgrove, and B. D. Clark, in M. J. Schlesinger, M. Ashburner, and A. Tissieres, Eds., *Heat Shock from Bacteria to Man*, Cold Spring Harbor Laboratory, Cold Spring Harbor, NY, p. 361 (1982a).

Brown, I. R., J. J. Heikkila, J. W. Cosgrove, in I. R. Brown, Ed., *Molecular Approaches to Neurobiology*, Academic, New York, p. 221 (1982b).

Brown, I. R., D. G. Lowe, and L. A. Moran, *J. Cell Biol.* **97,** 152 (1983).

Burdon, R. H., in M. Schlesinger, M. Ashburner, and A. Tissieres, Eds., *Heat Shock from Bacteria to Man*, Cold Spring Harbor Laboratory, Cold Spring Harbor, NY, p. 283 (1982).

Clark, B. D. and I. R. Brown, *Brain Res.* **247**, 97 (1982a).

Clark, B. D. and I. R. Brown, in A. M. Giuffrida Stella, G. Gombos, G. Benzi and H. S. Bachelard, Eds., *Basic and Clinical Aspects of Molecular Neurobiology*, Fondazione Internationale Meuarini, Milan, p. 274 (1982b).

Clark, B. D. and I. R. Brown, *Transactions of the American Society for Neurochemistry*, **15,** 248 (1984).

Cosgrove, J. W. and I. R. Brown, *J. Neurochem.* **36**, 1026 (1981).

Cosgrove, J. W. and I. R. Brown, *Proc. Natl. Acad. Sci. USA* **80**, 569 (1983).

Cosgrove, J. W. and I. R. Brown, *J. Neurochem.* **42,** 1420 (1984).

Cosgrove, J. W., B. D. Clark, and I. R. Brown, *J. Neurochem.* **36**, 1037 (1981).

Cosgrove, J. W., J. J. Heikkila, and I. R. Brown, *Neurochem. Res.* **7**, 505 (1982).

Currie, R. W. and F. P. White, *Science* **214**, 72 (1981).

Freedman, M. S., B. D. Clark, T. F. Cruz, J. W. Gurd, and I. R. Brown, *Brain Res.* **207**, 129 (1981).

Hall, C. and L. Lim, *Biochem. J.* **196**, 327 (1981).

Heikkila, J. J. and I. R. Brown, *Neurochem. Res.* **4**, 763 (1979a).

Heikkila, J. J. and I. R. Brown, *Life Sci.* **25,** 347 (1979b).

Heikkila, J. J. and I. R. Brown, *J. Neurochem.* **36**, 1219 (1981).

Heikkila, J. J., L. A. Holbrook, and I. R. Brown, *J. Neurochem.* **32**, 1793 (1979).

Hightower, L. E. and F. P. White, *J. Cell. Physiol.* **108,** 261 (1981).

Holbrook, L. A. and I. R. Brown, *J. Neurochem.* **27**, 77 (1976).

Holbrook, L. A. and I. R. Brown, *J. Neurochem.* **29**, 461 1977).

Holtzman, D., K. Obana, J. Olson, *Science* **213**, 1034 (1981).

Inasi, B. S. and I. R. Brown, *Biochem. Biophys. Res. Commun.* **106**, 881 (1982).

Kelley, P. M. and M. J. Schlesinger, *Cell* **15**, 1277 (1978).

Kelley, P. M. and M. J. Schlesinger, *Molec. and Cell. Biol.* **2**, 267 (1982).

Li, G. C., D. S. Shrieve, and Z. Werb, in M. J. Schlesinger, M. Ashburner, and A. Tissieres, Eds., *Heat Shock from Bacteria to Man*, Cold Spring Harbor Laboratory, Cold Spring Harbor, NY, p. 395 (1982).

Lowe, D. G., W. D. Fulford, and L. A. Moran, *Mol. Cell Biol.* **3**, 1540 (1983).

Mahony, J. B. and I. R. Brown, *Biochim. Biophys. Acta* **565**, 161 (1979).

Morrison, M. R., S. Pardue, and W. S. T. Griffin, *J. Biol. Chem.* **265**, 3550 (1981).

Schlesinger, M. J., M. Ashburner, and A. Tissieres, Eds., *Heat Shock from Bacteria to Man*, Cold Spring Harbor Laboratory, Cold Spring Harbor, NY (1982).

Singer, R. H. and D. C. Ward, *Proc. Natl. Acad. Sci. USA* **79**, 7331 (1982).

Szabo, P. and D. C. Ward, *Trends in Biochem. Sci.* **7**, 425 (1982).

White, F. P., *Neurosciences* **5**, 1793 (1980).

White, F. P., *J. Neurosci.* **1**, 1312 (1981).

# 7

# *The Molecular Genetic Analysis of sn-Glycerol-3-Phosphate Dehydrogenase Development in Mouse Cerebellum*

***Leslie P. Kozak and Marilyn Fisher***

The emerging molecular descriptions of the differential phenotypes of eukaryotic cells in higher organisms include two distinguishing patterns. One pattern is characterized by the production of a cell-specific protein at very high concentrations in a single cell type, e.g., hemoglobin in the red cell (Weatherall and Cleff, 1979) and ovalbumin in oviduct cells (Palmiter, 1975). In a second pattern a specific enzyme is expressed in a variety of cell types, although each cell type has a different but characteristic level of enzyme production, for example, lactate dehydrogenase (Markert, 1968) and aldolase (Rutter et al., 1968). Because of their abundance as cell-specific gene products, the hemoglobin and ovalbumin genes were among the first genes to be cloned with recombinant DNA techniques; as a result much of our current information on the structure and regulation of the gene is derived from the molecular analysis of these systems.

In designing a molecular genetic analysis of the structure and function of genes in the brain, one also confronts the fact that there are gene products that are neural specific and some that are non-specific. In contrast to some of the non-neural tissues, however, the proportion of the biosynthetic activity in a specific brain region which is devoted to synthesis of a cell-specific product may be very low compared to that found in the extraneural tissues. Whether this is a function of the quantitative aspects of gene expression in neural cells or a result of the heterogeneous composition of the cell populations is not certain. Whatever the underlying reason, the consequence of this situation in terms of the practical experimental aspects of pursuing a molecular biological analysis of gene expression in brain is that it may be more feasible to analyze the housekeeping genes rather than the neural-specific genes.

In part because of the above considerations, we have been analyzing the genetic expression of the enzyme *sn*-glycerol-3-phosphate dehydrogenase (EC 1.1.1.8; GPDH) as a model system to investigate mechanisms controlling developmental expression in the cerebellum of the mouse. However, in addition to the brain, GPDH has been detected in virtually every tissue of the body except blood (Lin, 1977). This feature of GPDH expression provides a useful tool that enables the investigator to determine not only what is specific to the brain with respect to the expression of GPDH but also to identify the characteristics the brain holds in common with other tissues. These common steps may turn out to be more easily analyzed in other tissues and thus accelerate our ability to understand brain development.

Another additional and important consideration in the development of the GPDH model is that it was originally selected because it expressed genetic variability (Kozak, 1972). Although recombinant DNA technology has recently provided us with a way to study structural genes without having to rely on allelic variation present in natural populations, the identification of regulatory loci is still dependent on Mendelian genetic variability. Even in the case of structural genes a mutation may be an aid in identifying which gene is functional in the presence of a gene family that also includes pseudogenes (Monk et al., 1981; Scarpulla et al., 1982).

In addition to the genetic variability at known structural and regulatory loci of GPDH, mutations have also been identified which affect various aspects of the normal organization and structure of the cerebellum (Green, 1981; Mallett, 1980). These mutations, some of which secondarily affect GPDH expression, can be used in a variety of experiments such as determining the cellular location of a gene product (Fisher et al., 1981) or testing the effect of cellular disorganization on gene expression. As a general experimental scheme we have used whatever genetic variation is available in inbred and wild populations of mice to study the regulation of GPDH in the developing brain. The results of the experiments have revealed that the GPDH genes *Gdc-1* and *Gdc-2* form a model system for analyzing several aspects of the development and differentiation of the cerebellum, which includes (1) mechanisms controlling gene switching of two structural genes at different phases of development; (2) the molecular basis for cell-specific gene expression; (3) the identification and mode of action of cerebellar-specific regulatory genes; and (4) the molecular basis for neuronal glial cell-to-cell interactions in determining glial cell gene expression. In this review we will describe in some detail the experimental data which has established this model gene system.

## 1. DEVELOPMENTALLY DEPENDENT EXPRESSION OF CEREBELLAR GPDH ISOZYMES

The specific activity of GPDH in the cerebellum of the newborn C57BL/6J mouse is approximately 5 nmol NADH oxidized/min/mg of protein (Kozak and Jensen, 1974). A slow, gradual increase in enzyme activity occurs until the ninth day. This is followed by an accelerated increase in enzyme activity that lasts until approximately 30 days of age to give a final enzyme-specific activity of approximately 300 nmol/min/mg protein. However, this apparently simple developmental increase in enzyme activity which resembles the pattern for many different enzymes and proteins is in fact much more complex. The reason is that the concentrations of three different isozymic forms of GPDH are changing during development.

The isozymic forms of GPDH were first detected by the presence of a heat sensitive GPDH enzyme activity which was gradually replaced by a heat stable form. These thermally sensitive GPDH isozymes can be separated by anion-exchange chromatography. The three peaks of activity observed in Figure 1 correspond to the following: (1) adult isozyme (peak I), the predominant form in adult tissues; (2) the embryonic isozyme (peak III), the only isozymic form detected in the 8-day embryo and the predominant form during fetal and early postnatal brain development; and (3) a hybrid form of the adult

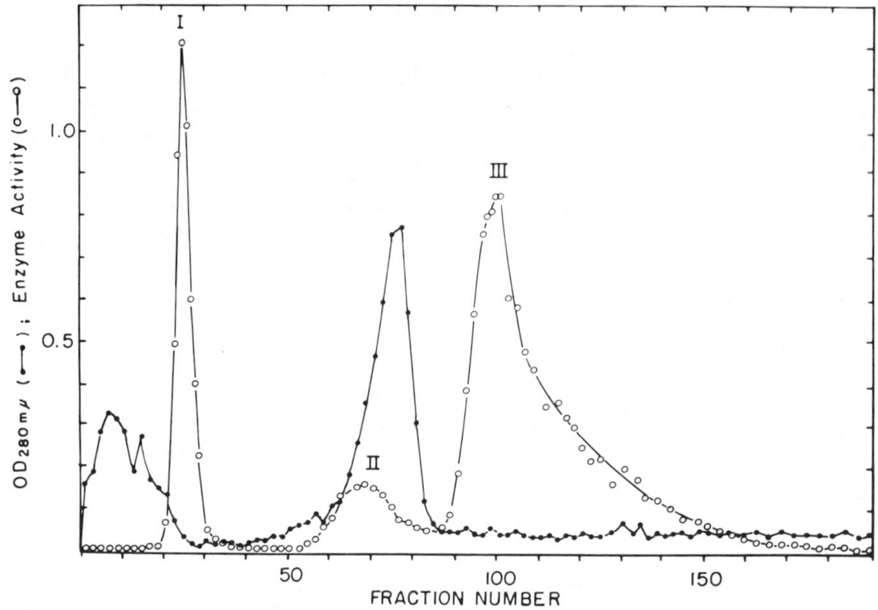


**Figure 1.** Separation of isozymes of GPDH from 11- to 13-day-old brains of C57BL/6J mice by DE52 ion exchange chromatography. Peak I isozyme is the product of the *Gdc*-1 locus and is the major isozyme found in all adult tissues, including muscle and brain; Peak III is the embryonic isozyme, while peak II is possibly a hybrid of peaks I and III. The enzyme from 128 g of tissue (400 mice) was first partially purified by affinity chromatography. The original homogenate contained 101 units of activity, while 79 and 51 units were recovered after affinity chromatography and ion exchange chromatography, respectively. Peak III in a volume of 84 mL contained 39 units which on rechromatography on blue Sepharose 6B to concentrate the protein was reduced to 26 units. This fraction contained significant quantities of lactate dehydrogenase activity which could be removed by chromatography on Sephadex G-100. Additional loss of activity continued to occur and this protein was still not pure. (From Kozak and Burkart, 1981.)

and embryonic isozyme (peak II) which is principally detected between 10 and 13 days of age. This time period is a transitional point in the development of the cerebellum when the structural genes for both isozymes are most likely being expressed in the same cell.

Although the adult and embryonic isozymes are both dimeric cytosolic proteins, the clear differences observed in thermal stability, isoelectric point, and kinetic constants (Kozak and Jensen, 1974; Wise and Green, 1979) suggest that the isozymes are distinct molecular species derived from separate structural genes and not simply epigenetic modifications of each other. On the other hand, structural similarities also exist; their molecular weights (68,000 Da) are indistinguishable and the subunits can reassociate to form hetero-

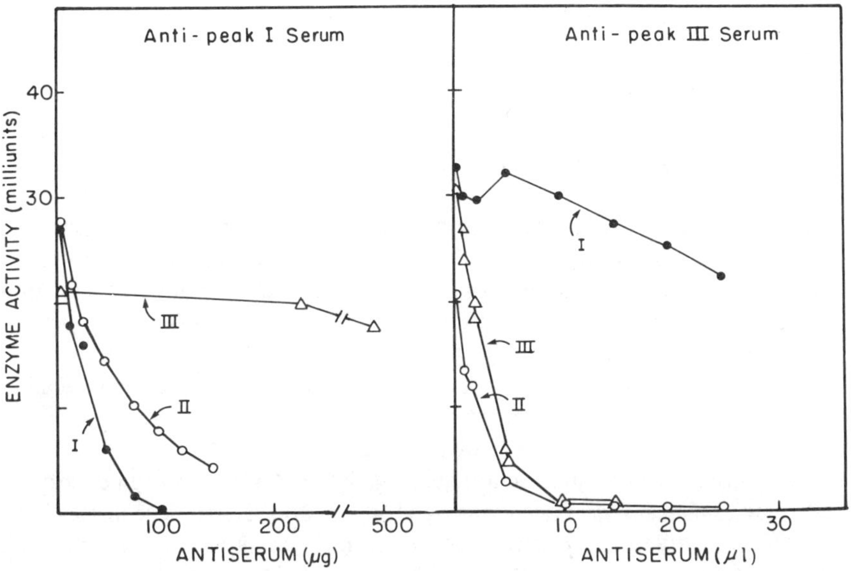


**Figure 2.** Immunoprecipitation of GPDH isoyzme by antiserum obtained from rabbits immunized with (A) purified peak I isozyme and (B) an anti-peak II–peak III antigen–antibody complex. Isozymes present in peaks I, II, and III were isolated from neonatal C57BL/6J mice by anion exchange chromatography on DE-52 resins (Whatman).

dimers both *in vivo* and *in vitro* (Kozak and Burkart, 1981). The ability to form heterodimers provided an additional means of comparing the structures of the isozymes by immunological methods (Kozak and Burkart, 1981). Such an immunological approach was essential for further structural analysis because the low levels of embryonic GPDH expression (approximately 0.003% of the soluble protein fraction of the cell) precluded purifying the isozyme to homogeneity by conventional methods. This approach made use of the observation that antibodies against highly purified adult GPDH cross-reacted with the hybrid isozyme. Immunization of rabbits with the immunoprecipitated hybrid species resulted in an antiserum which cross-reacted with all three isozymic species. A third cycle of immunization with an immunoprecipitate formed with the anti-hybrid GPDH antibody and partially purified embryonic isozyme provided a serum which reacted strongly with the embryonic and hybrid serum but very weakly with the adult isozyme (Fig. 2). This experimental protocol for producing specific antibody against embryonic GPDH not only provided a means to determine that the GPDH isozymes had similarities in their peptide structures but also demonstrated that they expressed distinct antigenic sites and were therefore derived from distinct genetic structural loci. Genetic evidence supporting this conclusion is summarized in the next section.

## 2. GENETICS OF GPDH ISOZYMES

Concurrently, in 1972, Kozak as well as Chapman and Ruddle identified genetic structural variants for GPDH that affected the thermal stability and ionic charge of the adult isozyme. The alleles determining this variation mapped to chromosome 15 at a position which was called *Gdc-1* for *g*lycerol-3-phosphate *d*ehydrogenase *c*ytoplasmic (Eicher, 1976). Each adult tissue which expressed GPDH activity also expressed the allelic variation indicating that only one major structural gene determined the GPDH phenotype in the various tissues of adult mice (Kozak and Erdelsky, 1975).

In contrast, when embryonic and fetal tissues taken from mice carrying alternate alleles at *Gdc-1* were analyzed, it was found that these alleles have no effect on structural properties such as thermal stability and isoelectric point of the embryonic isozyme (Kozak and Jensen, 1974). This observation was consistent with the hypothesis, derived from the protein and immunological studies, that separate structural genes encode the isozymes. Recently the hypothesis has been strengthened even further by the discovery in *M. m. castaneus* mice of a genetic variant which alters the electrophoretic migration of the embryonic isozyme without an effect on the adult isozyme (Kozak et al., 1982). The typing of both *Gdc-1* and *Gdc-2* in offspring produced by backcrossing (C57BL/6J × *M. m. castaneus*) $F_1$ mice to C57BL/6J gave a recombination frequency for the alleles at *Gdc-1* and *Gdc-2* of 0.425 ± 0.058 (31/73). This frequency indicates that these loci segregate independently and that the loci are probably not on the same chromosome.

The genetic linkage analysis of *Gdc-2* was determined with several reciprocal backcross matings made with $F_1$ offspring between *M. m. castaneus* mice and the inbred strains C57BL/6J, DBA/2J, and SEA/GnJ. *Gdc-2* shows linkage to three genes, *Mpi-1* (0.270 ± 0.033), dilute (0.197 ± 0.030), and *Mod*-1 (0.179 ± 0.032) which are known to be located on chromosome 9 (Nichols et al., 1973; Roderick et al., 1971). These recombination frequencies indicate a gene order *Mpi-1-d-Mod-1-Gdc-2*.

A summary of the alleles at the *Gdc-1* and *Gdc-2* loci and their phenotypes is given in Table 1.

*Drosophila* is the only other eukaryotic organism in which GPDH genetics has been investigated. Like the mouse, *Drosophila* has developmentally expressed larval and adult GPDH isozymes with differences in their physiochemical and kinetic properties which parallel those found for the mouse isozymes (Bewley et al., 1974; Wright and Shaw, 1969). In contrast to the

**TABLE 1. PROPERTIES OF ALLELES AT THE *Gdc-1* AND *Gdc-2* LOCI**

| Locus | | Allele | Heat Stability[a] $t_{1/2}$ at 50°C (min) | Electrophoretic[b] Migration | References[c] |
|---|---|---|---|---|---|
| *Gdc-1* | b | (most common) | 6 | ++++ | 1 |
| | c | (BALB/c, CBA) | 2 | ++++ | 1 |
| | d | (*M. m. castaneus*) | 6 | +++ | 2, 3 |
| | f | (DDS) | 12 | unknown | 4 |
| *Gdc-2* | b | (most common) | 1 | + | 5 |
| | d | (*M. m. castaneus*) | 1 | ++ | 5 |

[a]Heat stability may vary depending on the experimental conditions so these values are adjusted to reflect the relative differences observed for alleles of the two loci.

[b]The relative charges of the enzymes are indicated since pIs are not known for all the allozymes. At the two extremes the *Gdc-1*$^b$ enzyme has a pI of 6.7 and the *Gdc-2*$^b$ enzyme a pI of 4.8 (Kozak and Burkart, 1981).

[c]1. Kozak (1972). 2. Chapman and Ruddle (1972). 3. Kozak and Erdelsky (1975). 4. Chan and Thompson (1981). 5. Kozak et al. (1982).

mouse, however, the data strongly indicate that in *Drosophila* only one structural gene encodes the two isozymes (Bewley and Lucchesi, 1977; Grell, 1967; O'Brien and MacIntyre, 1972) and that the larval isozyme lacks the three c-terminal amino acids of the adult isozyme (Niesel et al., 1980).

The need for the brain to have two loci for GPDH expression, one active in development during periods of cell proliferation and migration and the other in its final differentiated state, is unclear. Although the isozymes have different kinetic properties, there is no obvious metabolic reason why one isozyme is more suitable than the other at any given stage of development. The most striking difference in the loci is that the adult locus, *Gdc-1*, is generally expressed at a level up to 1000-fold greater than *Gdc-2*. GPDH serves important metabolic functions for the biosynthesis of phospholipids and triglycerides as well as for the control of NADH levels and energy metabolism by means of the glycerol phosphate cycle (Lin, 1977). Since the highest levels of GPDH are found in those cells which are most active in phospholipid and triglyceride synthesis (Ratner et al., 1981; Wise and Green, 1979), it is possible that *Gdc-1* is regulated to produce the large amount of enzyme needed for these functions. By contrast, *Gdc-2* is generally expressed in embryonic cells and undifferentiated cultured cells, where the requirements for GPDH may be minimal even though an active synthesis of cell membranes is required for cell division. Perhaps the level of gene expression is best modulated by having multiple loci, each set at a specific level of expression rather than one locus which can be modulated over a broad range.

## 3. CELLULAR LOCALIZATION OF GPDH AND ITS EXPRESSION IN NEUROLOGICAL MUTANTS

Knowing which cells are involved is essential in understanding the factors involved in regulating GPDH expression in the developing cerebellum. That different regions of the brain varied up to fivefold in the level of enzyme activity initially suggested that GPDH was enriched in certain cell types, even though it could serve a metabolic function in virtually every cell of the brain. Indeed, the only tissue of the body in which we have failed to detect GPDH activity is in the circulating cells of the blood (unpublished observations).

Three lines of experimental evidence are pertinent to the question of cellular localization. One type comes from the biochemical determination of GPDH expression in normal brain cells and in nervous-tissue-derived tumor cells in tissue culture. Briefly stated (see also Section 6) culture conditions favorable to glial rather than neuronal survival favor the expression of high levels of GPDH activity (Kozak et al., 1977, 1978). Furthermore, cultured tumor cells derived from a glial cell line, C6 glioma, express high levels of GPDH (McGinnis and de Vellis, 1974), while a cultured neuronal tumor cell line, C1300 neuroblastoma, expresses only the embryonic GPDH isozyme (Kozak and Murphy, 1976). Thus, if the cultures are an accurate reflection of the *in vivo* situation, these results suggest that in the brain the population of cells enriched for GPDH is glial rather than neuronal.

The second line of experimental evidence relevant to cellular localization of GPDH is based on the biochemical determination of GPDH expression in cerebella of various neurological mutants of the mouse. Table 2 compares the activity levels of several enzymes of carbohydrate metabolism in 5 neurological mutants. Reeler (*rl/rl*), Purkinje cell degeneration (*pcd/pcd*), weaver (*wv/wv*), and staggerer (*sg/sg*) express severe defects of one or more neuronal populations while jimpy (*jp*/Y) mice show a severe reduction in differentiated oligodendroglial cells (Green, 1981). It is evident in Table 2 that while GPDH levels are significantly reduced in the mutants with serious neuronal defects, the level of activity in the glial mutant, jimpy, is apparently normal. These results suggest GPDH activity is enriched in neurons *in vivo*, particularly granule cells and/or Purkinje cells, the two neuron populations most severely affected in these mutants.

The third line of experimental evidence provides a definitive answer to the question of cellular localization, and when applied to the mutants reconciles the apparent conflict between the first two methods of localization. In normal mice direct visualization of the enzyme by immunohistochemistry using an antiserum specific for the adult isozyme of GPDH shows that two populations of cells in the cerebellum contain high levels of immunoreactive enzyme, Bergmann glia (Golgi epithelial cells) and oligodendrocytes (Fisher et al.,

**TABLE 2. ENZYME ACTIVITIES IN NEUROLOGICAL MUTANT MICE**

| | *pcd/pcd* | *wv/wv* | *sg/sg* | *Ta jp/Y* | *rl/rl* |
|---|---|---|---|---|---|
| | | | Percent of Littermate Controls | | |
| α-GPDH | 83 | 38 | 63 | 100 | 43 |
| G6PDH | 89 | — | 79 | — | 58 |
| LDH | 103 | 90 | 93 | — | 88 |
| TPI | 102 | 71 | 77 | — | 85 |

1981). Although oligodendrocytes stain positively with an intensity similar to that of the Bergmann glia, the mouse cerebellum has very few oligodendrocytes that stain relative to the number of Bergmann glia. This accounts for the difficulty in detecting a reduction in GPDH levels in the jimpy mutant. The staining pattern in jimpy cerebella shows a bleached-out white matter due to the absence of the oligodendrocytes but normal staining of the Bergmann glia (Fig. 7 of Fisher et al., 1981). A comparison of the GPDH staining in wild-type and jimpy mice to staining for the glial fibrillary acidic protein, which is present in both Bergmann glia and fibrous astrocytes but not in oligodendrocytes, indicates that if GPDH is present in astrocytes, it is too low to be detectable by immunoperoxidase (Figs. 4 and 7 of Fisher et al., 1981).

The detection of GPDH in the Bergmann glia by immunoperoxidase staining occurs at approximately 10 days of age (Fig. 3). Although the adult isozyme of GPDH is clearly increasing slowly between 4 and 9 days of age when analyzed by enzymatic methods, it is first detected in the cerebellum by immunoperoxidase localization methods at 10 days of age. Immunoreactive GPDH is detected along the entire Purkinje cell layer in the cell bodies of the Bergmann glial cells at a faint but uniform intensity (Fig. 3a). As the animal ages, the immunoreactive GPDH rapidly increases so that by 12 days of age an essentially adult pattern of GPDH staining is observed with staining of the cell bodies and the fibers which extend across the molecular layer to the surface of the cortex (Fig. 3). The most striking features of this pattern of GPDH expression are (1) the rapidity with which the adult pattern is established even though the tissue possesses less than 10% of the final adult level of enzyme protein and (2) the extraordinarily synchronous manner in which all the cells and fibers appear to be accumulating enzyme at the same time.

In trying to understand the mechanism regulating expression at the *Gdc-1* locus in the cerebellum of wild-type and mutant animals, the pattern revealed by immunohistochemistry is informative because such a synchronized pattern is not seen for GPDH expression during development in other tissues of the

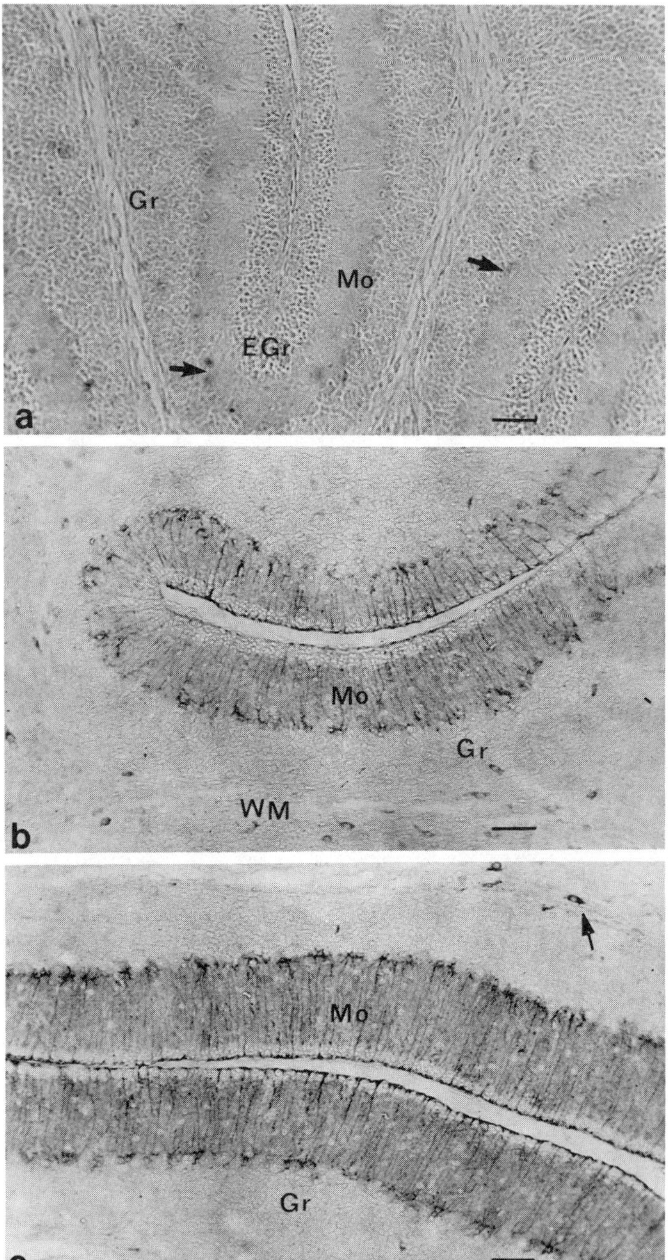
Gr
Mo
EGr
a
Mo
Gr
WM
b
Mo
Gr
c

mouse. For example, Figure 4 shows a similar experiment carried out on the developing epididymal fat pad (Cook and Kozak, 1982). In this tissue positive cells are first observed as isolated islands of cells; then as the tissue develops with age, these islands increase in size until all the fat cells are positive for enzyme. One also sees in this tissue a strong association between the presence of a lipid vacuole and the expression of GPDH. An even greater contrast is found in the entirely random fashion in which cells in tissue culture begin to express their differentiated functions. From studies of hemoglobin differentiation in cultured erythroleukemic cells, Gusella et al. (1976) postulated that the recruitment of cells into a differentiated pathway occurs in a stochastic manner. Nadal-Ginard (1978) arrived at a similar model for the differentiation of transformed myoblasts in culture.

An alternative to the stochastic model is that cells enter the pathway to terminal differentiation in proportion to the amount of organization that is present within the cell population. Accordingly, at one extreme tissue culture models are highly stochastic because they are essentially devoid of any organization. The epididymal fat pad shows clusters of cells differentiating at different times because the tissue has a low degree of organization derived from the arrangement of cells around the capillaries, while at the other extreme the cerebellum which is highly organized in time and space during development is highly synchronized with respect to the time at which it differentiates. The importance to development of interactions among cells and with their environment is fairly well established even if the mechanisms of these interactions are not presently understood. That Bergmann glia are differentiating in a highly synchronized manner and that the arrangement of these cells is highly organized suggests that specific cell contacts with neighboring cells is a necessary condition for normal development and continued expression of GPDH.

---

**Figure 3.** Light micrograph of a sagittal section through the cerebellum of a BALB/cBy mouse (a) 10 days of age, (b) 11 days of age, and (c) 12 days of age. Sections were treated with anti-GPDH serum (1 : 10,000) and stained with peroxidase as described by Fisher et al., 1981. In the 10-day-old mouse (a) very faint staining is detected in Purkinje cell layer (arrow), fibers are not yet detected in the molecular layer (Mo) or external granule layer (EGr). By 11 days of age (b) intensely staining fibers are clearly seen extending from the Purkinje cell layer through the molecular layer (Mo) and the external granule layer (EGr) to the surface of the cortex. Staining is less intense in areas on the outside of the folia (region at the upper right of b) as compared to the inner folds and this is generally seen at all ages of the mouse subsequent to 10 days of age. Positively stained cells in the white matter (WM) are oligodendroglia. On day 12 (c) a fully differentiated pattern of GPDH staining is observed. The arrow points to an intensely staining oligodendroglial cell. Calibration bar=50$\mu$m.

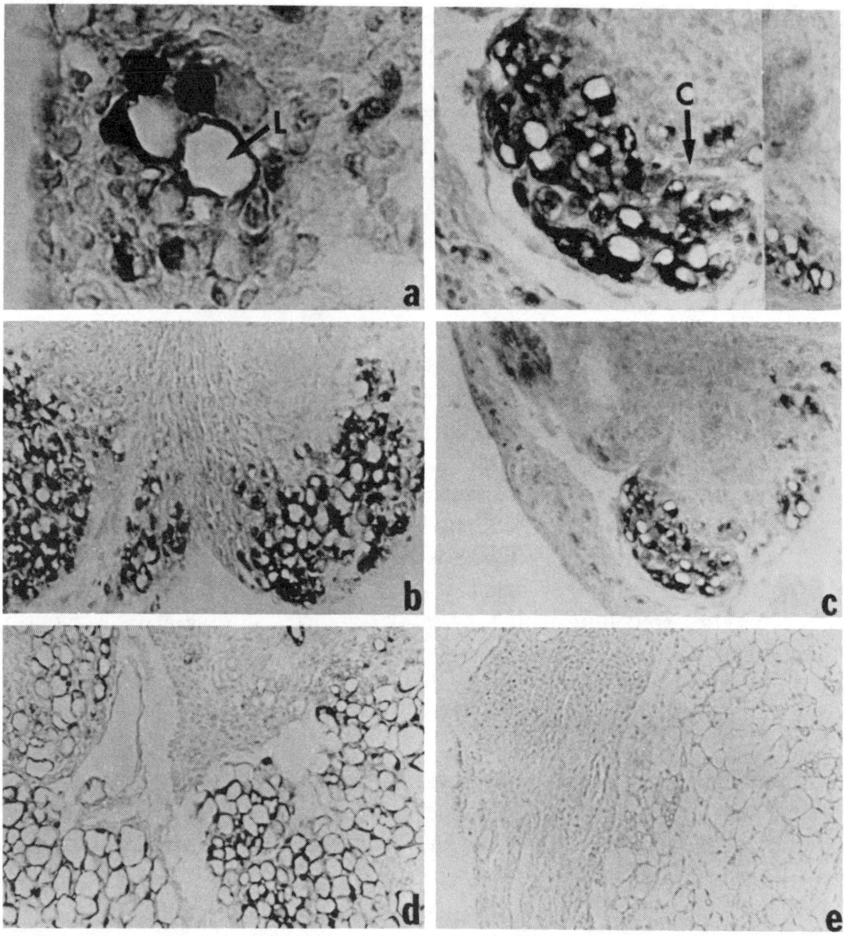


**Figure 4.** Immunohistochemical staining of adult GPDH in sectioned epidimyal fat pads of BALB/cByJ mice. (a) 3-day fat pad, L indicating lipid vacuole. (b,c) 4-day fat pads. Inset in c shows a capillary (C) penetrating the cluster of positive cells. (d) 5-day fat pad. (e) 5-day fat pad incubated wih control serum. (From Cook and Kozak, 1982.) With permission from *Dev. Biol.* Copyright: Academic Press, Inc.

The possibility that GPDH expression in Bergmann glia is influenced by their interaction with neighboring cells is strengthened by the pattern of enzyme expression seen in the mutants mentioned previously. When cerebellar sections from reeler, weaver, and staggerer mice were examined immunohistochemically, it was apparent that the reduced levels of GPDH measured biochemically correlated with a reduction in immunoreactive GPDH in the molecular layer and Bergmann glial cell bodies. Of the mutants found to have reduced GPDH levels, most have a reduced number of granule cells and all have some type of Purkinje cell defect. Specific Bergmann glial cell defects have been described only in weaver (Rakic and Sidman, 1973) where the defects seem to be relatively minor (Bignami and Dahl, 1974; Sotelo and Changeux, 1974) and in reeler where the defect is the apparent absence of immunoreactive glial fibrillary acidic protein (Bignami and Dahl, 1974). Thus, it would appear that the effects of these mutations on expression of GPDH in Bergmann glia might be secondary to their well-characterized effects on neuronal cells, most like Purkinje cells since this cell type is severely affected in each of the mutants. In *pcd/pcd* animals the Purkinje cell population dies around the end of the first month (Mullen et al., 1976). The staggerer mutation results in reduced numbers of Purkinje cells and abnormalities of dendritic arborization among those remaining (Bradley and Berry, 1978; Herrup and Mullen, 1979). The reeler and weaver mutations are characterized by malpositioning of the Purkinje cells and reduced arborization of their dendrites (Sidman, 1968). Thus, all these mutants have in common reduced expression of GPDH in the Bergmann glia and absent or grossly malformed Purkinje cells.

Certain observations suggest that these two conditions might be related. First, in addition to the radial Bergmann fibers, these glial cells produce elaborate membranous processes which ensheath the Purkinje dendrites (Palay and Chan-Palay, 1974). Second, GPDH is an important enzyme for lipid synthesis, and its presence in oligodendrocytes suggests its importance in myelination (de Vellis et al., 1977; Fisher et al., 1981). Finally, Aguayo et al. (1976a, 1976b) showed that in the peripheral nervous system myelination is specifically induced by the axons which the Schwann cells contact. Thus, it is conceivable that Bergmann glia are similarly "induced" to elaborate the membranous sheaths around Purkinje cell dendrites.

A number of explanations might be proposed regarding the common loss of GPDH expression in the Bergmann glia of several distinct mutants. For example, we cannot rule out the possibility that each of these mutations is characterized by an intrinsic defect of the Bergmann glia. This seems unlikely in the absence of descriptions of such defects in these mutations (with the possible exceptions of reeler and weaver as noted earlier). Furthermore, in some of the mutants (*pcd/pcd* and *sg/sg*) Purkinje cells have been shown to be a primary site of gene action (Herrup and Mullen, 1979; Mullen, 1977). It seems unlikely but not impossible that mutations having very different intrinsic effects on Purkinje cells would have a similar intrinsic, pleitropic effect on Bergmann glia.

All of these observations concerning GPDH expression in mutants as well as the known defects of these mutants are consistent with the following

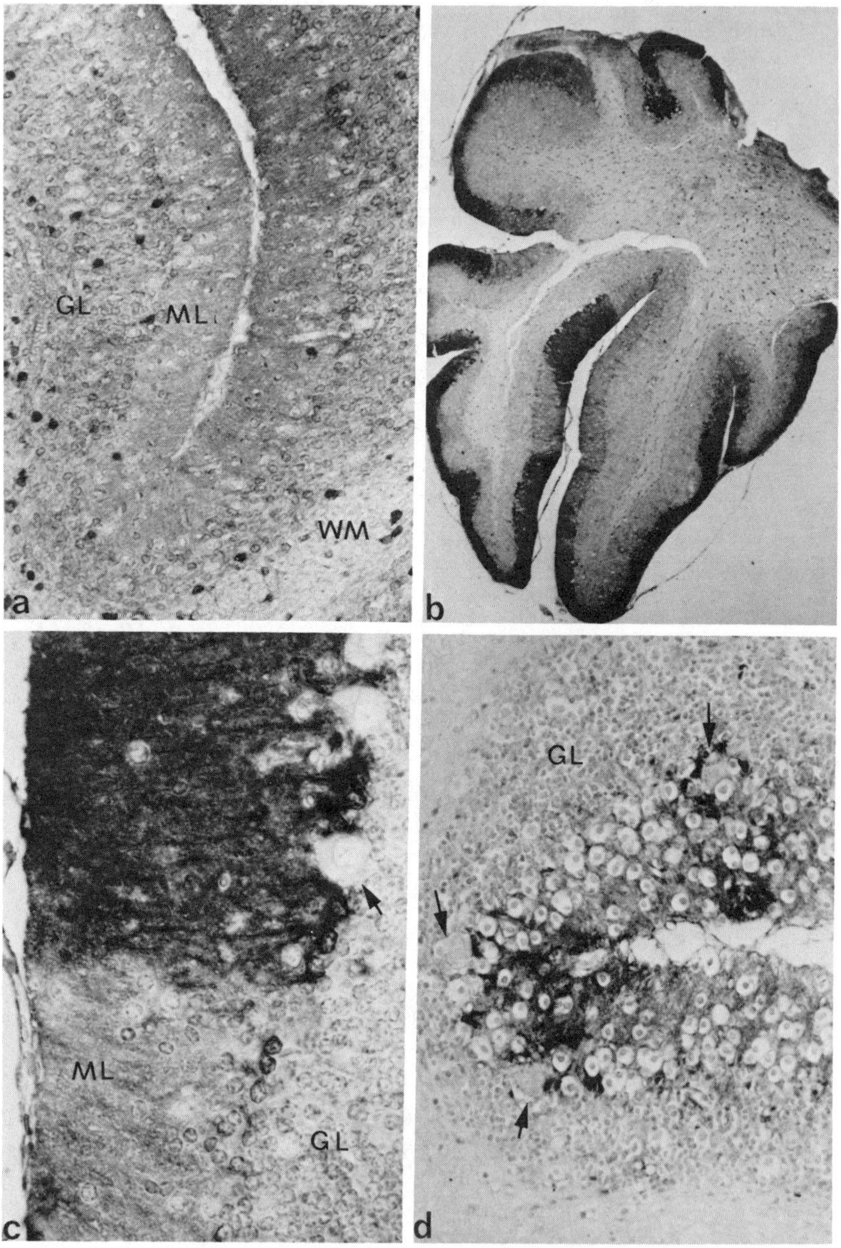

GL
ML
WM
a
b
ML
GL
c
GL
d

hypothesis: in the mouse cerebellum the induction and maintenance of a high level of GPDH activity in Bergmann glia is dependent upon their interactions with neighboring Purkinje cells.

More detailed studies of GPDH expression in the Purkinje-cell-deficient mutants, Lurcher (*Lc*/+), Purkinje cell degeneration (*pcd*/*pcd*), and nervous (*nr*/*nr*) (Fisher, unpublished results) and chimeric mice (Fisher and Mullen, unpublished observation) strongly support this hypothesis. Lurcher mutants, like *pcd*/*pcd* animals, lose their entire Purkinje cell population but the process starts earlier, around postnatal day 10 (Caddy and Biscoe, 1979). In nervous mice there is a regionally variable loss of Purkinje cells; up to 90% in the lateral hemispheres and 50% or more in the vermis disappear by two months of age (Sidman and Green, 1970). While all Purkinje cells in nervous mice show early ultrastructural anomalies, those that survive reacquire normal morphology (Landis, 1973). The nervous mutant is, therefore, unique in that it provides a system suitable for simultaneous, direct examination of enzyme expression in Bergmann glia in relation to the presence and absence of Purkinje cells in different regions of the same cerebellum.

Cerebella from *Lc*/+ and *pcd*/*pcd* mutants examined immunohistochemically show a progressive loss of GPDH staining with age following the loss of the Purkinje cells. Figure 5a shows a two-month-old *Lc*/+ cerebellum following immunostaining for GPDH. When this figure is compared with Figure 3c, a marked reduction of GPDH immunoreactivity in the Bergmann glial cell bodies and processes in the molecular layer is apparent. Most striking though is the staining pattern seen in the nervous cerebellum (Fig. 5b). Here there are obvious patches of normally stained Bergmann glial cell bodies and processes adjacent to patches of unstained molecular layer.

---

**Figure 5.** (a) Section through cerebellum of 2-month-old *Lc*/+mouse following immunostaining for GPDH. Purkinje cells are absent from the boundary between the molecular layer (ML) and granule cell layer (GL). There is little background staining in the molecular layer and the Bergmann glia are not visibly stained. Intensely stained oligodendrocytes appear in the granule cell layer and white matter (WM) as in controls. Magnification, 238×. (b) Section through cerebellum of 7-month-old *nr*/*nr* mouse following immunostaining for GPDH. The molecular layer shows regions of intense staining as well as regions with very light staining. Magnification, 48×. (c) Part of section shown in b shows that Purkinje cell somas (arrowheads) are associated with the intensely stained patch of molecular layer and Bergmann glia cell somas. No Purkinje cell somas are present underlying the adjacent, faintly stained patches of molecular layer. Magnification, 475× (d) Section through cerebellum of 9-month-old *pcd*/*pcd*←+/+ chimeric mouse following immunostaining for GPDH. Three Purkinje cell somas (arrowheads) are visible in the field. Adjacent to each Purkinje cell soma are several intensely stained Bergmann glia cell somas. Unstained patches of molecular layer not associated with a Purkinje cell are also apparent. Magnification, 238×.

Figure 5c shows that there are Purkinje cell somas associated with an intensely stained region of molecular layer while the adjacent, unstained patch is devoid of Purkinje cells. This patchy GPDH staining pattern is reproduced in chimeric mice made with either of the mutants *Lc/+* or *pcd/pcd*. The Purkinje cell death in each of those mutants is due to the intrinsic action of those genes within Purkinje cells (Mullen, 1977; Wetts and Herrup, 1982). In chimeric mice made with either of these mutants there are patches where mutant Purkinje cells have died and patches where normal Purkinje cells remain. As in the nervous mutant, in such chimeras there is always a positive correlation between the presence of Purkinje cells and intense GPDH staining in Bergmann glia (Fig. 5d). The patchy GPDH staining seen in the Bergmann glia of nervous mice and *Lc/+* or *pcd/pcd* chimeric mice contrasts with the uniform appearance of Bergmann fibers throughout the cortex in these mice as seen by immunostaining for the presence of glial filaments. Therefore, it is certain that Bergmann glia are still present in the patches that do not have intense GPDH staining, but their metabolic state has been altered so that they no longer have high levels of this enzyme activity.

Although evidence from the biochemical analysis of tissue culture cells and neurological mutants suggested different sites of GPDH localization, direct visualization of the the enzyme in wild-type, mutant, and chimeric mice by immunohistochemistry resolved the apparent conflict. Furthermore, this resolution as well as elucidating the cellular localization of GPDH also provided insights that explain why mutants with neuronal defects may be deficient in a glial enzyme.

The resolution of this apparent conflict of cellular localization of GPDH emphasizes a potential pitfall in the use of mutants with "known" cellular deficiencies in biochemical assays designed to determine the location, for example, of receptors or the source of a neurotransmitter since it is clear that a defect in one cell type can cause a profound change in the metabolism of another cell type that might appear by other criteria to be "normal."

## 4. GENETIC REGULATION OF CEREBELLAR GPDH EXPRESSION

The data presented in the preceding section indicate that the increase in GPDH expression in the cerebellum of the developing mouse is largely due to an increased expression of the *Gdc-1* locus in the Bergmann glial cells. Furthermore, the experiments with neurological mutants of the mouse suggest that the cellular environment is very important for normal expression and that GPDH expression in Bergmann glial cells is not autonomously regulated (i.e., interactions with other cells are required). In attempting to understand

*Gdc-1* regulation in the Bergmann glial cells, we have used an additional type of genetic variation besides that represented by neurological mutants, such as the reeler mouse which result in severe neuroanatomical abnormalities. This other genetic variation, found as the polymorphisms present in natural populations of mice, has no obvious effect on the normal functioning of the animal.

We have searched among inbred strains and wild stocks of mice for variation in the level of enzyme activity that is expressed in a cell-specific and developmentally dependent manner. We have attempted to determine what number of genes controls the variation in *Gdc-1* expression and how these genes exert their effects.

Among inbred strains of mice, BALB/cJ mice (as well as other BALB/c sublines) have an increased amount of GPDH in the cerebellum (Kozak, 1972). The magnitude of the difference between BALB/cJ mice and a low strain, like C57BL/6J mice, is age-dependent; a two- to threefold difference is found between 12 and 20 days of age. As the animals age, the difference in enzyme activity between strains decreases in magnitude. Adult BALB/cJ mice have 50% more activity than adult C57BL/6J mice (Fig. 6). Earlier than 10 days of age the difference is not observed unless the adult and embryonic isozymes are first separated by ion exchange chromatography (Kozak and Jensen, 1974), indicating that the difference in activity is associated only with the *Gdc-1* locus. Since 16 additional enzymes of carbohydrate metabolism, including glucose-6-P dehydrogenase, did not show any strain-dependent difference in activity levels, this indicates that the differences are not due to some generalized effect that depresses the number of GPDH-enriched cells or carbohydrate metabolism of the cerebellum of C57BL/6J mice. Moreover, other tissues of the animal including other regions of the brain do not show the difference in activity (Kozak, 1972). These observations together with the cell localization data suggest that this difference in GPDH expression between BALB/cJ and C57BL/6J mice is found specifically in the Bergmann glial cells.

The inheritance of the difference in GPDH activity between BALB/cJ and C57BL/6J is complex. To resolve this complexity and to determine the mechanisms by which the genes exert their effect, we have utilized conventional backcross analysis and recombinant inbred lines. For the backcross analysis, $F_1$ animals were crossed to each parent. The offspring were all typed at 20 days of age for the level of GPDH and the characteristics of the structural alleles. C57B/6J and BALB/cJ mice carry the *b* and *c* alleles at the *Gdc-1* locus, respectively. The expression of these alleles detected by differences in heat stability allows one to determine whether a particular allele at the structural locus is associated with the activity differences. The frequency distribution of the phenotypes in a backcross of the $B6CF_1$ to the BALB/cJ parent is

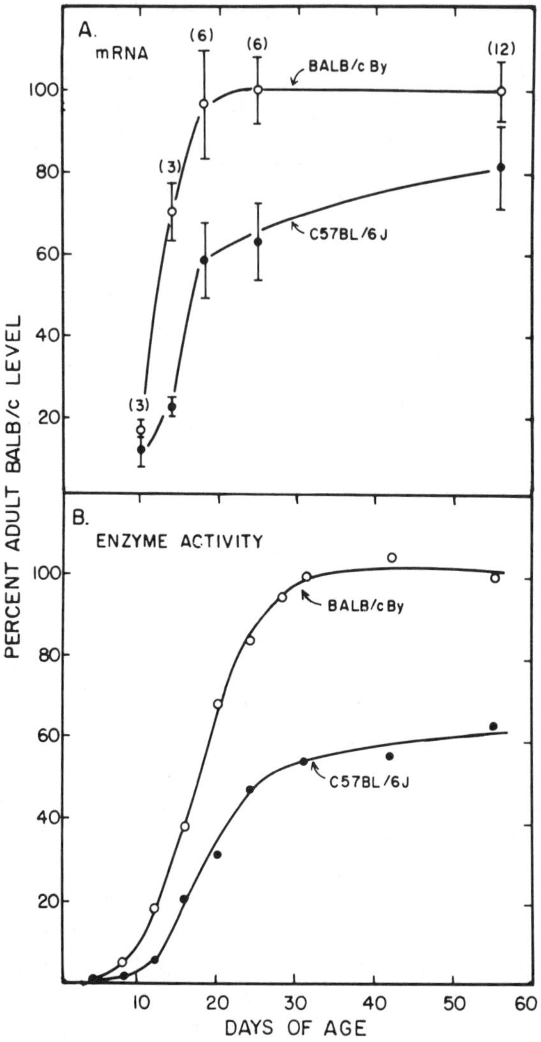


**Figure 6.** Developmental profile for the accumulation of translatable mRNA and enzyme activity for GPDH. (A) Poly (A+) mRNA was isolated from the cerebella of C57BL/6J (closed circles) and BALB/cBy (open circles) mice at the indicated ages and translated in the rabbit reticulocyte lysate system. The concentration of glycerol-P dehydrogenase mRNA is expressed as counts per minute in GPDH relative to counts per minute in trichloroacetate-precipitable protein and then graphed as a percentage of the adult BALB/cBy level. The number in parentheses refer to the number of assays for both BALB/cBy and C57BL/6J mice. Each assay was based on determinations with two or three levels of RNA isolated from 30 to 200 mice. The statistical significance of the difference between BALB/cBy and C57BL/6J for each age group was tested by the Student's *t* test. At 10 days of age the difference is not significant at $p=0.01$. (B) Glycerol-P dehydrogenase specific activity was determined on the 100,000 *g* supernatant fraction as described by Kozak and Jensen, 1974. Each point represents the mean of three to four animals. (From Kozak and Ratner, 1980.)

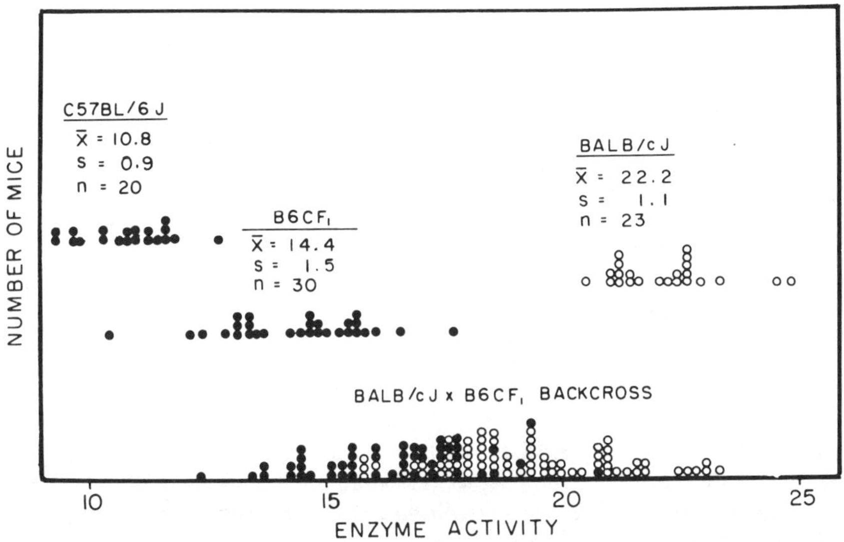


**Figure 7.** Frequency distributions of GPDH in the cerebellum in parental strains, $B6CF_1$s, and in the progeny of the backcross generation. Among the backcross progeny animals indicated by closed circles have a hybrid, BC, heat denaturation phenotype while those indicated by open circles have a BAlB/c CC, heat denaturation phenotype.

illustrated in Figure 7. Statistical analysis of the distribution suggests that two major loci are controlling the variance in the population. In addition, there is an association of the structural locus with the activity phenotype (i.e., all animals with an enzyme level similar to the BALB/cJ parent are homozygous for the heat-sensitive *c* allele) while animals that are heterozygous at the *Gdc-1* locus express enzyme at a level similar to the $B6CF_1$ parent. Some animals have a level of activity that is intermediate to the $B6CF_1$ and BALB/cJ parents. This intermediate group can be either homozygous, *c/c*, or heterozygous, *b/c*, at the *Gdc-1* locus. These data which implicate the structural gene and a second unlinked gene in determining the activity levels were verified by an analysis with the recombinant inbred lines constructed from the BALB/cBy and C57BL/6By progenitor strains.

Recombinant inbred lines are formed by crossing two unrelated inbred strains followed by 20 or more generations of brother × sister matings (Bailey, 1971). Since these lines carry genetically fixed recombinant chromosomes derived from both the BALB/c and B6 genotypes, it is possible to identify mice that have GPDH activity phenotypes due to the recombination of the two major chromosomal regions that determine GPDH expression. The phenotypes identified among the RI lines are shown in Table 3. As with the

**TABLE 3. CLASSES OF GPDH PHENOTYPES**

| Class | Enzyme Activity Phenotype | Structural Gene | Strains |
|---|---|---|---|
| I | High GPDH (0.242–0.273) | *c* | CXBG, CXBH, BALB/cBy (J) |
| II | Intermediate GPDH (0.153–0.192) | *b* or *c* | CXBD, CXBE, CXBI CXBJ, CXBK |
| III | Low GPDH (0.117–0.125) | *b* or *c* | C57BL/6By(J), BC61 Other inbred strains and wild stocks |

backcross analysis, high GPDH activity (class I) requires that the (*c*) allele at *Gdc-1* locus must be present. However, although the presence of a *c* allele at the *Gdc-1* locus is a necessary condition for high acitivity, it is not sufficient since it is possible to have an intermediate level of enzyme activity with a *c* allele at *Gdc-1*. We interpret this to mean that mice with a *c* allele at *Gdc-1*, but an intermediate level of activity, must have a *b* allele at another unlinked (regulatory) locus-controlling activity. Conversely, mice with a *b* allele at the *Gdc-1* locus and an intermediate level of activity must have a *c* allele at the unlinked (regulatory) locus. Accordingly, both regulatory and structural loci are controlling the enzyme activity differences present among inbred strains of mice. An intermediate level of expression in $F_1$ animals indicates that the alleles are codominantly expressed.

What is the mechanism by which these loci affect GPDH activity? The following data suggest that the difference in GPDH activity is determined by differences in the production or translatability of GPDH mRNA:

1. Kinetic analysis of GPDH enzyme purified from BALB/cJ and C57BL/6J mice indicates that the Michaelis constants, maximum velocities, and energies of activation are indistinguishable (Kozak, 1974).
2. Tissue extracts of BALB/cJ cerebellum have greater quantity of immunoreactive GPDH than similar extracts from C57BL/6J mice (Kozak and Ratner, 1980).
3. BALB/cJ mice have increased amounts of translatable GPDH mRNA than do C57BL/6J mice; this correlates well with the differences in enzyme activity observed during development (Kozak and Ratner, 1980; Fig. 6).

A difference in the level of translatable mRNA for GPDH does not by itself establish that this difference is controlled by the strain-specific alleles present in these inbred mice (Kozak, 1972). However, we must also note that the differences in mRNA levels are maximal between 14 and 25 days of age (Fig. 6) and specific for the cerebellum in a manner that parallels the enzyme activity levels which have previously been established as being under genetic control by testing the progeny from genetic crosses (Kozak, 1972). Therefore, we must consider the possibility that the observed differences in glycerol-P dehydrogenase mRNA levels are, in fact, controlled by the same strain-specific alleles that control GPDH activity levels.

## 5. MOLECULAR CLONING OF GPDH cDNA

The investigation of several problems of GPDH expression in mouse cerebellum would be aided by the availability of cDNA probles to the GPDH mRNA and its structural genes. They would allow us to measure the GPDH mRNA levels during development and among inbred strains of mice which differ in GPDH expression. They would also provide us with a reagent to isolate the genomic sequences that encode the adult and embryonic isozymes and to compare the nucleotide structures of these genes.

Recently we succeeded in isolating a bacterial clone carrying a plasmid with a GPDH cDNA insert from a cDNA plasmid library constructed from mRNA isolated from the brown adipose tissue, the tissue with the highest known level of GPDH (Kozak and Birkenmeier, 1983). The identification of the clone as a GPDH clone was achieved first by the characteristics of the signal obtained by colony hybridization with a series of cDNA probes made from mRNA with various amounts of translatable GPDH mRNA activity. Next the plasmid DNA from one of the clones identified as a candidate was shown to select by hybridization an mRNA which on translation *in vitro* gave immunoprecipitable GPDH. Finally, it was shown that the insert in the plasmid called C8 hybridized to a polymorphic restriction enzyme fragment which mapped to the *Gdc-1* locus on chromosome 15 (Fig. 8).

We have obtained some preliminary information on the characteristics of GPDH mRNA by a Northern blot experiment (Fig. 9). The most striking feature of the GPDH mRNA is that it is much larger than required to encode the GPDH protein. Mouse GPDH, which contains approximately 340 amino acids (Kozak and Burkart, 1981; Otto et al., 1980), requires about 1020 nucleotide bases for coding. However, the mRNA is approximately 3600 bases long or 3.5 times longer than required. The three adult tissues analyzed (liver, brown adipose tissue, and cerebellum) each have an mRNA of the

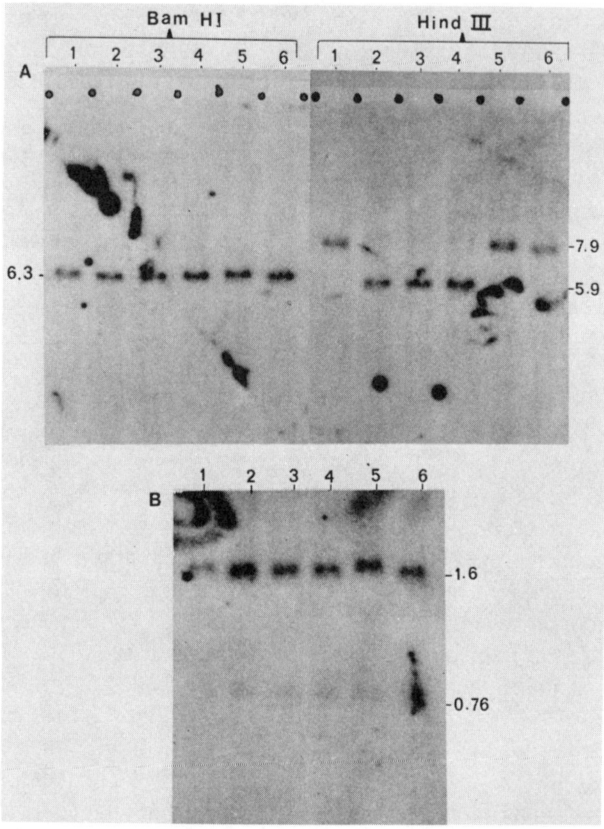


**Figure 8.** Southern blot analysis of cellular DNAs hybridized to C8. (A) Cellular DNAs from B6-*Gdc*-1[b] (1), B6. Cas-*Gdc*-1[d] (2), Cas-*Gdc*-1[d] (3), LT. Cas-*Gdc*-1[d] (4), LT-*Gdc*-1[b] (5), and BALB/c-*Gdc*-1[c] (6) mice were digested with *Bam HI* and *Hind III* restriction enzymes. The numbers in the margin indicate the size of the DNA fragments in kilobase pairs. (B) The same mouse DNAs were digested with *Eco RI* and the two fragments which hybridized to C8 are shown. (From Kozak and Birkenmeier, 1983.) With permission from *Proc. Natl. Acad. Sci. USA*. Copyright: National Academy of Sciences.

same size, which indicates that differences in tissue expression may not be due to differences in mRNA structure. The fourth mRNA analyzed in Figure 9 (lane 1) was isolated from neonatal brain. It showed no hybridization signal with the GPDH cDNA probe. This finding was consistent with data on GPDH mRNA expression obtained from *in vitro* translation of cerebellar mRNA preparations. Furthermore, the intensity of the signal seen in the other lanes of the Northern blot generally corresponds to the signal expected from enzymatic and immunological data (Ratner et al., 1981) and *in vitro* translation

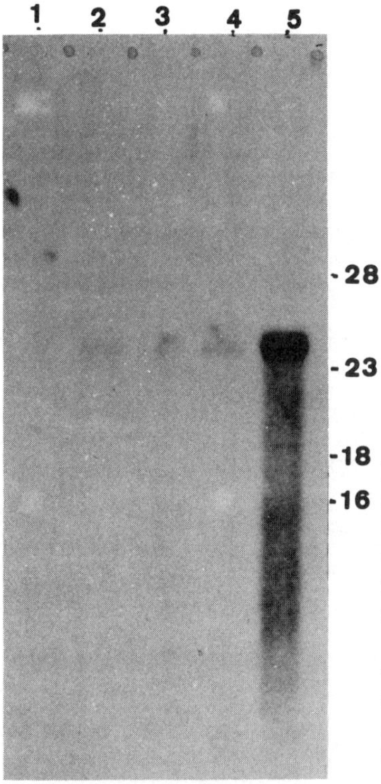


**Figure 9.** Northern blot analysis of poly (A+) RNA isolated from (1) neonatal brain, (2) C57BL/6J cerebellum, (3) BALB/cBy cerebellum, (4) liver, and (5) cold-adapted brown adipose tissue. Blots were hybridized to nick-translated C8 plasmid DNA. Each lane contained 5 $\mu$g of RNA, except lane 1 which contained 10 $\mu$g. Numbers along the side refer to S values of ribosomal RNA markers. (From Kozak and Birkenmeier, 1983.) With permission from *Proc. Natl. Acad. Sci. USA*. Copyright: National Academy of Sciences.

data of GPDH mRNA (Kozak and Ratner, 1980). In summary these data suggest that the increase in GPDH mRNA during development is controlled by the number of cytoplasmic transcripts of GPDH mRNA. Whether differences observed between strains are controlled by the mRNA levels remains to be determined since the experiment in Figure 9 was not designed to assess this particular question.

## 6. EXPRESSION IN CELL CULTURE

From the preceding discussion both qualitative and quantitative features of GPDH expression emerge as particularly prominent. The qualitative feature is observed as a switch in the structural gene used to encode the major isozyme. The quantitative feature is evidenced as genetic variations in the amount of enzyme that is present in the developing cerebellum. In considering cellular and genetic mechanisms to explain how these processes are controlled, one is compelled to consider that they are occuring in an organ that is in the process of acquiring a highly organized structure of glial and neuronal cells through a successive series of cell divisions, migrations, and interactions. How do these processes observed at the cellular level influence events that are occurring at the level of the *Gdc* genes?

In order to acquire information on this question, we adapted the existing techniques for culturing primary neural cells to the analyses of GPDH expression in reaggregating postnatal cerebellar cells (Moscona, 1961). Primarily, we wished to assess the effects of changes in the physical form of cell culture (i.e., aggregate culture vs monolayer culture) on the transition in isozyme expression and on the final level of GPDH expression. Secondarily, we wished to perturb interactions between cells in an aggregate culture and assess the effects of the perturbations on *Gdc* expression. The results of these experiments strongly indicate that cerebellar cells growing in a three-dimensional aggregate provide an environment essential for the transition from embryonic to adult isozyme expression. These results with primary normal cerebellar cells are corroborated by the information on GPDH in neuroblastoma cell lines (as well as other types of tumor cells) in culture and *in vivo*. In this case we detected the embryonic isozyme in cultured C1300 neuroblastoma cells and the adult isozyme in these cells growing as solid tumors.

Cerebellar cells taken from mice between birth and 7 days of age reaggregate when cultured in Erlenmeyer flasks in Eagle's basal medium wih 15% horse serum. In these cultures the expression of GPDH varies with respect to (1) the age of the animal from which the cerebellum is removed (Kozak, 1977), (2) the length of time the cells are in culture (Kozak, 1977), and (3) the composition of the medium (Kozak et al., 1978). We think that these variables affect the relative proportion of neurons to glial cells in the aggregate population and the extent to which cultures are in a gliotic state (Kozak et al., 1977, 1978). This interpretation is based on the analysis of GPDH and GFA expression (glial fibrillary acidic protein) in aggregate cultures and *in vivo*.

The general impression on the morphology of cerebellar aggregate culture obtained by microscopy (Kozak et al., 1977) is that for the first 10–12 days in culture the cell aggregates are composed of both glial cells and neuronal cells, principally granule neurons, in relative proportions that resemble the composition observed *in vivo*. Beyond 12 days in culture the granule neurons begin to degenerate with an accompanying glial reaction. The immunohistological analysis of GFA indicates that after 19 days in culture most aggregates resemble a glial nodule.

During this period of aggregate culture GPDH is first expressed as the embryonic isozyme; then at a time corresponding to that which would be expected to occur *in vivo*, the adult isozyme begins to be expressed (Kozak, 1977). Accordingly, during the first 10 days of age both the time of isozyme transition and the levels of GPDH activity resemble, to a remarkable degree, the characteristics observed *in vivo*. This also includes the differences in cerebellar GPDH observed between BALB/cJ and C57BL/6J mice (Table 4). Culturing the aggregates with radiolabeled amino acids indicates that the increase in adult isozyme activity is determined by increased rates of enzyme synthesis (Kozak, 1978). Furthermore, the difference between BALB/cJ and C57BL/6J mice is also determined by higher rates of enzyme synthesis in BALB/cJ mice (Kozak, 1978).

Beyond 12 days in culture, particularly in aggregates formed from cerebellar cells isolated from mice 6 days of age and older, a continuing increase in GPDH occurs to levels that exceed substantially those found *in vivo*. *In vivo* the specific activity of GPDH in adult C57BL/6J mice is approximately 0.25 μmol/min/mg protein whereas in aggregate culture it can be as high as 1.2. It is apparent from immunofluorescence studies with GFA that an increased proportion of fibrous astrocytes are present in response to the glial reaction (Kozak et al., 1977)., Accordingly, it is possible that GPDH is also increasing as a consequence of gliosis. This is unexpected because GPDH does not appear at high levels in the fibrous astrocytes of normal tissue, only in Bergmann glia and oligodendroglia (Fisher et al., 1981). It seems that in culture either the Bergmann glia are involved in the glial reaction or, alternatively, GPDH is able to be synthesized by fibrous astrocytes. The idea that

**TABLE 4. INFLUENCE OF AGE ON THE SPECIFIC ACTIVITY OF L-GLYCEROL-3-PHOSPHATE DEHYDROGENASE IN CEREBELLAR AGGREGATES FROM BALB/cBY AND C57BL/6J MICE**

| Age of Mouse | Days in Culture | α-GPDH (units/mg protein) BALB/cBy | C57BL6J | Ratio[a] | Significance[b] |
|---|---|---|---|---|---|
| *P*0 | 22 | 178.3 ± 14.2 (3) | 38.1 ± 6.2 (3) | 4.7 | $P < 0.01$ |
| *P*2 | 20 | 345.1 ± 69.7 (3) | 87.1 ± 6.4 (3) | 4.0 | $P < 0.05$ |
| | 20 | 255.1 ± 58.6 (6) | 111.4 ± 9.7 (4) | 2.3 | NS at $P = 0.05$ |
| *P*3 | 15 | 488.3 ± 77.0 (6) | 224.8 ± 36.4 (4) | 2.2 | $P < 0.05$ |
| | 20 | 706.6 ± 63.8 (5) | 233.6 ± 25.9 (4) | 3.0 | $P < 0.05$ |
| *P*4 | 12 | 131.6 ± 22.6 (3) | 75.7 ± 12.1 (3) | 1.8 | NS at $P = 0.05$ |
| | 17 | 493.0 ± 88.5 (3) | 206.5 ± 22.7 (3) | 2.4 | $P < 0.05$ |
| *P*6 | 8 | 197.7 ± 9.4 (4) | 192.1 ± 15.7 (4) | 1.0 | NS at $P = 0.05$ |
| | 15 | 974.1 ± 129.6 (5) | 825.0 ± 99.9 (3) | 1.2 | NS at $P = 0.05$ |

[a] (α-GPDH activity in BALB/cBy)/(α-GPDH activity in C57BL/6J).

[β] The significance of the differences in α-GPDH activity between BALB/cBy and C57BL/6J mice was tested by the Student's *t*- test.

**TABLE 5. GPDH EXPRESSION IN CEREBELLAR CELLS CULTURED *IN VITRO* AS MONOLAYER AND AGGREGATE CULTURES**

| Cell Type | Culture Conditions | GPDH (nmol/min/mg) | LDH (nmol/min/mg) |
|---|---|---|---|
| Experiment 1 | | | |
| BALB/c, 6 days of age | (a) Aggregation for 15 days | 779.6 ± 231.8 (3) | 1308.4 ± 287.8 (3) |
| | (b) Monolayer for 15 days | 86.7 ± 21.6 (3) | 1430.9 ± 58.6 (3) |
| Experiment 2 | | | |
| C57BL/6J 5 days of age | (a) Aggregation 14 days | 324.9 ± 29.9 | N.D. |
| | (b) Aggregation 1 day then monolayer 13 days | 119.7 ± 25.8 | N.D. |

GPDH is increasing in aggregate cultures as a consequence of gliosis is consistent with the observation that gliosis is repressed and GPDH levels are reduced in reaggregating cells cultured in medium containing 5 $\mu$g/mL polylysine (Kozak et al., 1978).

In contrast to the above results with aggregates, if cerebellar cells are cultured as a monolayer population on the surface of plastic dishes, the GPDH levels are much lower than those observed in aggregate cultures (Table 5, experiment 1). This result does not seem to be a consequense of selecting a different population of cells in monolayer than in aggregate cultures since cells which were initially selected to aggregate and then transferred to monolayers where they attached to the surface of the dish and spread out over the surface also had substantially reduced levels of GPDH (Table 5, experiment 2). This result suggests that the expression of the *Gdc-1* locus is dependent on interactions that occur between cerebellar cells when they are in direct contact in an aggregate.

A requirement for cell contact in the expression of *Gdc-1* is also apparent from the analyses of a series of murine tumors grown *in vivo* as a solid and ascites forms of growth (Kozak and Murphy, 1976). The general observation is that a tumor from a cell lineage which expresses the embryonic isozyme (*Gdc-2*) during embryonic and/or fetal development (e.g., brain, muscle, and fibroblasts) will express the embryonic isozyme when the tumor is grown in culture or in the ascites form *in vivo*. However, when the tumor is grown as a solid form, irrespective of its rate of growth or degree of differentiation, then the adult isozyme is expressed. A striking illustration of this reversible expression of two isozymic loci as a consequence of the form of tumor growth is depicted in Figure 10. In this experiment an anaplastic mammary carcinoma 15091A was transplanted on alternate generations either subcutaneously to produce solid tumor growth or interperitoneally to produce ascites tumor growth. At each transplant generation it was found that ascites tumors expressed low levels of activity and the embryonic isozyme while the solid tumors expressed high levels of activity and the adult isozyme. This sensitivity of the expression of the *Gdc* loci in tumor cells to an environment which maximizes cell contact also supports the hypothesis that the *Gdc* loci are sensitive to cellular signals that are acting at the cell surface.

The analysis of tumors and fetal tissues of the mouse revealed an additional important characteristic of *Gdc-2* expression. Those tumors derived from cell lineages that never express the embryonic isozyme (e.g., liver cells) will accordingly never express the embryonic isozyme but only the adult isozyme, whether the tumor is growing in culture or *in vivo*. Neuroblastoma, on the other hand, will express the adult isozyme if grown as a subcutaneous solid tumor but the embryonic isozyme when grown in tissue culture. We do not think this apparent reversibility is a consequence of cell selection since a

| Generation | SOLID TUMOR Specific Activity | SOLID TUMOR T½ in min. | ASCITES TUMOR Specific Activity | ASCITES TUMOR T½ in min. |
|---|---|---|---|---|
| 1 | 35.4 | 4.6 | 2.4 ± 0.1 | 0.6 |
| 2 | 145.0 | 3.6 | 3.8 ± 0.1 | 0.7 |
| 3 | 49.3 ± 12.4 | 3.9 | 7.9 ± 0.8 | 0.6 |
| 4 | not analyzed | | 9.9 ± 1.4 | 0.9 |
| 5 | 18.7 | 4.0 | 3.5 ± 0.3 | 0.5 |

**Figure 10.** Changes in GPDH expression during interconversion of 15091A tumor. Closed circles = solid tumors; open circles = ascites tumor. Specific activity is given in nmol/min $mg^{-1}$ protein. $t^{1/2}$ (in min) indicates the half-life of $\alpha$-GPDH at 50°C. (From Kozak and Murphy, 1976.) With permission from *Cancer Research*. Copyright: American Assoc. for Cancer Research.

cloned C1300 neuroblastoma line was established which expressed the embryonic isozyme in culture and the adult isozyme in a solid tumor. Since neuroblastoma cells will not grow as ascites tumors in mice, we have not been able to assess this particular form of growth on *Gdc* expression.

Several attempts have been made to determine whether agents such as cAMP, removal of serum, and bromodeoxyuridine, which promote differentiation of neuroblastoma cells as evidenced by morphological changes and increases in neural specific proteins (Blume et al., 1970; Schubert et al., 1969), could also induce expression of the *Gdc-1* locus. These manipulations all proved negative (unpublished observations). Since a 3T3 fibroblast cell line, 3T3-F442A, is able to differentiate in culture to adipocytes with the concomitant switch from *Gdc-2* to *Gdc-1* expression (Wise and Green, 1979), it is possible that conditions may yet be found that will promote the expression of *Gdc-1* in cultured neuroblastoma cells.

## 7. COMPARISON OF RAT AND MOUSE GPDH EXPRESSION

In an extensive series of experiments de Vellis and his colleagues have studied the expression of GPDH in the rat cerebellum (de Vellis and Inglish, 1968; de Vellis et al., 1977). The main focus of this work has been to investigate the induction of GPDH in the rat cerebellum by glucocorticoids. Hypophysectomy or adrenalectomy of the rat leads to about a 60% reduction in cerebellar and brain stem GPDH over a period of 30 days which can be restored by administration of ACTH or cortisol. Administration of cortisol to neonatal rats leads to a precocious development of GPDH in the brain stem and cerebellum (de Vellis and Inglish, 1973). In order to study the mechanism underlying this induction, tissue cultures have been studied quite extensively, particularly a glioma cell line (C6) originally developed by Davidson and Benda (1970). Strong evidence has been presented that the increases in GPDH activity in cultured cells is caused by an increase in the synthesis of enzyme molecules (McGinnis and de Vellis, 1978). No studies have been made at the RNA level.

In contrast, GPDH levels cannot be induced in the cerebellum of the mouse. This has been tested by experiments in which glucocorticoids were injected during early postnatal development and by the addition of hydrocortisone to reaggregating cerebellar cells cultured in the presence of horse serum depleted of steroid hormones (unpublished results). Since the rat has the same cellular localization as the mouse (i.e., Bergmann glial cells and oligodendroglial) (Fisher et al., 1981), it cannot be argued that the species differ with respect to glucocorticoid induction because the enzyme is located in different cell types.

The rat also differs from the mouse in that an embryonic isozyme does not appear to be expressed in the fetal and neonatal brain. In addition, both the induced and uninduced C6 glioma cells express the same GPDH protein as that expressed in the adult rat brain (McGinnis and de Vellis, 1974). Since genetic structural variants have not been reported for rat GPDH to test for genetic identity, it is still conceivable, although improbable, that different genetic loci encode fetal and adult enzymes in rat brain.

In spite of these major differences between rat and mouse GPDH expression, recent work by McGinnis and de Vellis suggests that interactions at the surface of both C6 glioma cells and primary rat cerebral cell cultures are involved in modulating the level of cytoplasmic glucocorticoid receptor and subsequently GPDH induction (McGinnis and de Vellis, 1981). Furthermore, the uncoupling of $\beta$-adrenergic receptor sites from adenylate cyclase in C6 glioma cells when they are grown in contact with neuroblastoma cells resembles somewhat the ability of neuroblastoma cells to block GPDH expression in mouse cerebellar cells when the two types of cells are reaggregated (Ciment and de Vellis, 1978; Kozak, 1979). Perhaps GPDH expression in mouse cerebellar cells is also regulated at the cell surface by $\beta$-adrenergic-like receptors but without the involvement of glucocorticoid hormones.

## 8. CONCLUDING REMARKS

A premise in developmental biology states that the determination and differentiation of cells in a developing organism are controlled in part by the nature of the cell-to-cell interactions that area available to those cells (Holtfreter, 1964). More recent experiments, consistent with this premise, have demonstrated that if an embryonal carcinoma cell is injected into a blastocyst so that it becomes integrated with the cells of the inner cell mass, then subsequent differentiation of the embryonal carcinoma cell is controlled in time and space by that of the developing embryo. We find that the embryonal carcinoma cell has contributed in mass to the normal form of the mouse (Mintz and Illmensee, 1975).

Although these experiments with embryonal carcinoma cells emphasize the importance of the cell environment in the control of development, we still understand very little about how this environment acts. Are there in fact ligands on one cell that interact with receptors on others that modulate gene activation much like a polypeptide hormone-cell surface? Or are these environmental interactions mediated by diffusible soluble substances including both macromolecules and low-molecular-weight substances like inorganic ions? We have presented evidence that the *Gdc* loci are acutely sensitive to the cerebellar environment, specifically the cellular interactions, for their normal expression. The cerbellum is a particularly good model system because this organ has a population of undifferentiated cells that are at their most dynamic stage of growth and present in sufficient quantities at a time convenient for biochemical analysis either *in vivo* or *in vitro*. In addition, many morphologically well-characterized mutations have been identified which perturb the interactions of cells in the developing cerebellum. The features of cerebellar GPDH expression described here and the availability of appropriate molecular genetic and cell biological methods offer the opportunity for a rigorous analysis of the mechanisms controlling this enzyme's expression in the developing brain.

## ACKNOWLEDGMENTS

This work would not have been possible without active collaboration and stimulating discussions from Paul Ratner and Jeffry Cook, Edward Birkenmeier, Richard Mullen, John Eppig, Amico Bignami, Doris Dahl, and Harriet Gray. We thank Donna Burkart for excellent technical assistance and Anne Kozak for assistance in preparing the manuscript. This work was supported by NIH grants HDO6712, HD80431, NS16156, AND NS17790.

## REFERENCES

Aguayo, A. J., L. Charron, and G. M. Bray, *J. Neurocytol.* **5,** 565 (1976a).

Aguayo, A. J., J. Epps, L. Charron, and G. M. Bray, *Brain Res.* **104,** 1 (1976b).

Bailey, D. W., *Transplantation* **11,** 325 (1971).

Bewley, G. C. and J. C. Lucchesi, *Biochem Genet.* **15,** 235 (1977).

Bewley, G. C., J. M. Rawls, Jr., and J. C. Lucchesi, *J. Insect. Physiol.* **20,** 135 (1974).

Bignami, A. and D. Dahl, *J. Comp. Neurol.* **155,** 219 (1974).

Blume, A., F. Gilbert, S. Wilson, J. Farber, R. Rosenberg, and M. Nirenberg, *Proc. Natl. Acad. Sci. USA* **65,** 786 (1970).

Bradley, P. and M. Berry, *Brain Res.* **142,** 135 (1978).

Caddy, K. W. T. and T. J. Biscoe, *Philos. Trans. Roy. Soc. Lond.* **287,** 167 (1979).

Chan, A. K. and E. A. Thompson, *Arch. Biochem. Biophys.* **207,** 96 (1981).

Chapman, V. M. and F. H. Ruddle, *Genetics* **70,** 299 (1972).

Ciment, G. and J. de Vellis, *Science* **202,** 765 (1978).

Cook, J. R. and L. P. Kozak, *Dev. Biol.* **92,** 440 (1982).

Davidson, R. L. and P. Benda, *Proc. Natl. Acad. Sci. USA* **67,** 1870 (1970).

de Vellis, J. and D. Inglish, *J. Neurochem.* **15,** 1061 (1968).

de Vellis, J. and D. Inglish, *Prog. Brain Res.* **40,** 321 (1973).

de Vellis, J., J. F. McGinnis, G. A. M. Breen, P. Leveille, K. Bennett, and K. McCarthy, in S. Fedoroff and L. Hertz, Eds., *Cell Tissue and Organ Cultures in Neurobiology,* Academic, New York, p. 485 (1977).

Eicher, E. M., *Mouse Newsletter* **54,** 41 (1976).

Fisher, M., D. A. Gapp, and L. P. Kozak, *Dev. Brain Res.* **1,** 341 (1981).

Green, M. C., in M. C. Green, Ed., *Genetic Variants and Strains of the Laboratory Mouse,* Gustav Fisher Verlag, Stuttgart/New York, p. 8 (1981).

Grell, E. H., *Science* **158,** 1319 (1967).

Gusella. J., R. Geller, B. Clarke, V. Weeks, and D. Housman, *Cell* **9,** 221 (1976).

Herrup, K. and R. J. Mullen, *Brain Res.* **178,** 443 (1979).

Holtfreter, J., in B. Willier and J. Oppenheimer, Eds., *Foundations in Experimental Embryology,* Prentice-Hall, Englewood Cliffs, NJ, p.186 (1964).

Kozak, L. P., *Proc. Natl. Acad. Sci. USA* **69,** 3170 (1972).

Kozak, L. P., *Biochem Genet.* **12,** 69 (1974).

Kozak, L. P., *Dev. Biol.* **55,** 160 (1977).

Kozak, L. P., *Dev. Biol.* **66,** 593 (1978).

Kozak, L. P., *Dev. Biol.* **68,** 407 (1979).

Kozak, L. P. and E. H. Birkenmeier, *Proc. Natl. Acad. Sci. USA* **80,** 3020 (1983).

Kozak, L. P. and D. Burkart, *J. Biol. Chem.* **256,** 5162 (1981).

Kozak, L. P. and K. J. Erdelsky, *J. Cell. Physiol.* **85,** 437 (1975).

Kozak, L. P. and J. T. Jensen, *J. Biol. Chem.* **249,** 7775 (1974).

Kozak, L. P. and E. D. Murphy, *Cancer Res.* **36,** 3711 (1976).

Kozak, L. P. and P. L. Ratner, *J. Biol. Chem.* **255,** 7589 (1980).

Kozak, L. P., J. J. Eppig, D. Dahl, and A. Bignami, *Dev. Biol.* **59,** 206 (1977).

Kozak, L. P., J. J. Eppig, D. Dahl, and A. Bignami, *Dev. Biol.* **64,** 252 (1978).

Kozak, L. P., D. L. Burkart, and J. P. Hjorth, *Dev. Genet.* **3,** 1 (1982).

Landis, S. C., *J. Cell Biol.* **57,** 782 (1973).

Lin, E. C. C., *Ann. Rev. Biochem.* **46,** 765 (1977).

Mallet, G., in R. K. Hunt, Ed., *Current Topics in Developmental Biology,* Vol. 15, Academic, New York, p. 41 (1980).

Markert, C. L., *Ann. NY Acad. Sci.* **151,** 14 (1968).

McGinnis, J. F. and J. de Vellis, *Nature* **250,** 422 (1974).

McGinnis, J. F. and J. de Vellis, *J. Biol. Chem.* **253,** 8483 (1978).

McGinnis, J. F. and J. de Vellis, *Proc. Natl. Acad. Sci. USA* **78,** 1288 (1981).

Mintz, B. and K. Illmensee, *Proc. Natl. Acad. Sci. USA* **72,** 3582 (1975).

Monk, R. J., O. Meyuhas, and R. P. Perry, *Cell* **24,** 301 (1981).

Moscona, A. A., *Exp. Cell Res.* **22,** 455 (1961).

Mullen, R. J., *Nature* **270,** 245 (1977).

Mullen, R. J., E. M. Eicher, and R. L. Sidman, *Proc. Natl. Acad. Sci. USA* **73,** 208 (1976).

Nadal-Ginard, B., *Cell* **15,** 855 (1978).

Nichols, E. A., V. M. Chapman, and F. H. Ruddle, *Biochem. Genet.* **8,** 47 (1973).

Niesel, D. W., G. C. Bewley, S. G. Miller, F. B. Armstrong, and C. Y. Lee, *J. Biol. Chem.* **255,** 4073 (1980).

O'Brien, S. J. and R. J. MacIntyre, *Genetics* **71,** 127 (1972).

Otto, J., P. Argos, and M. G. Rossmann, *Eur. J. Biochem.* **109,** 325 (1980).

Palay, S. L. and V. Chan-Palay, in S. L. Palay and V. Chan-Palay, Eds., *Cerebellar Cortex: Cytology and Organization,* Springer-Verlag, Berlin/New York, p. 61 (1974).

Palmiter, R. S., *Cell* **4,** 189 (1975).

Rakic, P. and R. L. Sidman, *Proc. Natl. Acad. Sci. USA* **70,** 240 (1973).

Ratner, P. L., M. Fisher, D. Burkart, J. R. Cook, and L. P. Kozak, *J. Biol. Chem.* **256,** 3576 (1981).

Roderick, T. H., F. H. Ruddle, and V. M. Chapman, *Biochem. Genet.* **5,** 457 (1971).

Rutter, W. J., T. Rjkumar, E. Penhoet, M. Kochman, and R. Valentine, *Ann. NY Acad. Sci.* **51,** 102 (1968).

Scarpulla, R. C., K. M. Agne, and R. Wu, *Proc Natl. Acad. Sci. USA* **79,** 739 (1982).

Schubert, D., S. Humphreys, C. Baroni, and M. Cohn, *Proc. Natl. Acad. Sci. USA* **64,** 316 (1969).

Sidman, R. L., in F. D. Carlson, Ed., *Physiological and Biochemical Aspects of Nervous Integration,* Prentice-Hall, Englewood Cliffs, NJ, p. 163 (1968).

Sidman, R. L. and M. C. Green, in M. Sabourdy, Ed., *Les Mutants Pathologiques chez L'animal,* Centre National de La Recherche Scientifique, Paris, p. 69 (1970).

Sotelo, C. and J. P. Changeux, *Brain Res.* **77,** 484 (1974).

Weatherall, D. J. and J. B. Cleff, *Cell* **16,** 467 (1979).

Wetts, R. and K. Herrup, *J. Embryol. Exp. Morph.* **68,** 87 (1982).

Wise, L. S. and H. Green, *J. Biol. Chem.* **254,** 273 (1979).

Wright, D. A. and C. R. Shaw, *Biochem. Genet.* **3,** 343 (1969).

# 8

# *Molecular Biology of Myelination: Gene Expression of Myelin Basic Protein*

***Anthony T. Campagnoni***

Relatively little is known about the mechanism and regulation of myelin assembly in the central nervous system (CNS). In view of the importance of the myelin sheath in the maintenance of a functionally normal brain and because of the large number of neurological diseases and environmentally induced conditions that result in myelin degeneration, an understanding of the mechanisms which regulate the formation and maintenance of myelin is particularly important. There are many features of myelinogenesis which make it an attractive neurological process to study from a molecular biological standpoint:

1. Myelin is a major brain membrane constituting 25–33% of the dry weight of adult brain depending upon the species.
2. Myelin has a relatively simply protein composition with the myelin proteolipid protein and basic protein(s) constituting over 70% of the protein content of the membrane. These proteins have been extensively studied and their chemical, physical, and biological properties are well known.
3. Myelination is a major, developmentally regulated event in the brain and the majority of the membrane is synthesized over a relatively short postnatal period in the mouse and rat ( ~ 10–45 days).
4. Myelin is the product of a single cell-type in the brain, the oligodendrocyte, which elaborates prodigious quantities of the membrane during myelinogenesis.
5. A number of genetically defined dysmyelinating mutant mice are available and these can be used as genetic probes of myelinogenesis.

These features make the synthesis of myelin-specific proteins and their assembly into the membrane highly amenable to investigation by molecular biological techniques. It is the purpose of this review to outline some of our contributions to recent progress in this area of research which is now only in its earliest stages.

## 1. MYELINOGENESIS

### 1.1. Myelin Formation

One of the major biological events to occur during the early postnatal development of the mouse brain is myelination. This process appears to occur in at least two stages commencing at about 5–10 days *postpartum*. The first stage ( ~ 5–15 days) involves the proliferation of oligodendroglial cells, cell enlargement, and plasma membrane formation and is characterized histologically by a distinct absence of myelin (Davison and Dobbing, 1968; McIlwain, 1959; Peters et al., 1970). Biochemically, the first accumulation of several myelin components such as cerebrosides, cholesterol, proteolipid proteins, and basic proteins are noted. During the second stage ( ~ 16–30 days) histologically recognizable myelin can be observed in the brain with the maximum rate of deposition occurring at about 20 days of age (Norton and Poduslo, 1973b). An increase in the number of oligodendroglial cells precedes the formation of myelin and it is generally believed that myelin results from outgrowths of oligodendroglial cell plasma membranes wrapping around axonal segments (Peters et al., 1970). Unlike the peripheral nervous system (PNS) where there is a close relationship between the Schwann cell and the myelin sheath which it produces, the oligodendrocyte in the CNS may be some distance from the myelin sheath which it elaborates, connected only by a slender cytoplasmic process (Fig. 1). The oligodendrocyte may myelinate multiple axons at varying distances from the cell body, and these axons may be part of different fiber tracts (Sternberger et al., 1978).

Myelin is formed as a specialized extrusion of the oligodendroglial cell plasma membrane in which individual bilayers interact to form the multilamellar structure so characteristic of myelin (Fig. 2). A large variety of

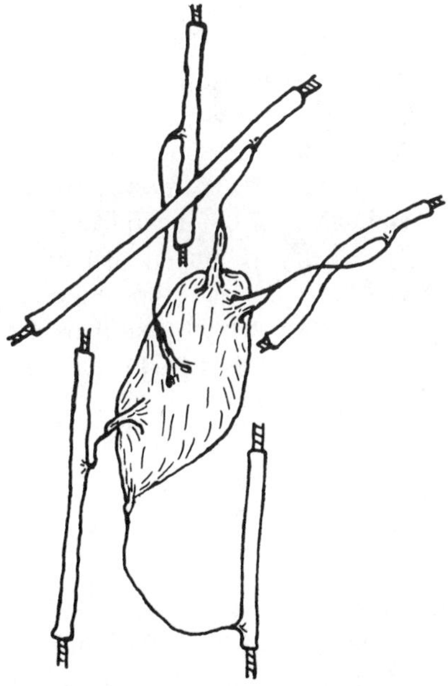

**Figure 1.** Diagrammatic representation of an oligodendrocyte myelinating five axons in two fiber tracts. Note that each myelin sheath is connected to the cell body of the oligodendrocyte by a slender cytoplasmic process of variable length depending upon the distance of the axon being myelinated from the oligodendrocyte.

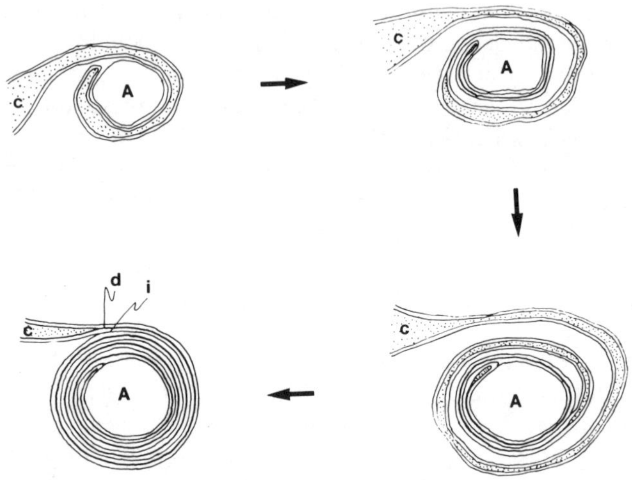


**Figure 2.** Cross-sectional representation of an oligodendrocyte process engulfing an axon and subsequent stages in the myelination process. After the cytoplasmic process reaches the axon (A) it begins to whorl loosely around the axon as the oligodendrocyte elaborates myelin. Eventually cytoplasm (c) is squeezed out and the points at which the cytoplasmic surfaces of the oligodendroglial cell plasma membrane make contact form the major dense line (d) of myelin where the MBP is localized. Apposition of the extracytoplasmic surfaces of the oligodendroglial cell plasma membrane results in the formation of the intraperiod line (i). Eventually the compact, multilamellar structure typical of mature myelin is formed.

evidence suggests that the myelin basic protein (MBP) is localized on the side of the "unit bilayer" which corresponded originally to the cytoplasmic surface of the oligodendrocyte plasma membrane (Dupouey et al., 1979; Omlin et al., 1982; Poduslo and Braun, 1975; Sternberger et al., 1978). In contrast the proteolipid protein (PLP) is believed to be embedded within the lipid bilayer (Boggs and Moscarello, 1978; Braun, 1977). When myelinogenesis begins, the bilayers are rather loosely whorled around the axon, and as myelination proceeds, these bilayers interact and compact to form a tight, multilamellar structure. The cytoplasmic surfaces of these bilayers interact to form the major dense line of myelin where the MBP is located and the extracytoplasmic surfaces interact to form the intraperiod line (See Fig 2).

### 1.2. Myelin Composition

Early studies on the composition of myelin relied heavily on indirect methods of analysis such as a comparison of the lipid components of white (myelin-rich) and gray (myelin-poor) matter as well as data obtained on lipid formation during the period of rapid myelination (Norton, 1977; Rossiter,

1962). For obvious reasons conclusions about the chemical composition of myelin were only tentative until techniques became available for its isolation in pure form (DeRobertis et al., 1962; Gray and Whittaker, 1962; Laatsch et al., 1962; Norton and Autilio, 1965). Since these methods have been developed, the composition of purified myelin has been investigated in several species (Cuzner et al., 1965; Dickerson, 1968; Gerstl et al., 1967; Horrocks, 1968; Norton and Poduslo, 1973a). "Mature" or adult myelin consists of about 70–75% lipid and 25–30% protein and represents one of the most lipophilic membranes known.

The macromolecular constituents of myelin are proteins and the two major components are the myelin PLP and the MBP. A number of other proteins have been identified in lesser amounts in the membrane, including a glycoprotein (Quarles et al., 1972, 1973) and a minor proteolipid called the DM-20 protein (Agrawal et al., 1972) which may bear a structural relationship to the major PLP (Agrawal and Hartman, 1980; Stoffel et al., 1982; Vacher-Lepretre et al., 1976). Several enzymatic activities are also associated with myelin (Braun and Brostoff, 1977); and a group of high-molecular-weight polypeptides called "Wolfgram" proteins (Wolfgram, 1966) have been observed upon electrophoresis of myelin in SDS-containing polyacrylamide gels.

Both of the major proteins of myelin (i.e., the PLP and MBP) have unusual properties. The major PLP is an integral membrane protein and constitutes about 50% of the total protein content of myelin. It is very hydrophobic and it easily aggregates under a variety of conditions (Agrawal and Hartman, 1980; Lees et al., 1979) making it a difficult protein to work with. The PLP apoprotein itself is extremely hydrophobic and the presence of 2–4% fatty acids covalently linked to the protein further contributes to its hydrophobicity (Folch-Pi and Stoffyn, 1972; Gagnon et al., 1971; Jolles et al., 1977). When myelin is analyzed on SDS–polyacrylamide gels, the PLP migrates with an apparent molecular weight between 23,000 and 28,000 (Agrawal and Hartman, 1980; Agrawal et al., 1972; Magno-Sumbilla and Campagnoni, 1977; Nicot et al., 1973). The minor PLP, designated DM-20, migrates with an apparent molecular weight of approximately 20,000 (Agrawal et al., 1972). The relationship between the major PLP and DM-20 protein is not clear at the present time. They appear to have identical amino acid compositions (Chan and Lees, 1978; Nicot et al., 1973) and N-terminal sequences (Vacher-Lepretre et al., 1976). In addition, antisera prepared against the major PLP often cross-reacts with the DM-20 band (Macklin et al., 1982). Since performic acid oxidation of a mixture of the major PLP and DM-20 results in the production of a single band on SDS gels ($M_r$ = 28 to 30,000), Stoffel et al. (1982) have suggested that the DM-20 protein represents a stable conformational isomer of the major PLP. However, the exact structural relationship between the two proteins still remains to be elucidated.

The myelin basic protein was first isolated from guinea pig by Laatsch et al. (1962) and its localization in the myelin sheath subsequently established by its isolation from purified myelin and by immunocytochemical techniques. The protein has been isolated from a number of species and the complete primary sequence is known in many cases (Carnegie and Moore, 1980; Mendz et al., 1982). The protein has several regions which have been conserved through evolution, especially in the central section from residues 86–109 and the C-terminal region from residues 130–149 (Mendz et al., 1982). The MBP from most species contains approximately 170 amino acid residues and has a molecular weight of 18,400. The basic proteins from a number of species in the mammalian suborders *Myomorpha* and *Sciumorpha* (which include the rat and mouse) can be resolved into two major bands when examined by either acid/urea or SDS–polyacrylamide gel electrophoresis (Martenson et al., 1971). These two MBPs have been designated either L (large) or S (small), or 18.5K and 14K, reflecting their apparent molecular weights. The rat 18.5K protein has been found to be similar in its size, chromatographic properties, amino acid composition, and encephalitogenic activity to the single MBP found in other species, such as man and ox (Kies et al., 1972). The rat 14K protein has been "fingerprinted" (Martenson et al., 1972) and sequenced (Dunkley and Carnegie, 1974) and appears to be identical to the rat 18.5K protein except for a deletion of 40 amino acid residues (i.e., 117–157) within the interior of the molecule. Although the mouse proteins have not yet been completely sequenced, on the basis of their amino acid compositions (Campagnoni et al., 1978a), electrophoretic properties (Magno-Sumbilla and Campagnoni, 1977), and peptide maps (Barbarese et al., 1977) the 14K and 18.5 K mouse proteins would appear to bear the same structural relationship to each other as do the rat proteins. In these rodents the 14K MBP predominates over the 18.5K MBP by a ratio of 2–3 : 1 (Martenson, 1980).

Barbarese et al. (1977) first identified and characterized two additional forms of the MBPs present in minor amounts in myelin isolated from mouse brain. These two proteins are immunologically related to the two major forms of the mouse basic proteins (i.e., 14K and 18.5K) and have apparent molecular weights of approximately 21,500 and 17,000. They appear to be structurally related to the 14K and 18.5K basic proteins having an extra sequence of 3000 Da in the N-terminal half of the molecule. The exact position of this sequence has not yet been determined although Barbarese et al. (1977) initially placed it at the N-terminus. The structural relationship among the four mouse MBPs appears to be as follows: (a) the 21.5K protein possesses the complete primary sequence without deletions and contains two small sequences of 3000 and 4500 Da, one of which is localized in the N-terminal half of the molecule (3000 Da) and the other of which is localized in the C-terminal half (4500 Da); (b) the 18.5K protein is missing the N-terminal half sequence; (c) the 17K protein is missing the C-terminal half sequence; and (d) the 14K protein is missing both sequences.

### 1.3. Developmental Aspects of Myelination

Stimulated by the results of Smith (1968) which indicated that myelin was not the inert membrane that it was originally thought to be, many studies were performed to measure the metabolic activity of the membrane and its components (D'Monte et al., 1971; Fischer and Morell, 1974; Hennacy and Horrocks, 1978; Jungalwala, 1974; Sabri et al., 1974; Smith, 1972). Although there is now abundant evidence to show that myelin is not totally inert and that at least a portion of it turns over at a measurable rate, it is much more stable than other cell membranes. In many ways this makes it an ideal membrane for assembly studies since when examined at the appropriate point in development, its synthesis greatly exceeds its breakdown. Thus, rates of entry of components into the membrane can be measured without resorting to excessively short labeling periods and these rates can be compared without the additional complication of rapid rates of loss of these components from the membrane.

An important consequence of these early studies on the metabolic activity of myelin were a number of investigations into compositional changes in the membrane during the period of active myelination in rat and mouse brain (1–9 weeks). These studies on the "maturation" of myelin revealed that the composition of myelin changes with age in a relatively predictable fashion with respect to certain components. For example, the proportion of cerebrosides and sulfatides increases in the membrane with age, with a corresponding decrease in the amount of phospholipid and cholesterol (Cuzner and Davison, 1968; Eng and Noble, 1968; Fishman et al., 1975; Horrocks, 1968; Norton and Poduslo, 1973b). In addition, in rats and mice the ratio of the 14K : 18.5K basic protein increases as myelin becomes enriched in the 14K protein with age (Adams and Osborne, 1973; Giorgi, 1976; Magno-Sumbilla and Campagnoni, 1977). Thus, using such indices it is possible to infer the relative degree of "maturation" of the membrane from its protein and lipid composition.

Because it was noted that the ration of 14K : 18.5K basic protein increased significantly in myelin during development, and because no studies had been performed on the synthesis of these proteins with age, we initiated such a study in the developing mouse brain (Campagnoni et al., 1978b). In addition, we also examined the *in vivo* synthesis of the major myelin PLP (Campagnoni and Hunkeler, 1980) over the same period. The purpose of these studies was to define the developmental pattern of synthesis of the major myelin proteins in mouse brain and to correlate the relative rates of synthesis of the two basic proteins with their proportions as they existed in intact myelin as a function of development. While the synthesis of all three proteins peaked generally around the age of maximum myelin deposition in the brain, maximal synthesis of the PLP appeared to lag behind that of the MBPs by several days (Fig. 3). The peaks of synthesis of the 14K and 18.5K MBPs were developmentally synchronous at 18 days.

Shown also in Figure 3 are the ratios of the rates of synthesis of the 14K and 18.5K mouse MBPs with development. It is clear that the rate of synthesis of the 14K MBP relative to the 18.5K MBP increases steadily from 14 to 32 days; at later ages this ratio declines. Thus, the composition of myelin with respect to the 14K and 18.5K basic proteins probably reflects the relative rates of synthesis of the proteins in the brain at any given time.

### 1.4. Dysmyelinating Mouse Mutations

A number of mouse neurological mutants have been characterized which are abnormal in the structure, composition, and/or metabolism of myelin (Baumann, 1980; Hogan, 1977). These include quaking (*qk*), myelin deficient (*mld*), shiverer (*shi*), which are autosomal recessive, and jimpy (*jp*), a sex-linked recessive mutation. The mutants *mld* and *shi* have been shown to be alleles (Lachapelle et al., 1980). All four of these dysmyelinating mutants are characterized by severe reductions in the levels of myelin in their brains although the reasons for the myelin deficit are probably different in each case (Baumann, 1980; Hogan, 1977). The *qk* mutation has been the most well studied of the dysmyelinating mutants. In this mutation both CNS and PNS myelin are affected. Myelin composition is abnormal and most of the enzymatic activities associated with myelin lipid synthesis are low. Evidence has been presented that synthesis of myelin proteins is close to normal in quaking (Brostoff et al., 1977; Campagnoni et al., 1980a; Greenfield et al., 1977) but that myelin assembly is defective. The *jp* mutation differs from *qk*, not only in its chromosomal location, but also in the fact that the PNS does not appear to be affected. Furthermore, while *qk* mice have nearly normal life spans, *jp* mice die by 3–4 weeks of age. Jimpy CNS contains very little myelin of apparently abnormal composition and many enzymatic activities associated with myelin lipid synthesis have been found to be reduced. It has been suggested that *jp* represents a specific defect of oligodendroglial cell metabolism (Norton, 1980) which is expressed as an arrest of myelinogenesis (Burkart et al., 1981; Matthieu et al., 1973) possibly because of a defect in myelin assembly (Braun et al., 1980).

The *shi* and *mld* mutations have been characterized to a lesser extent than either *jp* or *qk*. These mutants have extremely low levels of myelin and MBP

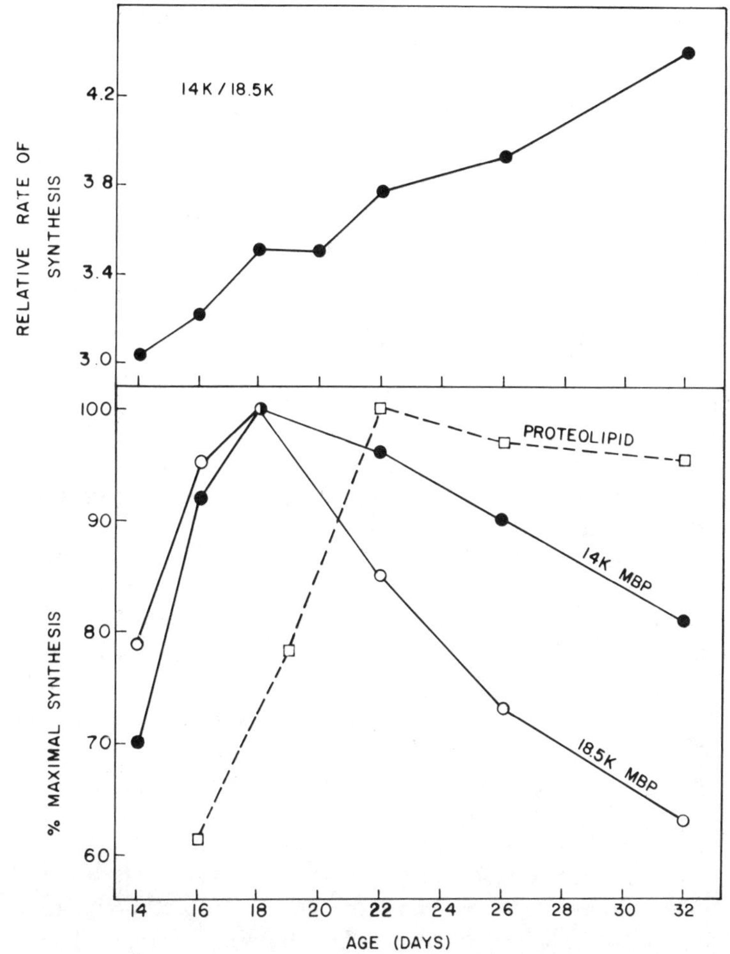


**Figure 3.** *In vivo* patterns of synthesis of the 14K and 18.5K MBPs and the major myelin proteolipid during early postnatal brain development in the mouse. Synthesis of the two MBPs peaks at 18 days and correlates with the peak in MBP mRNA concentration in the brain (see Section 3.1). Synthesis of the myelin PLP peaked somewhat later at 22 days. In the upper portion of the figure the ratio of the rate of synthesis of the 14K basic protein is plotted relative to the rate of synthesis of the 18.5K protein. This value increases by over 25% between 14 and 32 days and correlates with an increasing enrichment of the 14K basic protein in myelin over the same developmental period. Data from Campagnoni et al. (1978b) and Campagnoni and Hunkeler (1980). With permission from *Journal of Neurobiology* (Copyright: John Wiley & Sons) and *Archives of Biochemistry and Biophysics* (Copyright: Academic Press, Inc.).

in the CNS. There is also a lack of compaction at the major dense line of CNS myelin in both mutations. Ganser and Kirschner (1980) have shown that even though PNS myelin appears normal in *shi*, it contains very low levels ( $< 0.4\%$ of controls) of immunoreactive basic protein. In contrast to the low levels of MBP and myelin-associated lipids (Bird et al., 1978), the levels of several enzyme activities associated with myelin lipid synthesis and 2′, 3′-cyclic nucleotide phosphodiesterase have been shown to be normal in *shi* and *mld* CNS (Bird et al., 1980; Matthieu et al., 1981).

## 2. SYNTHESIS OF MYELIN PROTEINS

### 2.1. General Approach

The overall goal of our research has been to elucidate the mechanisms by which the synthesis of the major myelin proteins and their incorporation into the membrane is regulated. Initially we developed several cell-free systems capable of synthesizing myelin proteins (Campagnoni et al., 1980b; Carey and Campagnoni, 1979; Matthees and Campagnoni, 1980; Yu and Campagnoni, 1982) because of the potential importance they might have in studies on the general problem of myelin assembly. In addition to providing fundamental information about the early events in the synthesis and possible processing of myelin proteins, we felt these *in vitro* systems might also be used to identify specific regulatory mechanisms governing the synthesis of these proteins and ultimately their assembly into the membrane. Each *in vitro* system, however, possesses its own set of limitations in terms of the specific problems which it may be used to solve and the extent to which results obtained with it can be extrapolated to what occurs *in vivo*. Therefore, we felt that it would be advantageous to have the use of as many *in vitro* systems as possible. The two cell-free protein synthesizing systems which have proven to be most useful have been a homologous brain polysome "run-off" system and a heterologous mRNA-stimulated reticulocyte lysate. Our approach has been to use these systems (a) as a means of identifying mRNA for the MBPs, (b) to establish the site of synthesis of the basic proteins in brain, (c) to elucidate metabolic precursor–product relationships among four forms of the MBP, and (d) to examine possible mechanisms of translational control over synthesis of the MBPs.

Another important objective of this research has been to isolate, characterize, and study the metabolism of the mRNAs for the MBPs. Even before the discovery of the two minor MBPs (i.e., the 17K and 21.5K proteins) the structural relationship between the 14K and 18.5K MBPs suggested the possibility that their mRNAs might be derived from the same gene by different splicing mechanisms. Toward this end it has been possible to use brain mRNA to prepare a cDNA clone of part of the MBP gene(s). Using this probe it is now possible to hybrid-select MBP mRNAs and study their structure and metabolism in normal and mutant mice in order to determine directly those mechanisms which govern the expression of the genes for the MBPs.

## 2.2. Sites of Synthesis of Myelin Proteins

One of the first steps in elucidating the molecular mechanisms of the early events of myelin assembly is the determination of the sites of synthesis of the major myelin proteins. Since the basic proteins and the proteolipid protein differ so considerably in their biochemical properties and in their topographical location in the membrane, it might be expected that they are synthesized on different classes of ribosomes and thereby incorporated into the growing membrane via different mechanisms. Benjamins and co-workers were the first to investigate the kinetics of entry of proteins into myelin (Benjamins and Morell, 1978; Benjamins et al., 1975, 1976, 1978). They showed, using animals injected with radioactive amino acids *in vivo*, that label incorporated into the major PLP of isolated myelin followed a different time course from that of the 14K and 18.5K basic proteins. Their data suggested that PLP addition to the myelin membrane occurs subsequent to the MBPs. They then observed in a brain slice system that the entry of the major PLP into myelin seems to be delayed compared to the basic and Wolfgram proteins. Furthermore, in pulse-chase experiments where the slices were incubated with radioactive amino acids for a period of time followed by inhibition of further protein synthesis with cycloheximide, there was an abrupt inhibition of label into the Wolfgram and basic proteins but not into the major PLP in which label continued to accumulate for a period of time. These results are consistent with a delayed entry of the major PLP into the membrane relative to the basic proteins and, in addition, with separate routes of assembly.

We were the first laboratory to attempt to define the site of synthesis of the MBPs on either free or bound ribosomes. Using conventional biochemical means of identification of the products of *in vitro* synthesis (i.e., differential solubility, ion exchange chromatography, and two-dimensional gel electrophoresis), we observed predominant synthesis of the two major mouse basic proteins (14K and 18.5K) on free ribosomes (Campagnoni et al., 1980b). Subsequent experiments using immunological methods to identify the products of synthesis have shown that all four basic proteins are synthesized on free ribosomes (Colman et al., 1982; Hall et al., 1982; Yu and Campagnoni, 1982). Others using the procedure of Ramsey and Steele (1977) to separate free and bound ribosomes from brain have reported the synthesis of the basic proteins on both classes of polyribosomes (Hall et al., 1982). However, it has been observed that there is a class of MBP-synthesizing free ribosomes which sediment with crude myelin fractions (Colman et al., 1982). It is likely, therefore, that synthesis of MBP on "bound" ribosomes prepared by the Ramsey and Steele (1977) procedure is due to contamination of this fraction with MBP-synthesizing free ribosomes. Colman et al. (1982) have studied the synthesis and incorporation of the basic proteins and the major PLP into rat brain myelin. Their results confirm the earlier work of Benjamins and co-workers that newly synthesized basic proteins enter the membrane rapidly and prior to the entry of the PLP. They observed synthesis of the PLP on bound ribosomes and the basic proteins on free ribosomes. They have interpreted their results to indicate that the free ribosomes synthesizing the MBPs are concentrated in the oligodendrocyte processes involved in myelination. Immunohistochemical studies have shown differences in the immunofluorescent patterns of oligodendroglial cells treated with fluorescent antibodies to several myelin proteins (Agrawal and Hartman, 1980; Sternberger et al., 1978, 1979). During early myelinogenesis antibodies to the MBP are scattered diffusely throughout the cytoplasm, whereas antibodies against the PLP and glycoprotein are concentrated in clumps of immunoreactive material. Thus, the majority of the biochemical and immunohistochemical data favor the notion that the MBPs are synthesized on free polysomes and transported quickly to the growing membrane, whereas the major PLP is synthesized on polyribosomes bound to the endoplasmic reticulum and transported more slowly via a membrane-mediated process to the site of assembly of myelin in the oligodendrocyte process.

## 2.3. Metabolic Interrelationships among the Four Mouse Myelin Basic Proteins

In 1977 Barbarese et al. characterized two additional forms of the MBP present in minor amounts in myelin isolated from mouse brain. These two proteins were immunologically related to the two major forms of the mouse MBPs and had apparent molecular weights of 21,500 and 17,000. Barbarese et al. (1977) isolated the 17K and 21.5K MBPs from mouse myelin and interpreted their peptide maps to indicate that these two proteins were identical to the 14K and 18.5K MBPs except for an additional sequence of about 3000 Da at the N-termini of these proteins, similar to the "leader" sequence observed in the precursors to many proteins. Other evidence also supported the notion that precursors to the 18.5K and 14K MBPs might exist. For example, Shapira et al. (1978) detected the synthesis of a highly labeled 20,000-Da protein in chopped rabbit brain that could possibly have been an

MBP precursor; and our laboratory had obtained kinetic data on the synthesis of the 14K and 18.5K MBPs in mouse brain homogenates (Carey and Campagnoni, 1979) which were consistent with the notion that these two proteins might be synthesized as precursors. These studies prompted us to investigate in detail the question of precursors to the MBPs using a brain polysome protein-synthesizing system and a reticulocyte lysate system programmed with brain mRNA (Yu and Campagnoni, 1982). Incorporation of [$^{35}$S]-methionine into the four MBPs (14K, 17K, 18.5K, and 21.5K) was detected by immunoprecipitation of the *in vitro* products of synthesis followed by SDS–polyacrylamide gel electrophoresis. The four MBPs were identified by cross-reactivity with purified anti-MBP antibodies and their apparent molecular weights in SDS gels. Typical results illustrating the synthesis of the four MBPs and the immunospecificity of the reactions are shown in Figure 4.

All four MBP bands were competed out with unlabeled MBP but not with proteins of comparable size and charge, such as cytochrome *c* and histones. Furthermore, non-immune prebleed rabbit serum failed to precipitate any of the four proteins. We have consistently noted that the 17K and 21.5K MBPs are heavily labeled in the *in vitro* translation systems when [$^{35}$S]-methionine is used as a precursor. This is probably due to the presence of an additional methionine in these MBPs over that present in the 14K and 18.5K proteins (Agrawal et al., 1981, 1982; P. R. Carnegie, personal communication). In order to determine whether a precursor–product relationship existed among the MBPs (e.g., 21.5K → 18.5K and/or 17K → 14K) the kinetics of incorporation of [$^{35}$S]-methionine into the four proteins was examined. In the brain

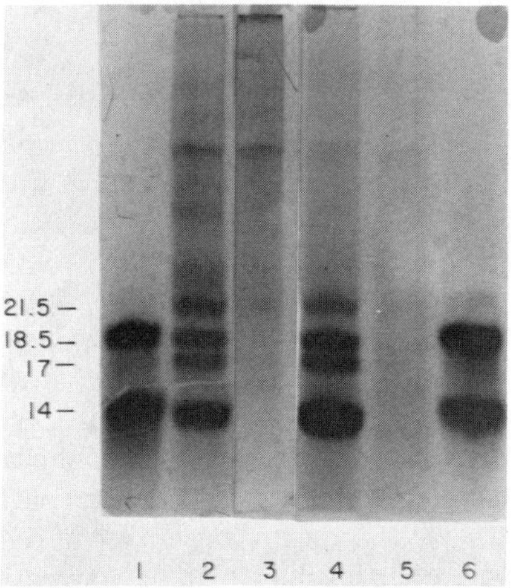


**Figure 4.** Fluorograms of MBPs synthesized *in vitro* using free polyribosomes in a homologous mouse brain system. [$^{35}$S]-methionine-labeled proteins were immunoprecipitated and separated by SDS–PAGE. Lanes 2–5 illustrate the cell-free synthesis of the four MBPs in the system and the immunospecificity of the precipitation reaction. Lanes 1 and 6 are [$^{125}$I]-labeled 14K and 18.5K mouse MBP standards. Lane 4 shows the four MBPs synthesized in the standard incubation. Lane 3 shows the results of immunoprecipitation in the presence of 200 $\mu$g/mL unlabeled mouse MBP. Lane 2 shows the results of immunoprecipitation in the presence of 200 $\mu$g/mL cytochrome *c*. Lane 5 shows the results of immunoprecipitation of the incubation products with non-immune rabbit serum.

polysome run-off system synthesis of all four proteins could be detected as early as 1.25 min after the incubation was begun and label increased linearly into the four bands for up to 10 min. If puromycin or unlabeled methionine was added 2.5 min after synthesis was begun, further incorporation of [$^{35}$S]-methionine into all four MBP bands stopped abruptly (Fig. 5D). During the chase period there was no evidence that the 17K and 21.5K proteins were being converted into the 14K and 18.5K proteins or that any precursor–product relationship existed among any of the four proteins. Thus, the system itself contained "processing" activity, effectively eliminating the possibility that conversion of the 21.5K and 17K proteins to the 18.5K and 14K MBPs was occurring *in vitro* at times faster than our ability to detect it.

To eliminate the possibility that the results with the homologous polysome system were not due to "run-off" of nascent MBP chains that had been processed *in vivo* on the ribosomal surface or during the isolation of the polyribosomes, we isolated brain mRNA from free polyribosomes and translated it in a reticulocyte lysate (Fig. 5A and 5B). Synthesis of all four MBPs was observed in this "initiating" system indicating that the proteins were the products of four separate mRNAs. To be absolutely sure that this cell-free system was not processing the 21.5K and 17K MBPs to the 18.5K and 14K MBPs, time course and puromycin inhibitor experiments were carried out. There was no evidence of a precursor–product relationship among the four MBPs. Thus, these data suggest the existence of multiple messages for the MBPs and they also indicate that basic proteins are not synthesized with a "leader" sequence. Colman et al. (1982) and Dautigny et al. (1983) have observed that the major myelin PLPs also do not appear to be synthesized as a precursor with a leader sequence.

### 2.4. Synthesis of the Myelin Basic Proteins in Mutant Mice

Over the past two decades a number of dysmyelinating mouse mutants have become available for study, and like many others we have felt that they might be useful for (a) elucidating the biochemical steps involved in myelin assembly, (b) discovering important regulatory mechanisms which govern myelinogenesis, and (c) serving as useful models of human dysmyelinating diseases. Therefore, we undertook an examination of the *in vitro* protein synthetic capacity of polyribosomes isolated from the brains of C57B1/6 mice and the dysmyelinating mutants *qk, jp, shi*, and *mld*. One aim of these experiments was to determine if the overall rates of protein synthesis were altered in any or all of the mutant mice. A second aim was to examine directly the *in vitro* rates of synthesis of the individual mouse MBPs to determine if all four proteins were affected to the same extent in the mutations. We hoped the study might be used to help classify these dysmyelinating mutations into

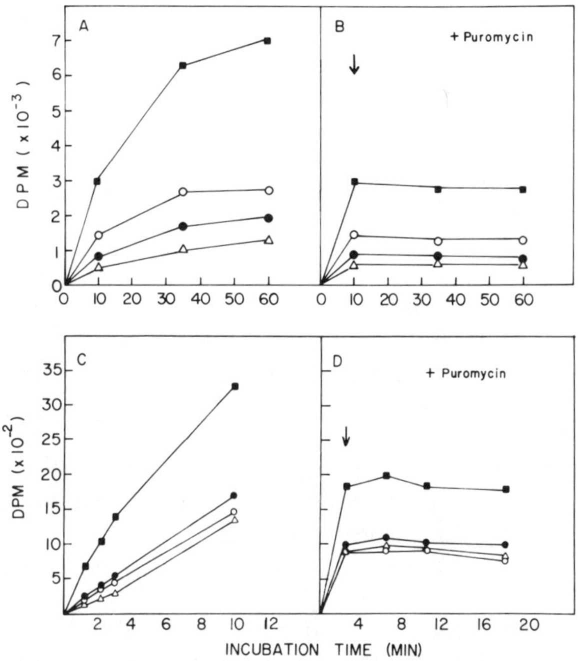


**Figure 5.** Incorporation of [$^{35}$S]-methionine into the four mouse MBPs as a function of time and treatment with puromycin in a reticulocyte lysate programmed with brain poly (A+) mRNA (A and B); and in a brain homologous "run-off" system using free polysomes (C and D). Incorporation into each of the four MBPs was measured by cutting out the appropriate bands on SDS–polyacrylamide gels (after fluorography), solubilizing the gel slices, and counting them in a scintillation counter. For the inhibition experiments incorporation was allowed to proceed for 2.5 min (brain, "run-off" system, D) or 10 min (reticulocyte lysate, B) before the addition of puromycin (arrow). The systems were then incubated for up to 50 min longer and further incorporation into each of the four MBPs determined as described. Figure redrawn from data published by Yu and Campagnoni (1982) where complete methodological details may be found. 14K MBP, closed squares; 18.5K MBP, closed circles: 17K MBP, open circles; 21.5K MBP, open triangles.

those capable of synthesizing the MBPs, but in which the newly synthesized MBPs are not incorporated into myelin (i.e., "assembly" or "post-translational" defects) and those which cannot synthesize MBP (i.e., "synthesis" defects).

The levels of MBP in all four mutants have been measured by radioimmunoassay (Barbarese et al., 1979, 1983; Delassalle et al., 1981; Ganser and Kirschner, 1980; Matthieu et al., 1980; Zimmerman and Cohen, 1979). Using this technique the CNS of the mutants has been determined to contain the following levels of MBP as a percentage of controls: *shi*, 0.1–0.4%; *mld*,

2–3%; *jp*, 3–12%; and *qk* 5–20%. These presumably represent steady-state levels of MBP wherein the protein is associated with myelin or some myelin-like structure, but they cannot be used as an indication of the rates of synthesis of the proteins. *In vivo* studies have suggested that MBP synthesis is normal in *qk* (Brostoff et al., 1977; Greenfield et al., 1977) and in *jp*, (Carson et al., 1975) and incorporation of newly synthesized MBP into the membrane may possibly be defective in these animals.

For these studies we chose to use a "run-off," homologous brain polysome system because [$^{35}$S]-methionine incorporation into each of the four MBPs should be proportional to the numbers of active mRNAs for those proteins being translated at the time the polysomes were isolated. In these studies polyribosomes were isolated from the brains of 16- to 20-day-old C57B1/6 mice, the four neurological mutants, and four heterozygote or littermate controls. There were no significant differences observed in either the yields, purity, or *in vitro* protein synthetic capacity among the nine genotypes examined. However, when the four MBPs were isolated from the *in vitro* incubation mixture by immunoprecipitation and analyzed by SDS–PAGE and fluorography, significant differences in their synthesis were observed (Fig. 6).

In order to obtain a semiquantitative estimate of the relative levels of incorporation of [$^{35}$S]-methionine into each of the four MBPs as a function of the mutation, after fluorography each band was cut out from the dried gel, hydrolyzed with solubilizer, and counted. Incorporation into the mutant or

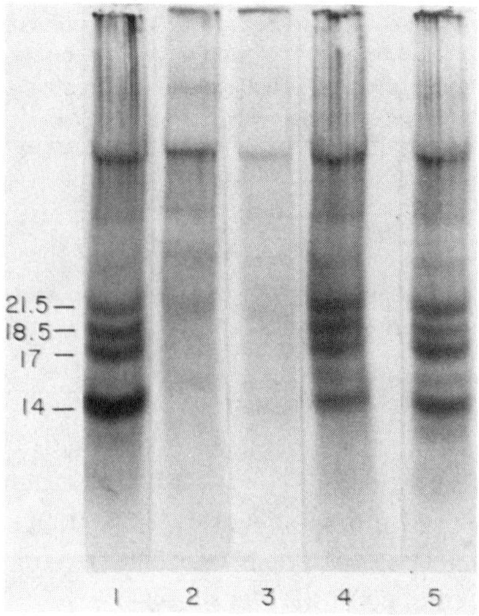


**Figure 6.** Cell-free synthesis of four mouse MBPs using polysomes isolated from 16- to 20-day-old C57B1/6 (lane 1), *shi/shi* (lane 2), *mld/mld* (lane 3), *jp*/Y (lane 4) and *qk/qk* (lane 5) mice. The MBPs were synthesized *in vitro* in a homologous brain system. They were immunoprecipitated with immunopurified anti-MBP antibodies and separated on SDS–polyacrylamide gels (Weber and Osborne, 1969).

**TABLE 1. RELATIVE SYNTHESIS OF BASIC PROTEINS IN SEVERAL MOUSE GENOTYPES**

| | Major MBPs | | Minor MBPs | | Steady State Levels from RIA Data[a] |
|---|---|---|---|---|---|
| Genotype | 14K | 18.5 K | 17K | 21.5K | |
| C57B1/6 | 1.00 | 1.00 | 1.00 | 1.00 | 1.00 |
| *qk/qk* | 0.70 | 0.87 | 1.15 | 1.48 | 0.05–0.20 |
| *jp/Y* | 0.07 | 0.43 | 0.48 | 0.46 | 0.03–0.12 |
| *mld/mld* | <0.05 | <0.05 | <0.05 | <0.05 | 0.02–0.03 |
| *shi/shi* | <0.05 | <0.05 | <0.05 | <0.05 | 0.001–0.004 |
| *qk/+* | 1.02 | 0.97 | 1.29 | 1.51 | — |
| *jp* littermates | 0.66 | 0.83 | 0.90 | 0.86 | — |
| *mld* littermates | 0.49 | 0.53 | 0.70 | 0.29 | — |
| *shi/+* | 0.42 | 0.51 | 0.64 | 0.42 | 0.5 |

[a] See text for appropriate references (Section 2.4).

littermate MBPs was always measured relative to a control (C57B1/6) incubation done at the same time in order to eliminate variations in total incorporation which occurred from experiment to experiment. These data are presented in Table 1 where the synthesis data may be compared to RIA data obtained from the literature on the steady-state levels of MBP present in the brains of the mutants. From Figure 6 and Table 1 it is clear that there was essentially no incorporation of [$^{35}$S]-methionine into the four MBPs by either the homozygous *shi* or *mld* polyribosomes, whereas significant incorporation of label was observed into the four MBPs by *jp*/Y and *qk*/*qk* polysomes. The inability to detect synthesis of the MBPs in *mld* and *shi* homozygotes does not mean that the mRNAs for these proteins are not there but that they are present at very low levels ( $<$ 5% of controls). These studies (and the data shown in the Northern blots of Fig 8) indicate that the MBP defect in *shi*/*shi* and *mld*/*mld* lies prior to the translation of the mRNA indicating that the MBP mRNAs do not reach the ribosome. The reason for this is not clear but it could involve either the lack of transcription of the MBP gene(s) or altered processing of the MBP mRNAs. Note that the heterozygous shiverer mice synthesize approximately half the levels of the four MBPs as the controls and this correlates well with the levels of MBP found in their brains (Barbarese et al., 1983). Thus, the shiverer and *mld* mutation fall into the category of having an "MBP synthesis" defect which probably occurs at the level of transcription or post-transcriptional processing of the MBP mRNAs. Furthermore, all four MBPs appear to be affected.

The *qk* mutation appears to have little effect on the synthesis of the MBPs. The synthesis of the major MBPs occurred at close-to-normal levels and the synthesis of the minor forms of the MBP (i.e., 17K and 21.5K) appeared to be elevated, yet the steady-state level of MBP in the brain of the *qk* mice is only 5–20% of normal mouse brain. This indicates that most of the newly synthesized basic proteins are unable to be incorporated into the myelin membrane. Thus, with respect to MBP metabolism, the defect in this mutation occurs post-translationally and should be useful in studies on the assembly of the membrane.

The *qk* mutation has been described as an "arrest" of myelin maturation because the ultrastructure and composition of adult *qk* myelin resembles "immature" normal myelin (Hogan and Joseph, 1970; Wisniewski and Morell, 1971). One of the characteristics of *qk* myelin and "immature" myelin is a lower 14K : 18.5K MBP ratio (Druse and Hogan, 1974; Greenfield et al., 1971; Magno-Sumbilla and Campagnoni, 1977; Matthieu et al., 1978; Morell et al., 1972; Nussbaum and Mandel, 1973). Since normal levels of all the MBPs are being synthesized in the mutant, then proportionately less of the 14K MBP must be entering the membrane relative to the 18.5K protein.

The jimpy mutation appears to be more complicated with respect to the metabolism of MBPs than the other three mutations and, therefore, it is not easy to categorize as either a "synthesis" or "assembly" defect. In this mutation the synthesis of the 14K MBP appears to be more reduced than the other three basic proteins. While synthesis of the 17K, 18.5K, and 21.5K basic proteins is lower than normal, it is still substantially greater than the steady-state levels of MBP found in *jp* brains relative to controls (3–12%). Several groups have attempted to examine membrane fractions which might contain or correspond to jimpy "myelin" and there is a striking reduction in amounts of the 18.5K and 14K basic proteins in these preparation (Matthieu and Waehneldt, 1978; Matthieu et al., 1974; Nussbaum and Mandel, 1973). This is a much greater reduction than can be explained by the reduction in the synthesis of the proteins. Thus, it appears that the jimpy mutation is unable to incorporate the newly synthesized MBP into myelin. The reduction of the synthesis of the 14K MBP observed in this study is in accord with the findings of Carnow et al. (1982). They concluded that there was very little mRNA for the 14K MBP in the brains of *jp* mice based on translation of mRNA in a reticulocyte protein-synthesizing sytem.

It is clear that these four dysmyelinating mutations represent a spectrum of genotypes capable of synthesizing vastly different levels of each of the four MBPs. Now that the phases of MBP metabolism which are affected in these mutants have been defined in general terms, it should be possible to focus future work on the mutations to specific aspects of MBP synthesis and assembly into myelin.

## 3. CHARACTERIZATION OF THE MYELIN BASIC PROTEIN mRNA

An important prerequisite for the isolation of a specific mRNA is an assay system to detect the message, such as an *in vitro* protein-synthesizing system. We were able to use the reticulocyte lysate system to help characterize a cDNA clone to a portion of the MBP message which is common to all four proteins. In this section the use of the clone to detect and isolate MBP mRNA and the characteristics of the message will be discussed.

### 3.1. Preparation and Identification of an MBP cDNA Clone

Zeller et al. (1983 and 1984) used two families of tetradecamer deoxyribo-oligonucleotides to prepare a cDNA clone of part of the MBP gene. These two deoxyribo-oligonucleotide families were complementary to all possible ribonucleotide sequences (in mRNA) corresponding to amino acid residues 82–86 and 89–92 of the MBP. Since these two sequences in the protein possess minimal degeneracy, the number of potential deoxyribo-oligonucleotide sequences required for one to be a correct complementary sequence could be limited to eight and sixteen, respectively. Using the family of tetradecamers corresponding to MBP 89–92 as primers, cDNA reverse transcripts of poly(A+) mRNA were prepared. They were made double-stranded and cloned into pBR322 by standard procedures (Zeller et al., 1984).

Of 600 recombinant clones examined, two clones hybridized with both the [$^{32}$P] (89–92) and [$^{32}$P] (82–86) tetradecamer families. One of these clones, pNZ111, contained 94 nucleotides and its sequence corresponds to residues 60–91 of the myelin basic protein. Although the complete sequence of the mouse MBP has not yet been determined, Burgess et al. (1978) have sequenced a small region from residues 74–83 and the nucleotide sequence of pNZ111 in this region corresponds exactly to the amino acid sequence. The amino acid sequence of the mouse protein deduced from the DNA sequence of the clone is identical to that of the rat 18.5K protein, except that the mouse MBP contains the His-Gly dipeptide at residues 78–79 which is missing from the rat protein. In this respect the mouse protein is similar to the human MBP.

Figure 7 shows an SDS–polyacrylamide gel of all the products synthesized in a reticulocyte lysate programmed with hybrid-selected MBP mRNA. The mRNA was prepared by immobilizing denatured pNZ111 DNA on nitrocellulose and hybrid-selecting MBP-specific mRNA from mouse polysomal

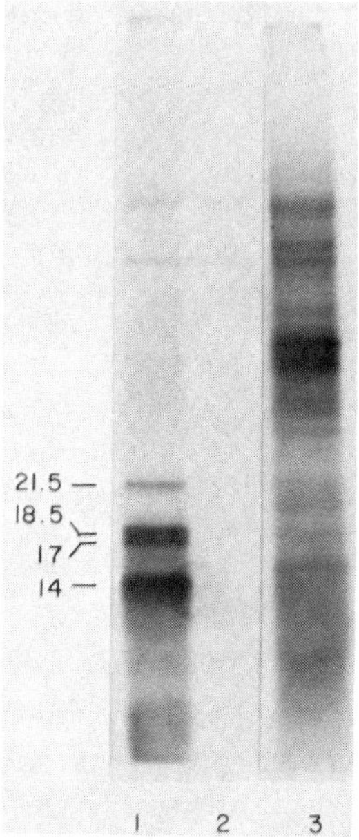


**Figure 7.** Fluorogram illustrating the total products of synthesis of hybrid-selected MBP mRNA and brain poly (A+) mRNA translated in a reticulocyte lysate. Using denatured DNA from pNZ-111 immobilized on filters, MBP-specific mRNA was prepared by hybrid selection and translated in a reticulocyte lysate. Total products of synthesis were precipitated with trichloroacetic acid and separated on SDS–polyacrylamide gels (Pearson et al., 1983). Lane 1, hybrid-selected MPB mRNA; lane 2, minus mRNA; lane 3, brain poly(A+) mRNA before hybrid selection. Note that the only products of synthesis of the hybrid-selected mRNA are the four MBPs (lane 1) and that these represent only a small proportion (1–2%) of the products synthesized using brain poly(A+) mRNA (lane 3).

poly(A+) RNA. The predominant products of synthesis were the four mouse MBPs. These were identified by their comigration with authentic [$^{125}$I]-14K and [$^{125}$I]-18.5K MBP standards, their apparent molecular weights, and their immunoprecipitation by purified anti-mouse MBP antibodies (Zeller et al., 1984). Further characterization of the clone was performed by using "dot blots" to measure the levels of MBP mRNA in mouse brain between 4 and 33 days (Zeller et al., 1983, 1984). In agreement with our previous studies (Campagnoni et al., 1978b), the concentration of MBP mRNA increased sharply after 12 days, peaked at 18 days, and then declined (see Fig 3). It was possible to detect small amounts of MBP mRNA in mouse brain four days after birth.

### 3.2. Characteristics of MBP mRNA in Normal and Mutant Mice

Figure 8 (lane 2) shows a Northern blot of normal mouse poly(A+) mRNA separated on formaldehyde agarose gels, transferred to a nylon membrane, and hybridized with the [$^{32}$P]-labeled MBP probe. One of the most surprising features of the mouse MBP mRNA is its size. The mRNA for the mouse MBP migrates as a broad band of RNAs ranging in length from approximatley 2100 to 2400 bases. Although they are not evident in Figure 8, minor hybridization occurs to RNAs of 1200, 1500, 1900, and 4100 nucleotides in length (Zeller et al., 1983, 1984). Since only 600 nucleotides are required to code for the

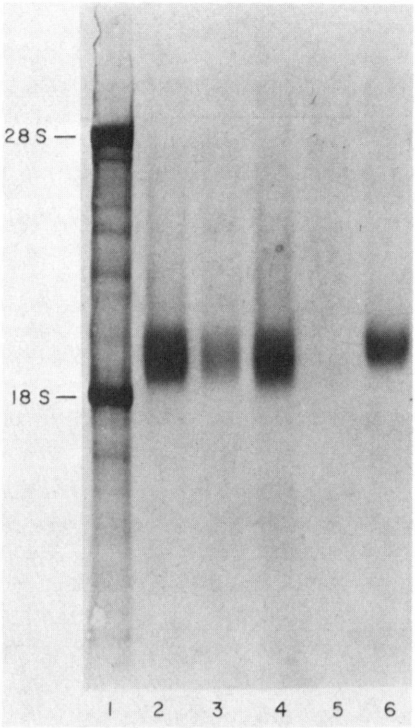


**Figure 8.** Northern blot analysis of several poly(A+) mRNA preparations hybridizing to the [$^{32}$P]-labeled cloned MBP probe. Poly(A+) mRNA was prepared from brain polyribosomes isolated from 18-day-old C57B1/6J (lane 2); *jp*/Y (lane 3), *qk*/*qk* (lane 4), and *shi*/*shi* (lane 5) mice; and fetal guinea pig (lane 6). These poly(A+) mRNA preparations were separated along with a mixture of [$^{14}$C]-18S and -28S rRNA standards (lane 1) on formaldehyde–agarose gels. The MBP mRNA migrated as a broad band just behind 18S rRNA which corresponds to a length of 2100–2400 bases.

largest MBP (i.e., the 21.5K protein) then the MBP mRNA must contain large untranslated regions. All of the *in vitro* protein synthetic work has indicated that multiple mRNAs exist for the four MBPs. At the present time it seems likely that the broad MBP mRNA band of 2100–2400 nucleotides is heterogeneous and contains these multiple mRNA species. Consistent with this interpretation is the result of the Northern blot with guinea pig MBP mRNA (Fig 8, lane 6). Polyribosomes were isolated from fetal guinea pigs at an age when active myelination was occurring and poly(A+) mRNA prepared. As shown in Figure 8 the guinea pig MBP mRNA hybridized well with the [$^{32}$P]-labeled mouse MBP probe and the width of the band was significantly reduced, corresponding to the higher molecular weight half of the mouse MBP mRNA band. This is compatible with the fact that the guinea pig only has a single major MBP (the 18.5K protein) and possibly the 21.5K protein but it is lacking the 14K and 17K basic proteins. Thus, it is probable that the mRNAs for the four mouse MBPs are of similar length and encompassed within a range of 2100–2400 nucleotides.

Figure 8 also shows Northern blots of poly(A+) mRNA isolated from jimpy (lane 3), quaking (lane 4), and shiverer (lane 5) polyribosomes. The results are consistent with the *in vitro* protein synthetic data (Campagnoni et al., 1984). Quaking polysomes appear to contain close-to-normal levels of MBP mRNA whereas jimpy polysomes contain substantially reduced levels of these mRNAs. Very little MBP mRNA is present on shiverer polysomes. Thus, quaking represents a clear example of a mutant able to synthesize MBP but unable to incorporate it into the newly synthesized membrane. It is also clear that the shiverer defect occurs prior to translation of the MBPs and that the jimpy defect involves both the transcription (or processing) of MBP mRNA and the assembly of the MBPs into myelin.

## 4. CONCLUDING REMARKS

One of the most important questions to be answered in the near future relates to the number of MBP genes. Do individual genes exist for the four MBPs or are two or more MBPs derived from a single gene? What is the structure of the MBP gene(s), and what are the locations of any introns? There is a good possibility that there exist two classes of MBP genes, one class of which codes for the 21.5K and 18.5K proteins and the other class of which codes for the 17K and 14K proteins. There is growing evidence, for example, that those species which contain a single major basic protein (i.e., 18.5K) may also contain an MBP comparable to the 21.5K protein. We have noted such a protein in bovine brain (Reidl et al., 1981). Kerlero de Rosbo et al. (1983) have reported the existence of such a protein in other species (e.g., rabbit and sheep). Those genes which code for the 17K and 14K MBPs in rats and mice could have arisen by gene duplication and partial deletion of this presumptive (21.5K + 18.5K) MBP gene.

At the present time very little is known about factors which regulate transcription of MBP genes or the processing of MBP mRNA. The *shi* and *mld* mutations may prove to be useful for determining whether there exist common controlling elements for the expression of the four basic proteins since synthesis of all of them is severely depressed in these mutations. Presumably this could be accomplished by isolating and sequencing the MBP genes and their flanking regions from these two mutations and comparing them with the structures of similar DNA regions from normal mice. The characteristics of the *jp* mutation also support the notion of multiple MBP genes since there is a greater reduction in the level of the 14K MBP mRNA than in the levels of mRNAs for the other three proteins.

Another important area of future investigation will be the processing of the MBP mRNA. If individual genes for all four MBPs do not exist and some of the mRNAs are derived from common genes, then it wil be important to elucidate the route by which the primary gene products are "processed" to the mature MBP mRNAs. On the basis of the scanty data available, we would like to propose the working hypothesis that there are two classes of MBP genes in the mouse. The 21.5K and 18.5K MBP mRNAs may be derived by splicing mechanisms from the same precursor transcript and the 17K and 14K MBP mRNAs may be derived from splicing of another precursor transcript. In the case of the *mld* and the *shi* mutations we would speculate that a structural element which governs the expression of both genes has been altered, thereby causing a dramatic reduction in the mRNAs for all four MBPs. In the case of the *jp* mutation not only has there been an alteration (though not as severe) in a controlling element governing expression of both genes but also an alteration in the splicing mechanism which governs production of the 14K MBP mRNA.

We have noted on innumerable occasions that in "initiating" *in vitro* systems the rate of synthesis of the 14K MBP relative to the 18.5K MBP is almost twice that in "run-off" systems (see Matthees and Campagnoni, 1980). We have also noted a similar phenomenon for the 17K and 21.5K MBP rates of synthesis (see Fig. 5). These data suggest that the 14K and 17K MBP mRNAs are more efficiently translated than the 18.5K and 21.5K MBP mRNAs and that the relative rates of synthesis of the MBPs may be regulated by differences in the rates of initiation of their mRNAs. The regulation of protein synthesis at the level of translation by such a mechanism has been extensively studied in rabbit reticulocytes (Lodish, 1976). Althogh the reticulocyte synthesizes equal numbers of $\alpha$- and $\beta$-hemoglobin chains, it contains 40% more $\alpha$-globin mRNA than $\beta$-globin mRNA (Lodish, 1971; Lodish and Jacobson, 1972). This capacity for the $\beta$-globin chain to be initiated more frequently seems to be a consequence of the mRNA structure itself and its ability to interact more favorably with initiation factors and ribosomal subunits (Lodish, 1976). It is possible that a similar situation exists with respect to the mRNAs for the MBPs, and translational regulation may also exist over the expression of MBP in the brain. At the present time all that is known about the structure of MBP mRNAs is that they are polyadenylated (Lim et al., 1974) and that they are very large (Zeller et al., 1983, 1984). It will be important, therefore, to examine their structure in more detail to determine if they are capped, for example, and to determine what function, if any, the non-MBP-coding regions of these large messages may have. In addition, a more detailed analysis of the specific steps in the translation mechanism (e.g., initiation, elongation, etc.) which acount for the apparent differential efficiency with which the MBP mRNAs are translated will need to be performed.

The majority of available evidence favors the view that the MBPs are synthesized on free ribosomes and that the major PLP is synthesized on bound polysomes. It will now be important to determine the routes by which these proteins are transported through the cell and assembled into the membrane.

Colman et al. (1982) have suggested that MBP-synthesizing free ribosomes become concentrated in oligodendrocytic processes near the site of assembly of myelin and that the proximity of these ribosomes accounts for the rapidity with which MBP becomes incorporated into the membrane, presumably by diffusion. If this is true, then it must be determined how MBP-synthesizing ribosomes are selectively transported from the oligodendrocyte cell body to the process. Immunohistochemical data at young ages ( ~ 5 days in the rat) show MBP scattered throughout the cytoplasm and at later ages in myelin. At the present time there are no reports of immunohistochemical localization of MBP solely in the oligodendrocyte processes. The results of Braun et al. (1980) suggest that MBPs become associated with vesicles shortly after synthesis. It is not yet clear whether these vesicles may be transport vesicles or whether they are derived from myelin. Thus, the mechanism by which the MBPs are transported to the growing myelin membrane has not yet been resolved and represents an important area of investigation in the near future. In this regard the quaking mutant may prove to be valuable since synthesis of the MBPs is essentially normal in this mutant, but their incorporation into the membrane is defective.

Although transport of the major PLP undoubtedly involves the Golgi bodies in the oligodendrocyte, elucidation of the factors which govern the transport of this myelin protin to the far-reaches of the oligodendrocytic processes are bound to be important, not only for understanding of myelinogenesis but also for contributing to our knowledge of how the cell "sorts" out the ultimate destination of membrane proteins.

Even though our understanding of the molecular biology of myelination is very incomplete, important advances are being made which should increase our knowledge rather quickly. With the availability of a probe of the MBP gene it should not be long before the gene is isolated and sequenced; and we can begin to understand what regulates its expression. For the first time, it will also be possible to examine the molecular genetics of the dysmyelinating mouse mutants and discover if they can help us learn what regulates myelinogenesis in the brain. The study of myelinogenesis has entered into a new molecular biological phase, and it will be very interesting to see what future discoveries will be made and how they will contribute to our understanding of brain function.

## NOTE ADDED IN PROOF

After this review was submitted for publication the isolation and sequence of a cDNA clone corresponding to most of the rat 14K MBP was reported (Roach, A., K. Boylan, S. Howath, S. B. Prusiner and L. E. Hood, *Cell* **34,**

799 [1983]). The size of the rat and mouse MBP mRNAs appear to be very similar, each containing large untranslated regions. In addition, another report appeared identifying the site of insertion of the extra N-terminal sequence in the sheep 21.5K MBP to be between residues 57 and 58 of the 18.5K protein (Carnegie, P. R. and C. A. Dowse, *Science* **223,** 936 [1984]).

## ACKNOWLEDGMENTS

This effort was supported in part by Grants NS-13934 and NS-15469 from the National Institutes of Health. The author wishes to thank his many students and collaborators for their contribution to this work. He also wishes to thank Celia W. Campagnoni for her support and valuable comments on the manuscript.

## REFERENCES

Adams, D. H. and J. Osborne, *Neurobiology* **3**, 91 (1973).

Agrawal, H. C. and B. K. Hartman, in R. A. Bradshaw and D. M. Schneider, Eds., *Proteins of the Nervous System*, 2nd Ed., Raven Pres, New York, p. 145 (1980).

Agrawal, H. C., R. M. Burton, M. A. Fishman, R. F. Mitchell, and A. L. Prensky, *J. Neurochem.* **19**, 2083 (1972).

Agrawal, H. C., C. L. Randle, and D. Agrawal, *J. Biol. Chem.* **256**, 12243 (1981).

Agrawal, H. C., K. O'Connell, C. L. Randle, and D. Agrawal, *Biochem. J.* **201**, 39 (1982).

Barbarese, E., P. E. Braun, and J. H. Carson, *Proc. Natl. Acad. Sci. USA* **74**, 3360 (1977).

Barbarese, E., J. H. Carson, and P. E. Braun, *J. Neurochem.* **32**, 1437 (1979).

Barbarese, E., M. L. Nielson, and J. H. Carson, *J. Neurochem.* **40**, 1680 (1983).

Baumann, N., Ed., *Neurological Mutations Affecting Myelination*, INSERM Symposium, No. 14, Elsevier-North Holland, Amsterdam, (1980).

Benjamins, J. A. and P. Morell, *Neurochem. Res.* **3**, 137 (1978).

Benjamins, J. A., M. Jones, and P. Morell, *J. Neurochem.* **24**, 1117 (1975).

Benjamins, J. A., M. Gray, and P. Morell, *J. Neurochem.* **27**, 571 1976).

Benjamins, J. A., R. Iwata, and J. Hazlett, *J. Neurochem.* **31**, 1077 (1978).

Bird, T. D., D. F. Farrell, and S. M. Sumi, *J. Neurochem.* **31**, 387 (1978).

Bird, T. D., D. F. Farrell, S. Stranahan, and E. Austin, *Neurochem. Res.* **5**, 885 (1980).

Boggs, J. M. and M. A. Moscarello, *Biochim. Biophys. Acta* **515**, 1 (1978).

Braun, P. E., in P. Morell, Ed., *Myelin*, Plenum, New York, p. 91 (1977).

Braun, P. E. and S. W. Brostoff, in P. Morell, Ed., *Myelin*, Plenum, New York, p. 201 (1977).

Braun, P. E., P. M. Pereyra, and S. Greenfield, in N. Baumann, Ed., *Neurological Mutations Affecting Myelination*, INSERM Symposium, No. 14, Elsevier/North Holland, Amsterdam, p. 413 (1980).

Brostoff, S. W., S. Greenfield, and E. L. Hogan, *Brain Res.* **120**, 517 (1977).

Burgess, K. T., C. C. A. Bernard, and P. R. Carnegie, *Adv. Exp. Med. Biol.* **100**, 303 (1978).

Burkart, T., U. N. Wiesmann, H. P. Siegrist, and N. N. Herschkowitz, *Biochim. Biophys. Acta* **673**, 351 (1981).

Campagnoni, A. T. and M. J. Hunkeler, *J. Neurobiol.* **11**, 355 (1980).

Campagnoni, A. T., D. L. Whitehead, and R. Rowan, III, in P. F. Agris, R. N. Loeppky, and B. O. Sykes, Eds., *Biomolecular Structure and Function*, Academic, New York, p. 413 (1978a).

Campagnoni, A. T., G. D. Carey, and Y. T. Yu, *J. Neurochem.* **34,** 677 (1980a).

Campagnoni, A. T., J. Matthees, and G. D. Carey, in N. Baumann, Ed., *Neurological Mutations Affecting Myelination,* INSERM Symposium, No. 14, Elsevier-North Holland, Amsterdam, p. 299 (1980b).

Campagnoni, A. T., C. W. Campagnoni, J. M. Bourre, C. Jacque, and N. Baumann, *J. Neurochem.* **42,** 733 (1984).

Campagnoni, C. W., G. D. Carey, and A. T. Campagnoni, *Arch. Biochem. Biophys.* **190**, 118 (1978b).

Carey, G. D. and A. T. Campagnoni, *J. Neurochem.* **33**, 81 (1979).

Carnegie, P. R. and W. J. Moore, in R. A. Bradshaw and D. M. Schneider, Eds., *Proteins of the Nervous System*, Raven Press, New York, p. 120 (1980).

Carnow, T. B., J. H. Carson, E. L. Hogan, and S. W. Brostoff, *Trans. Amer. Soc. Neurochem.* **13**, 256 (1982).

Carson, J. H., N. N. Herschkowitz, and P. E. Braun, *Trans. Amer. Soc. Neurochem.* **6**, 207 (1975).

Chan, D. S. and M. B. Lees, *J. Neurochem.* **30**, 983 (1978).

Colman, D. R., G. Kreibich, A. B. Frey, and D. D. Sabatini, *J. Cell Biol.* **95**, 598 (1982).

Cuzner, M. L. and A. N. Davison, *Biochem. J.* **106**, 29 (1968).

Cuzner, M. L., A. N. Davison, and N. A. Gregson, *Ann. NY Acad. Sci.* **122**, 86 (1965).

Dautigny, A., P. M. Alliel, J. L. Nussbaum, and P. Jolles, *Biochem. Biophys. Res. Comm.* **110**, 432 (1983).

Davison, A. N. and J. Dobbing, in A. N. Davison and J. Dobbing, Eds., *Applied Neurochemistry*, Blackwell Scientific, Oxford, England, p. 253 (1968).

Delassalle, A., B. Zalc, F. Lachapelle, M. Raoul, P. Collier, and C. Jacque, *J. Neurosci Res.* **6**, 303 (1981).

DeRobertis, E., A. Pelligrino de Iraldi, G. Rodriquez de Lores Arnaiz, and L. Salgonicoff, *J. Neurochem.* **9**, 23 (1962).

Dickerson, J. W. T., in A. N. Davison and J. Dobbing, Eds., *Applied Neurochemistry*, Blackwell Scientific, Oxford, England, p. 48 (1968).

D'Monte, B., P. Mela, and N. Marks, *Eur. J. Biochem.* **23**, 355 (1971).

Druse, M. J. and E. L. Hogan, *Proc. Soc. Exp. Biol. Med.* **145**, 747 (1974).

Dunkley, P. R. and P. R. Carnegie, *Biochem J.* **141**, 243 (1974).

Dupouey, P., C. Jacque, J. M. Bourre, F. Cesselin, A. Privat, and N. Baumann, *Neurosci. Lett.* **12**, 113 (1979).

Eng, L. F. and E. P. Noble, *Lipids* **3**, 157 (1968).

Fischer, C. A. and P. Morell, *Brain Res.* **74**, 51 (1974).

Fishman, M. A., H. C. Agrawal, A. Alexander, J. Golterman, R. E. Martenson, and R. F. Mitchell, *J. Neurochem.* **24**, 689 (1975).

Folch-Pi, J. and P. J. Stoffyn, *Ann. NY Acad. Sci* **195**, 86 (1972).

Gagnon, J., P. R. Finch, D. D. Wood, and M. A. Moscarello, *Biochemistry* **10**, 4756 (1971).

Ganser, A. L. and D. A. Kirschner, in N. Baumann, Ed., *Neurological Mutations Affecting Myelination*, INSERM Symposium, No. 14, Elsevier/North Holland, Amsterdam, p. 171 (1980).

Gerstl, B., L. F. Eng, R. B. Hayman, M. G. Tavaststjerna, and P. R. Bond, *J. Neurochem.* **14**, 661 (1967).

Giorgi, P. P., *Biochem. Soc. Trans.* **4**, 742 (1976).

Gray, E. G. and V. P. Whittaker, *J. Anat. (London)* **96**, 79 (1962).

Greenfield, S., W. T. Norton, and P. Morell, *J. Neurochem.* **18**, 2119 (1971).

Greenfield, S., S. Brostoff, and E. Hogan, *Brain Res.* **120**, 507 (1977).

Hall, C., L. C. Mahadevan, S. A. Whatley, T. S. Ling, and L. Lim, *Biochem. J.* **202**, 407 (1982).

Hennacy, D. M. and L. A. Horrocks, *Bull. Mol. Biol. Med.* **3**, 207 (1978).

Hogan, E. L., in P. Morell, Ed., *Myelin*, Plenum, New York, p. 489 (1977).

Hogan, E. L. and K. C. Joseph, *J. Neurochem.* **17**, 1209 (1970).

Horrocks, L. A., *J. Neurochem.* **15**, 483 (1968).

Jolles, J., J. Nussbaum, F. Schoentgen, P. Mandel, and P. Jolles, *FEBS Lett.* **74**, 190 (1977).

Jungalwala, F. B., *Brain Res.* **78**, 99 (1974).

Kerlero de Rosbo, N., P. R. Carnegie, and C. C. A. Bernard, *J. Neurochem.* **41**, S129 (1983).

Kies, M. W., R. E. Martenson, and G. E. Deibler, in A. N. Davison, P. Mandel, and I. G. Morgan, Eds, *Functional and Structural Proteins of the Nervous System*, Plenum, New York, p. 201 (1972).

Laatsch, R. H., M. W. Kies, S. Gordon, and E. C. Alvord, Jr., *J. Exp. Med.* **115**, 777 (1962).

Lachapelle, F., C. de Baecque, C. Jacque, J. M. Bourre, A. Delassalle, D. Doolittle, J. J. Hauw, and N. Baumann, in N. Baumann, Ed., *Neurological Mutations Affecting Myelination*, INSERM Symposium, No. 14, Elsevier/North Holland, Amsterdam, p. 27 (1980).

Lees, M. B., J. D. Sakura, V. S. Sapirstein, and W. Curatolo, *Biochim. Biophys. Acta* **559**, 209 (1979).

Lim, L., J. O. White, C. Hall, S. Berthold, and A. N. Davison, *Biochim. Biophys. Acta* **361**, 241 (1974).

Lodish, H. F., *J. Biol. Chem.* **246**, 7131 (1971).

Lodish, H. F., *Ann. Rev. Biochem.* **45**, 39 (1976).

Lodish, H. F. and M. Jacobson, *J. Biol. Chem.* **247**, 3622 (1972)

Macklin, W. B., P. E. Braun, and M. B. Lees, *J. Neurosci. Res.* **7**, 1 (1982).

Magno-Sumbilla, C. and A. T. Campagnoni, *Brain Res.* **126**, 131 (1977).

Martenson, R. E., in S. Kumar, Ed., *Biochemistry of the Brain*, Pergamon, New York, p. 49 (1980).

Martenson, R. E., G. E. Deibler, and M. W. Kies, *J. Neurochem.* **18**, 2427 (1971).

Martenson, R. E., G. E. Deibler, M. W. Kies, S. S. McKneally, R. Shapira, and R. F. Kibler, *Biochim. Biophys. Acta* **263**, 193 (1972).

Matthees, J. and A. T. Campagnoni, *J. Neurochem.* **35**, 867 (1980).

Matthieu, J. M. and T. V. Waehneldt, *Brain Res.* **150**, 307 (1978).

Matthieu, J. M., S. Widmer, and N. Herschkowitz, *Brain Res.* **55**, 403 (1973).

Matthieu, J. M., R. H. Quarles, H. DeF Webster, E. L. Hogan, and R. O. Brady, *J. Neurochem.* **23**, 517 (1974).

Matthieu, J. M., B. Koell-Reutter, and M. L. Joyet, *J. Neurochem* **30**, 783 (1978).

Matthieu, J. M., H. Ginalski, R. L. Friede, S. R. Cohen, and D. P. Doolittle, *Brain Res.* **191**, 278 (1980).

Matthieu, J. M., H. Ginalski-Winkelmann, and C. Jacque, *Brain Res.* **214**, 219 (1981).

McIlawain H., *Biochemistry and the Central Nervous System*, Little Brown, Boston, MA (1959).

Mendz, G. L., W. J. Moore, and P. R. Carnegie, *Aust. J. Chem.* **35**, 1979 (1982).

Morell, P., S. Greenfield, E. Constantino-Ceccarini, and H. Wisniewski, *J. Neurochem.* **19**, 2545 (1972).

Nicot, C., T. Nguyen Le, M. Lepretre, and A. Alfsen, *Biochim. Biophys. Acta* **332**, 109 (1973).

Norton, W. T., in P. Morell, Ed., *Myelin*, Plenum, New York, p. 161 (1977).

Norton, W. T., in N. Baumann, Ed., *Neurological Mutations Affecting Myelination*, INSERM Symposium, No. 14, Elsevier/North Holland, Amsterdam, p. 545 (1980).

Norton, W. T. and L. A. Autilio, *Ann. NY Acad. Sci.* **122**, 77 (1965).

Norton, W. and S. Poduslo, *J. Neurochem.* **21**, 749 (1973a).

Norton, W. T. and S. E. Poduslo, *J. Neurochem.* **21**, 759 (1973b).

Nussbaum, J. L. and P. Mandel, *Brain Res.* **61**, 295 (1973).

Omlin, F. X., H. DeF Webster, C. G. Palkovits, and S. R. Cohen, *J. Cell. Biol.* **95**, 242 (1982).

Pearson, W. R., J. J. Windle, J. F. Morrow, A. M. Benson, and P. Talalay, *J. Biol. Chem.* **258**, 2052 (1983).

Peters, A., S. L. Palay, and H. DeF Webster, in A. Peters, S. L. Palay, and H. DeF Webster, Eds., *The Fine Structure of the Nervous System. The Cells and Their Processes*, Harper & Row, New York (1970).

Poduslo, J. F. and P. E. Braun, *J. Biol. Chem.* **250**, 1099 (1975).

Quarles, R. H., J. L. Everly, and R. O. Brady, *Biochem. Biophys. Res. Comm.* **47**, 491 (1972).

Quarles, R. H., J. L. Everly, and R. O. Brady, *J. Neurochem.* **21**, 1177 (1973).

Ramsey, J. C. and W. J. Steele, *J. Neurochem.* **28**, 517 (1977).

Reidl, L. S., C. W. Campagnoni, and A. T. Campagnoni, *J. Neurochem.* **37**, 373 (1981).

Rossiter, R. J., in K. A. C. Elliot, I. N. Page, and J. H. Quastel, Eds., *Neurochemistry*, 2nd Ed., Thomas, Springfield, p. 870 (1962).

Sabri, M. I., A. H. Bone, and A. N. Davison, *Biochem J.* **142**, 499 (1974).

Shapira, R., F. C. H. Chou, M. R. Wilhelmi, and R. F. Kibler, *Fed. Proc.* **37**, 1784(1978).

Smith, M. E., *Biochim. Biophys. Acta* **164**, 285 (1968).

Smith, M. E., *Neurobiology* **2**, 35 (1972).

Sternberger, N. H., Y. Itoyama, M. W. Kies, and H. DeF Webster, *J. Neurocytol.* **7**, 251 (1978).

Sternberger, N. H., R. H. Quarles, Y. Itoyama, and H. DeF Webster, *Proc. Natl. Acad. Sci. USA* **76**, 1510 (1979).

Stoffel, W., W. Schroder, H. Hillen, and R. Deutzmann, *Hoppe-Seyler's Z. Physiol. Chem.* **363**, 1117 (1982).

Vacher-Lepretre, M., C. Nicot, A. Alpen, J. Jolles, and P. Jolles, *Biochim. Biophys. Acta* **420**, 323 (1976).

Weber, K. and M. Osborne, *J. Biol. Chem.* **244**, 4406 (1969).

Wisniewski, H. and P. Morell, *Brain Res.* **29**, 63 (1971).

Wolfgram, F., *J. Neurochem.* **13**, 461 (1966).

Yu, Y. T. and A. T. Campagnoni, *J. Neurochem.* **39**, 1559 (1982).

Zeller, N. K., J. A. Sprague, R. A. Lazzarini, Y. T. Yu, and A. T. Campagnoni, *Trans. Amer. Soc. Neurochem.* **14**, 255 (1983).

Zeller, N. K., M. J. Hunkeler, A. T. Campagnoni, J. Sprague, and R. A. Lazzarini, *Proc. Natl. Acad. Sci. USA* **81**, 18 (1984).

Zimmerman, T. R., Jr. and S. R. Cohen, *J. Neurochem.* **32**, 817 (1979).

# 9

# *The Bag Cell Neurons of Aplysia as a Possible Peptidergic Multitransmitter System: From Genes to Behavior*

**Barry S. Rothman, Earl Mayeri, and Richard H. Scheller**

The marine mollusk *Aplysia* is a convenient experimental system for investigating at the cellular and molecular levels the roles of peptides as neurotransmitters and their roles in the regulation of behavior. These problems can be studied in considerable detail in the peptidergic bag cell neurons and their target neurons located in the abdominal ganglion (Mayeri, 1979). This preparation is accessible to detailed studies at the electrophysiological, biochemical, morphological, and behavioral levels. Recently this system has also been studied at the molecular genetic level.

As evidenced by this book, molecular genetic techniques are now being used in neurobiology as a powerful tool for investigating how gene products such as peptides, receptors for transmitters, and ion channel proteins determine the individual properties of neurons. Because peptides, as a rule, are synthesized as parts of larger precursor proteins (Herbert et al., 1981), their structure and organization on the precursor are directly approachable by the use of molecular genetic techniques. In *Aplysia* a variety of recombinant DNA techniques have been employed to isolate a family of genes that encode the precursors for the neuroactive peptides used by the bag cells, other neurons of the central nervous system (CNS), and a non-neuronal tissue, the atrial gland. By determining the nucleotide sequences of three of these genes the complete amino acid sequences of three precursor proteins have been determined. The genes encode related sets of peptides which are expressed specifically in the bag cells or the atrial gland. Three previously described peptides, two of which are neuroactive, were found to be cleaved from a common precursor in the bag cells. The protein precursors were also found to contain new potentially active peptide sequences. Electrophysiological, biochemical, immunocytochemical, and behavioral studies are delineating the actions of these peptides as neurotransmitters in the CNS and as hormones in the periphery. The actions of these peptides are thought to mediate the stereotyped behavior accomopanying egg laying in *Aplysia*.

In this chapter a brief historical perspective is presented of the bag cells as a system of peptidergic neurons that mediate egg laying. The atrial gland is also reviewed historically. Molecular biological data recently obtained by the authors concerning a family of genes encoding peptides in the bag cells and atrial gland are presented followed by data on the products of transcription and translation of these genes. The bag cell peptides are then discussed as candidate neurotransmitters and as mediators of the behavior associated with egg laying. Finally, an overview of the bag cells with regard to their possible advantages as a multitransmitter system is presented.

## 1. ABDOMINAL GANGLION AND BAG CELLS

The CNS of *Aplysia* contains about $10^4$ neurons and consists of several ganglia linked by connective nerve trunks. The abdominal ganglion has been studied extensively at the electrophysiological, biochemical, and behavioral levels and is known to control several visceromotor functions such as egg laying, gill and siphon withdrawal, circulation, defensive release of ink from the purple gland, and kidney function (Kandel, 1976). The ganglion is a simple and isolatable preparation that contains only about 2000 neurons. These neurons are large (up to 500 $\mu$m in diameter); they can be identified from preparation to preparation and monitored with intracellular microelectrodes for many hours (Figs. 11A and B). Many of these target cells have been extensively studied with regard to their electrophysiological properties and roles in control of behavior. Located in the rostral margin of the abdominal ganglion are two clusters of 400 cells each known as the bag cells (Frazier et al., 1967). The bag cells constitute a large and apparently homogeneous population of peptide-secreting neurons that can be surgically isolated from the rest of the ganglion. The *Aplysia* central neurons are highly polyploid and thereby contain large amounts of mRNA and protein (Coggeshall et al., 1971; Lasek and Dower, 1971). Therefore, while only relatively small numbers of cells are available, these properties allow the bag cells to be used effectively in biochemical and molecular genetic studies.

### 1.1. Bag Cells as Mediators of Egg Laying

*Aplysia* are non-self-fertilizing hermaphrodites which culminate their annual life cycle with the laying of many large egg masses each containing up to $10^6$ fertilized oocytes. As is the case with other reproductive processes in both vertebrates and invertebrates, egg laying is associated with a stereotyped pattern of behavior which lasts as long as several hours. As an egg string emerges from the external genital groove located near the mouth, it is deposited onto an appropriate substrate as an irregularly shaped mass due to a complex motion of the head (Arch and Smock, 1977; Cobbs and Pinsker, 1982a). While these events occur, there is increased respiratory pumping (Brownell and Schaefer, 1982) and inhibition of locomotion and feeding (Cobbs and Pinsker, 1982b; Stuart and Strumwasser, 1980).

The first indications of the central role of bag cells in reproduction came from experiments in which sea water extracts of bag cells caused egg laying when injected into sexually mature animals (Kupfermann, 1967, 1970; Strumwasser et al., 1969; Toevs and Brackenbury, 1969). Initial work also demonstrated that the active material was released by the bag cells when they were stimulated electrically (Kupfermann, 1970) suggesting that the active substance(s) was a neurohormone. Later experiments in which the bag cells were monitored electrically in otherwise intact animals revealed that spontaneous egg laying is always preceded by a burst discharge of neuronal activity in the bag cells (Dudek et al., 1979; Pinsker and Dudek, 1977). The burst discharge lasts for about 20 min and occurs in near synchrony among all bag cells because the cells are electrically coupled (Blankenship and Haskins, 1979; Haskins and Blankenship, 1979; Kupfermann and Kandel, 1970). These properties of the bag cell system presumably allow it to release sufficient quantities of neurohormone to cause egg laying.

The active component of the bag cell extract was thought to be a protein because the egg-laying activity comigrated with a bag-cell-specific staining band on polyacrylamide gels, and the activity and band disappeared after treatment with proteases (Toevs and Brackenbury, 1969). Further chemical characterization of the active component showed that it was approximately a 6000-Da peptide with an isoelectric point of about 9.3 and that it was synthesized by the bag cells from a precursor of about $M_r$ of 25–29 kDa (Arch, 1972a, 1972b; Arch et al., 1976a, 1976b; Loh et al., 1975). The active peptide, egg-laying hormone (ELH) was purified to homogeneity from bag cell extracts and its amino acid sequence determined (Chiu et al., 1979). It is a 4400-Da peptide that is sufficient to cause egg laying when injected into sexually mature animals, although it is not known whether injection of pure ELH will elicit the full behavioral program associated with egg laying (Brownell and Schaefer, 1982). ELH is released from bag cells when they are electrically stimulated, and this release is dependent upon the presence of calcium in the medium (Stuart et al., 1980). Thus, the bag cells are a neuroendocrine system that release a blood-borne hormone, ELH, that causes egg laying. ELH also causes changes in the electrical activity of neurons in the abdominal (see below) and buccal ganglia (Stuart and Strumwasser, 1980) and releases eggs from the ovotestis (Rothman et al., 1983b). Bag cell extract which contains ELH reversibly suppresses serotonin-induced locomotion (Mackey and Carew, 1983). Our working hypothesis is that the bag cells induce egg-laying behavior by the action of their released products on the CNS and peripheral organs (Mayeri et al., 1979b).

### 1.2. Bag Cells as Model Peptidergic Neurons

In the last several years more than 30 biologically active peptides have been discovered in the CNS of vertebrates and invertebrates. Combined immunocytochemical, pharmacological, and electrophysiological studies suggest that these peptides act as neurotransmitters. However, at present little is known about the normal functioning of peptides in the CNS at the cellular and subcellular levels. The bag cells and their target neurons in the abdominal ganglion are a useful system for such studies because they represent one of the relatively few experimental systems where both the peptide-secreting cells and their targets are morphologically defined and accessible at the cellular level. The bag cells have profusely branched processes which terminate on blood spaces in the ganglionic sheath (Coggeshall, 1967) and do not make anatomically identifiable synapses with any neuron. However, electrical stimulation of the bag cells causes specific, reproducible changes in the electrical activity of certain identified neurons in the abdominal ganglion (Mayeri et al., 1979a, 1979b). These interactions are likely to be mediated by a non-synaptic mode of communication similar to the release of transmitters at autonomic nerve endings (Branton et al., 1978a, 1978b; Mayeri and Rothman, 1982). The transmitters involved are thought to be two or more peptides, namely, ELH (Branton et al., 1978b; Mayeri and Rothman, 1982) and $\alpha$-bag cell peptide (Rothman et al., 1983a).

### 1.3. Atrial Gland

The atrial gland may also play a role in reproduction since extracts of this gland initiate egg laying when injected into mature animals (Arch et al., 1978). The atrial gland is situated at the distal end of the genital tract where it forms a continuous lumen with the large hermaphroditic duct. It consists of two major cell types, large exocrine cells which contain electro-dense granules and ciliated capping cells. The dense core granules fuse prior to release into the lumen of the duct, identifying the secretory modality as exocrine (Arch et al., 1980). Three biologically active peptides, A and B and ERH, each 34 residues long were the first to be isolated from the atrial gland (Heller et al., 1980; Schlesinger et al., 1981). Peptides A and B are identical except for four amino acids in their $NH_2$-terminal halves and are not structurally related to ELH (Fig. 10). ERH is a hybrid peptide consisting, essentially, of the $NH_2$-terminal half of ELH and COOH-terminal half of peptides A or B. Each of the three native peptides initiates a burst discharge when applied to the bag cells *in vitro*. Three additonal peptides have been recently isolated from the atrial gland (Rothman et al., 1982, 1984; see below), two closely resemble ELH and one resembles an acidic bag cell peptide. Although the function of the atrial gland is still unknown (Blankenship et al., 1983) the fact that it is located in the reproductive tract and contains large amounts of biologically active peptides suggest that it is likely to play an important role in reproduction.

## 2. GENERATION OF GENE LIBRARY AND SELECTION OF CLONES

An *Aplysia* gene library was constructed using lambda phage Charon 4 (CH4) as a vehicle (Scheller et al., 1982, 1983a) by the procedure of Maniatis et al. (1978). Sperm was collected from the small hermaphroditic duct of a single *Aplysia californica* and several hundred micrograms of DNA isolated. The

DNA was partially digested with the restriction enzyme *Eco R1* to generate fragments of DNA about 15–20 kb in length. These fragments were ligated to the arms of the lambda phage derivative CH4. The resultant concatenate was packaged *in vitro* and the mixture used to infect *E. coli* DP50 SupF. This procedure yielded about $10^6$ independent clones, a number sufficiently high to assure representation of the entire *Aplysia* genome of $8.4 \times 10^5$ kb (Angerer et al., 1975).

The library was then amplified and $6 \times 10^5$ independent clones screened by filter hybridization using cDNA probes radiolabeled to high specific activity. The cDNA was synthesized *in vitro* from poly(A+) RNA purified from the bag cells, the abdominal ganglion, and the hepatopancreas using reverse transcriptase in the presence of [$^{32}$P]-labeled nucleotides. This procedure detects only the more prevalent species of mRNA and is not an exhaustive analysis. Clones were selected that hybridized to bag cell cDNA but not to the cDNA from the other two tissues. These clones were rescreened against cDNA from the bag cells, the abdominal ganglion, the hepatopancreas, and the atrial gland. Six clones were selected that hybridized to both the bag cell cDNA and atrial gland cDNA but not to cDNA from other tissues. One clone (ELH-1) was further characterized by restriction enzyme mapping and DNA sequence analysis and shown to contain a sequence coding exactly for ELH (Scheller et al., 1982) as purified from the bag cells (Chiu et al., 1979).

Further analysis was necessary to deduce the tissue specificity of the six clones. The six clones were completely digested with *Eco R1* which does not cleave in the ELH coding region of ELH-1 but does separate the recombinant DNA from the lambda phage arms. Three clones (ELH-1, -3, and -8) each contained two fragments that annealed to the cDNA probe, and three clones each contained a single hybridizing fragment. Blot hybridization of *Eco R1* digests of the clones performed in 30% formamide, 0.5 *M* NaCl, 48°C, showed that all fragments annealed both to bag cell cDNA and atrial cDNA (Fig. 1A and B). As expected, no significant hybridization occurred with cDNA derived from other parts of the *Aplysia* nervous system or other *Aplysia* tissues (not shown). When the stringency of the reassociation reaction was increased by raising the concentration of formamide to 70%, the tissue-specific expression of the gene family was seen (Fig. 1C and D). Each fragment hybridized to cDNA from either bag cells or atrial gland but not both. Three clones, ELH-1, ELH-3, and ELH-8, each contained one fragment specific to the bag cells and one specific to the atrial gland. The six clones contain a family of as few as three and as many as nine independent genes depending on allelic relationships.

Using immunocytochemical and *in situ* hybridization techniques it has recently been shown that in addition to the bag cells (Fig. 6) and atrial gland, a small network of neurons throughout the *Aplysia* CNS (Chiu, 1981; L. B. McAllister, et al., 1983) express members of the gene family. At present the

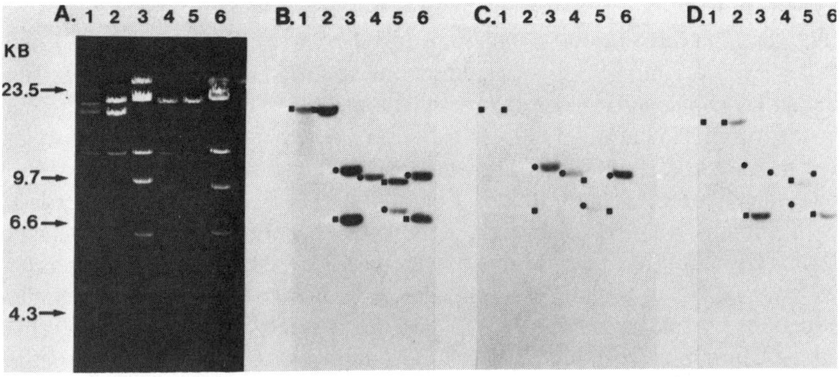


**Figure 1.** Southern blot analysis of ELH clones. A. Six different ELH-homologous recombinant clones were cleaved with restriction enzyme *Eco R1* and subjected to electrophoresis on a 0.7% agarose gel. The gel was stained with ethidium bromide and photographed. A *Hind III* digest of wild-type lambda phage was used as a size marker. Lane 1, ELH-18; lane 2, ELH-13; lane 3, ELH-8; lane 4, ELH-7; lane 5, ELH-3; and lane 6, ELH-1. B. The gel shown in A was transferred to nitrocellulose and hybridized with [$^{32}$P]-labeled cDNA from either bag cells or atrial gland. Hybridization was carried out in 30% formamide-TNE [TNE is 0.1 *M* Tris (pH 7.5), 0.5 *M* NaCl, 5 m*M* EDTA, and 0.1% SDS] at 48°C. C. A gel identical with that shown in A was hybridized with [$^{32}$P]-labeled cDNA from bag cells in 70% formamide-TNE at 48°C. The buffer is the same as in B. D. A third gel was hybridized to [$^{32}$P]-labeled cDNA from atrial gland under conditions described in C. The squares indicate the positions of restriction enzyme fragments that hybridize specifically to atrial gland cDNA under stringent hybridization conditions. The dots indicate the positions of fragments that hybridize specifically to bag cell cDNA under the same conditions. (From Scheller et al., 1983 a; Copyright 1983 M.I.T.)

function of these other cells is not known nor precisely which genes of the family are expressed in them. It is possible, however, that other members of the gene family that are not expressed in the bag cells and atrial gland are expressed in these cells. It is often the case that individual members of a multigene family are expressed either at different times in development or in different adult tissues, presumably satisfying the functional requirements of the tissue in which they are expressed. In contrast to the pro-opiomelanocortin system (Herbert et al., 1981) in which a single precursor undergoes different processing pathways in different tissues, the ELH gene family appears to use tissue-specific expression of related genes to generate different sets of peptides.

## 2.1. Restriction Mapping of the Clones

The organization of sequences in the clones has been deduced by restriction enzyme mapping and heteroduplex analysis. Restriction mapping of the six clones revealed that they were distinct from each other although clones

ELH-1 and ELH-7 may contain 5′ and 3′ extentions of the same fragment (Fig. 2). The three clones containing linked genes (ELH-1, -3, and -8) had these genes oriented in opposite directions about an axis of symmetry. The linked genes appear to arise from duplications of at least a 6.5-kb fragment of the chromosome. Two closely associated *Eco R1* sites flank the axes of symmetry for the inversions. However, as revealed by the restriction mapping, the symmetry is not perfect. For example, ELH-1, ELH-3, and ELH-8 each contain a 1.7-kb insert not present on the other side of the axis of symmetry (Fig. 2). This finding is supported by heteroduplex analysis in which the DNA of each clone was denatured, allowed to form intramolecular structures by folding back on itself, and then visualized by electron microscopy. The results for clone ELH-1 show a duplex structure resulting from the inverted duplication that is interrupted several times by loops, two about 250 bases in length and one 1.7 kb long (Scheller et al., 1982; Fig. 6). Heteroduplex analysis of the other clones was also consistent with their restriction maps (Scheller et al., unpublished data).

Recent genomic blotting experiments have demonstrated that the *Eco R1* fragments in clone ELH-3 are not present in the genome and suggest that during cloning the 1.7-kb element has moved from the other side of the inverted repeat to its present location (not shown). Thus, ELH-3 may be identical to ELH-1 and, therefore, not a unique clone. It is possible that the 1.7-kb element is mobile within the *Aplysia* genome as well during cloning. The fact that the 1.7-kb element is present at or near the 5′ end of some genes (e.g., ELH-1) but not others (e.g., ELH-18) suggests that this element may play an important role in governing the tissue-specific expression of the gene family. The resolution of this issue awaits further experimentation.

### 2.2. DNA Sequence Analysis of Three Genes

The nucleotide sequences of three DNA fragments that hybridize with cDNA from the atrial gland and the bag cells have been determined. Two fragments, one specific to the atrial gland and one specific to the bag cells, were derived from clone ELH-1, and a second atrial-specific fragment was derived from clone ELH-18. Two kilobases of nucleotide sequence were determined for the bag-cell-specific fragment and 0.6 and 0.8 kb determined for the two atrial fragments. These fragments were chosen because R-loop and Southern blot analyses indicated that they contained DNA sequences that were transcribed (see below). Sequencing of 5′-labeled restriction fragments was performed by the method of Maxam and Gilbert (1980).

Each of the three fragments contains a gene that appears to code for a precursor protein (Fig. 3). The bag-cell-specific gene codes for the precursor to ELH. One atrial gene codes for the peptide A precursor and one for the

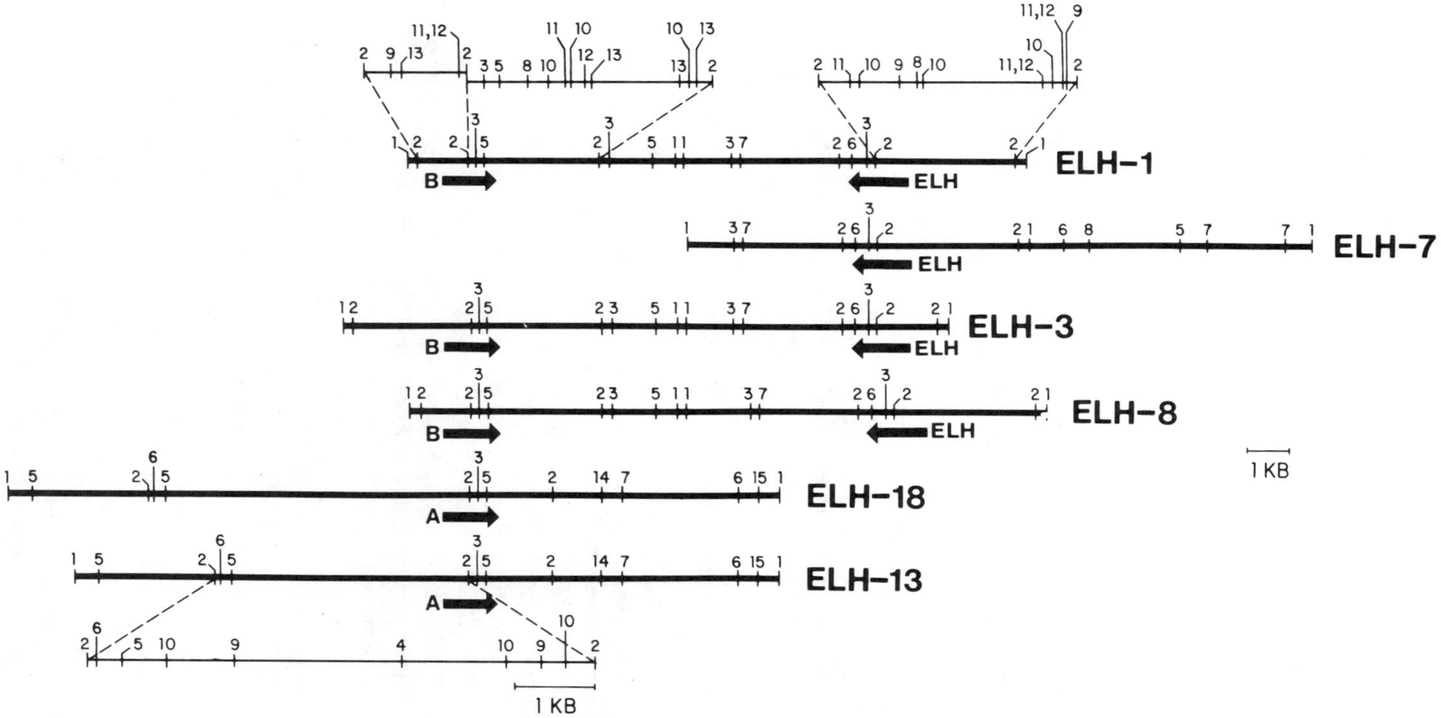


**Figure 2.** Restriction enzyme maps of ELH recombinant clones. Maps were determined by a combination of a single, partial, and double digests on intact clones and isolated fragments. The arrows indicate the position of mRNA homologous sequences and point in the direction of transcription as determined from DNA sequencing. Numerical code for restriction enzymes: 1, *Eco RI*; 2, *Pst I*; 3, *Xho I*; 4, *Stu I*; 5, *Pvu II*; 6, *Hind III*; 7, *Bgl I*; 8, *Xba I*; 9, *Ava II*; 10, *Hinc II*; 11, *Hae II*; 12, *Hha I*; 13, *Hpa II*; 14, *Bam HI*; and 15, *Sal I*. (From Scheller et al., 1983a. Copyright 1983 M.I.T.)

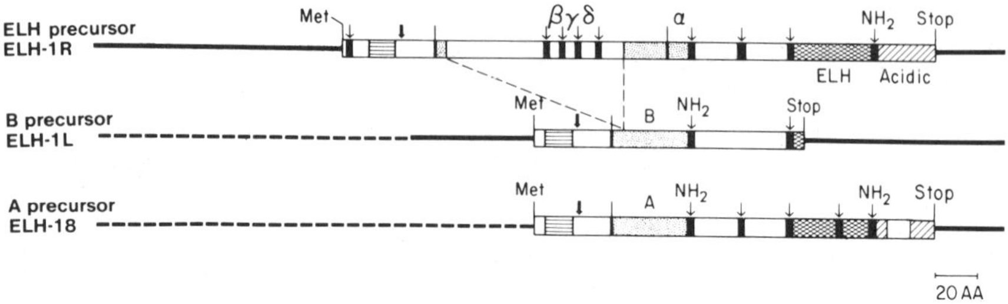


**Figure 3.** Comparison of the precursors and signal sequences encoding the ELH, A, and B peptides. The ELH precursor encodes ELH, α-BCP and acidic peptide (AP), and the potential cleavage products β-, γ-, and δ-BCP. The peptide B precursor encodes peptide B and an $NH_2$-terminal fragment of ELH. The peptide A precursor encodes peptide A, an ELH-like peptide, and an acidic peptide. Each of the three precursor proteins is initiated by a methionine followed by a hydrophobic region (▤). Thick arrows represent the putative cleavage site of the signal sequence (↓). Line above the sequence represents potential cleavages at single arginine residues (|), while arrows (↓) represent potential or known cleavages at dibasic, tribasic, or tetrabasic residues. If COOH-terminal amidation is believed to occur, an $NH_2$ appears above the arrow. The sequence homology to peptides A or B is represented by stippled boxes (░), the homology to ELH by cross-hatched boxes (▩), and the homology to acidic peptide by parallel lines enclosed in boxes (▨). Solid lines symbolize sequenced non-coding regions, and dashed lines depict regions not sequenced. (Adapted from Scheller et al., 1983a. Copyright 1983 M.I.T.)

peptide B precursor. As will be detailed below, ELH and three other peptides encoded on the ELH precursor have been purified from bag cell extracts, and peptides A and B and three other peptides have been purified from atrial gland extracts. On the precursor the sequences of the known peptides are flanked by basic amino acids which are presumed cleavage sites (Steiner et al., 1980). Other potential cleavage sites are present, and cleavage at these sites would yield peptides that have not yet been observed in exract of the bag cells and atrial gland. All three genes share a 90% homology of DNA sequence and, therefore, code for homologous precursors.

**2.2.1. The ELH Precursor Expressed in the Bag Cells.** The nucleotide sequence of the bag-cell-specific fragment predicts the synthesis of a precursor that contains 271 amino acids and has a molecular weight of 31 kDa (Fig. 3). Several domains of interest are defined by the potential cleavage sites on this precursor. As many as 11 peptides could be generated if cleavage occurred at all potential sites. Starting at the $NH_2$-terminus there is a hydrophobic region (horizontal lines) of about 20 amino acids that are thought to be necessary for attachment of the nascent protein chain to the endoplasmic reticulum and ultimately for vesiculation (Blobel and Dobberstein, 1975). The existence of such a "signal" sequence is supported by the observation that when dog pancreas microsomes are added to an *in vitro* translation of bag cell poly(A+) RNA, the precursor is reduced in apparent molecular weight (Fig. 7C). Past the hydrophobic region are the first six amino acids of peptide A/B (stippled area, Fig. 3), followed by 80 amino acids that are not present in either of the atrial genes. The base sequence coding for the 80-amino-acid segment contains a direct repeat, TCATCA, at each end. Such repeated sequences are commonly found at the ends of insertions or deletions. The insertion contains sequences for a 47-residue peptide followed by three peptides 5–7 residues long, $\beta$-bag cell peptide (BCP), $\gamma$-BCP, and $\delta$-BCP. The three small peptides are each flanked by pairs of basic residues. After 10 more residues the inserted segment ends and a 29-amino-acid sequence homologous to the COOH-terminus of peptides A and B occurs (stippled area). This includes a sequence of nine amino acids coding for $\alpha$-BCP, a neuroactive peptide purified from the bag cells (see below). The $\alpha$-BCP sequence is flanked on the $NH_2$-terminus by a single arginine residue and on the COOH-terminus by the cleavage sequence Arg-Lys-Arg. A single arginine residue is an unusual cleavage site for peptide precursors but does occur for peptides A and B (see below) and for the vertebrate magnocellular hormones oxytocin and vasopressin (Land et al., 1982). Cleavage at a single arginine residue may also generate a novel opiod peptide (E. Weber, personal communication). Further on there are two predicted cleavage products. Each is 19 residues long and is flanked by a sequence of 2–4 basic residues. These peptides are followed by the sequence of 36 amino acids coding for ELH (cross-hatching). The ELH sequence ends with the amino acids Gly-Lys-Arg which may signal a cleavage point at which the glycine residue is destroyed to yield an amidated COOH-terminus (Bradbury et al., 1982). ELH is followed by a 27-amino-acid sequence coding for an acidic peptide (diagonal lines) which is synthesized by the bag cells and present in bag cell extract (Scheller et al., 1983a) and may be identical to the acidic peptide observed by Arch et al. (1976a) and Berry (1981). The peptide and the precursor are ended by a stop codon.

**2.2.2. The Peptide A and B Precursors Expressed in the Atrial Gland.** The nucleotide sequences of the two atrial fragments predict the synthesis of a 19.8-kDa precursor to peptide A that contains 177 amino acids and the synthesis of a 13.6-kDa precursor to peptide B that contains 122 amino acids (Fig. 3). As a result of internal cleavage sites, the A precursor has the potential to yield seven peptides and the B precursor a potential to yield four. Eight of the first sixteen residues of each precursor are hydrophobic and appear to represent signal sequences (horizontal lines). Seventeen residues further down, a single arginine residue is found on each precursor marking the beginning of the A or B peptide sequence each 34 residues long (stippled area). The A and B peptide sequences end with the residues Gly-Lys-Arg, signaling cleavage and possible amidation of their COOH-termini. The B precursor continues for 42 residues to another diabasic cleavage site encoding the first six amino acids of ELH (cross-hatching), then ends at a stop codon. In contrast the A precursor has the 42-residue sequence interrupted by a tribasic sequence creating another potential cleavage site. The A precursor continues for 36 residues to code for an ELH-like peptide (cross-hatching) that is interrupted by another potential tribasic cleavage site. As in the ELH precursor, this peptide ends with the sequence Gly-Lys-Arg which signals a cleavage site at which amidation may occur. The ELH-like peptide is followed by a 27-residue sequence homologous to the bag cell acidic peptide (diagonal lines). The peptide and the precursor end at a stop codon.

**2.2.3. Conservative Changes in the Precursors.** Although the sequenced genes show 90% sequence homology, they have small differences, many of which alter the proteolytic processing of the precursor proteins, thus generating a different set of peptides by each precursor. This implies that these changes have become fixed in the population because of the selective advantage they provide. It also argues that the different peptide components are functionally important to the organism.

The creation of the single basic cleavage site at the amino terminus of $\alpha$-BCP and the removal of the Gly residue signaling possible amidation at its COOH-terminus each result from single base changes in the ELH precursor relative to the A and B presursors. Four other base changes occur within the $\alpha$-BCP sequence and these account for the three amino acid substitutions that occur in $\alpha$-BCP relative to peptides A and B (Fig. 10). The creation of a potential cleavage site, Arg-Arg-Arg, in the middle of the ELH-like region of the A precursor (Fig. 3) results from a single base change in a sequence coding for Arg-Gln-Arg, a site unlikely for cleavage in the bag cell precursor. Similarly the cleavage site that exists halfway between the A/B peptide region and the ELH-like region of the precursors shows two small changes that are potentially important. The ELH precursor contains a sequence of four basic residues, Arg-Arg-Lys-Arg (black bar). A single base change in the A precursor changes this sequence to Trp-Arg-Lys-Arg (black bar), and a second base change in the B precursor changes the region to Trp-Arg-Glu-Arg (no black bar), a site unlikely for cleavage. Finally, the early termination of the B precursor at the sixth residue of the ELH-like sequence results from a frame shift caused by the insertion of a single base.

While the nature of the events that generate different sets of peptides from the family of related genes is understood, the molecular mechanisms utilized to generate the tissue-specific expression of the genes is less clear. Although the chromosomal position of the gene may be important, it is also likely that the small number of substitutions, insertions, and deletions in the coding and flanking regions of the genes may play a role in tissue-specific expression.

## 3. TRANSCRIPTS

Only indirect evidence is available for the identification of mRNA coded by the cloned genes. RNA gel blot analysis (Fig. 4) reveals that five species of poly(A+) RNA from the bag cells hybridize with the fragment of clone ELH-1 that codes for ELH. The species are 1.60, 1.45, 1.20, 0.7, and 0.4 kb in size. Similarly, four species of atrial poly(A+) RNA 1.40, 1.15, 9.7, and 0.4 kb in size hybridize to this probe (Fig. 4). The diffuse nature of the two smallest bands is unexplained and may reflect multiple transcripts or a variety of degradation products of the larger RNA species.

R-loop analysis by electron microscopy has allowed the tentative identification of a transcript from the bag cells. DNA from clone ELH-1 was allowed to hybridize with bag cell poly(A+) RNA and the RNA-DNA complex viewed by electron microscopy. Where the double-stranded DNA was interrupted by the binding of a complementary RNA molecule, a loop (R-loop) was seen (Fig. 5). The loop, 1.2 kb long, indicates the site of transcription. Its size is in good agreement with one bag cell RNA species seen in hybridization blots (Fig. 4). The location of the R-loop matches the region that codes for ELH (Fig. 2, arrow at left on clone ELH-1). Thus, the mRNA coding for the ELH precursor is expected to be 1.2 kb in length. *In situ* hybridization of the abdominal ganglion with the DNA fragment coding for ELH shows that an ELH-like transcript is expressed in the bag cells as expected (Fig. 6B).

While the precise nature of the various mRNA species is at present unclear, recent studies of bag cell cDNA clones have revealed several interesting features. Four different cDNA clones have been characterized (Scheller et al., unpublished data). DNA sequence analysis of the clones suggests that two different 5′ ends are generated via the splicing of either one or two donor sites to a unique acceptor. In addition, different 3′ ends arise by adding poly A sequences to two unique sites. The combination of two possible 5′ and two possible 3′ ends generates four different mRNA species. The differences among these species occur in untranslated regions and as a result the mRNA species appear to encode the same precursor protein. Therefore, the functions of the different transcripts are at present unclear. For the atrial gland transcripts, the working hypothesis is that the two largest RNA species, 1.40 and 1.15 kb in size, are mRNAs that code for the A and B precursors, respectively.

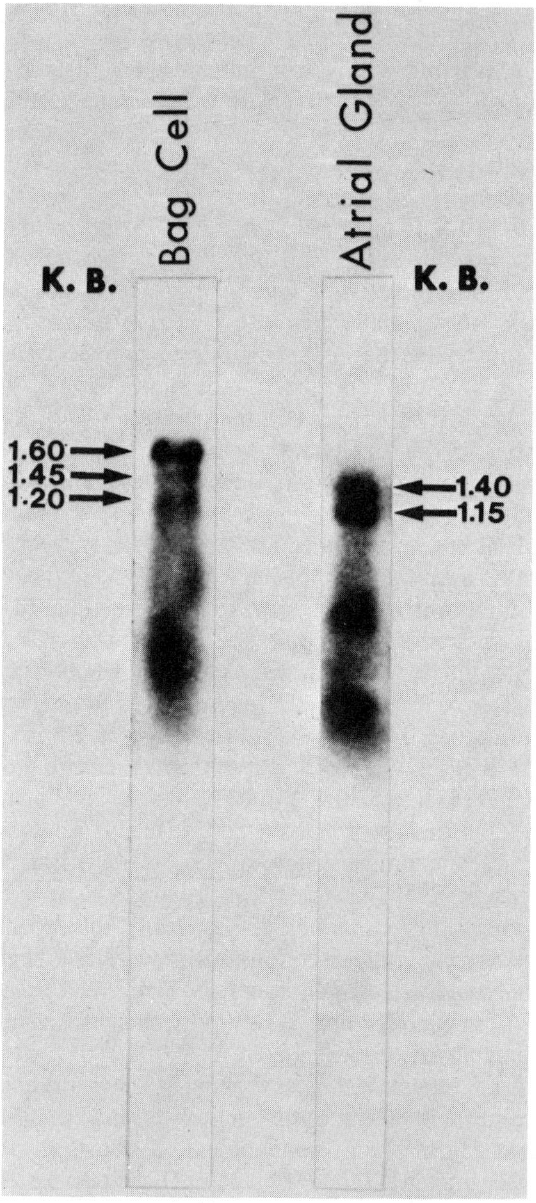


**Figure 4.** RNA gel blot analysis of bag cell and atrial gland RNA. One microgram of poly(A+) RNA from the bag cells and the atrial gland was fractionated on a 1.0% agarose formaldehyde gel and transferred to nitrocellulose. Nick-translated DNA of the ELH-coding region of clone ELH-1 ($10^6$ cpm) was hybridized to the blot in 50% formamide at 42°C. (From Scheller et al., 1982. Copyright 1982 M.I.T.)

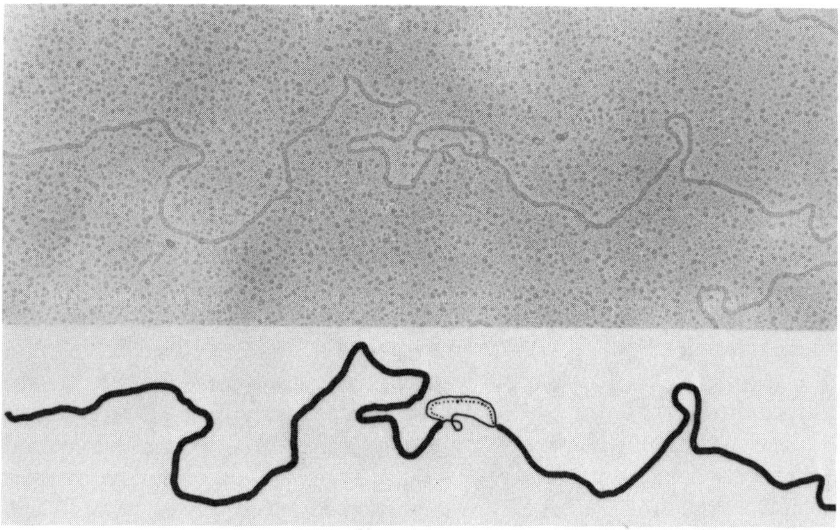

**Figure 5.** R-loop and interpretive tracing of the ELH gene on clone ELH-1. Top: R-loops were formed using whole-phage DNA and spread for examination in the electron microscope in 55% formamide. Loops are regions to which poly(A+) RNA from the bag cells is hybridized to the recombinant clone. Bottom: Interpretive tracing of micrograph above. Thick lines: double-stranded DNA. Thin lines: single-stranded DNA. Dotted line: RNA. (From Scheller et al., unpublished.)

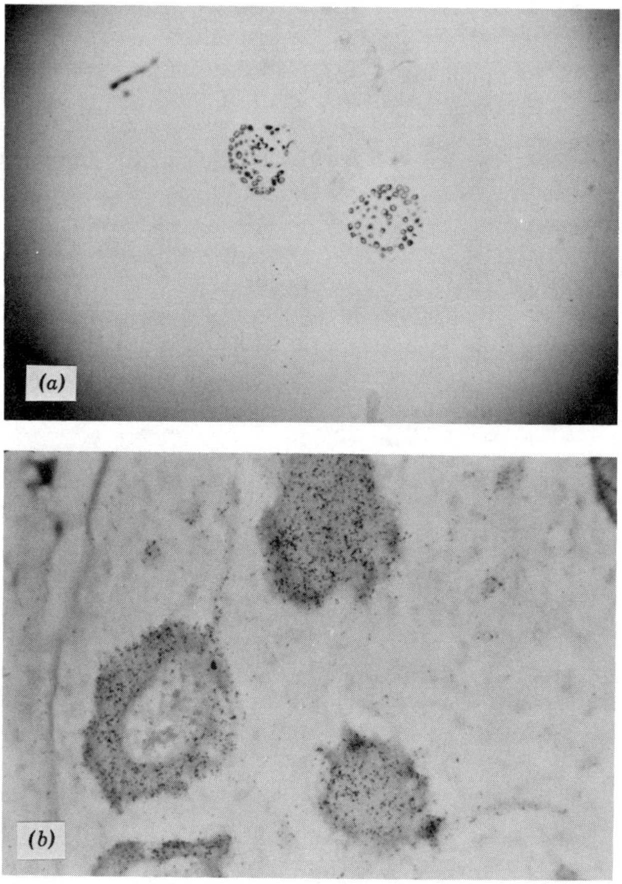

**Figure 6.** Immunocytochemical and *in situ* hybridization labeling of the bag cells. A. Fluorescence micrograph of immunocytochemical staining of bag cells in a 45-day postmetamorphic animal using rabbit anti-ELH as a primary antibody and goat anti-rabbit IgG coupled to rhodamine as a secondary antibody. 20×. B. [$^{125}$I]-labeled cloned DNA encoding ELH was hybridized to glutaraldehyde-fixed 5-$\mu$m sections through the abdominal ganglion. Slides were dipped in Kodak NTB-2 emulsion (diluted 1 : 1 with water), exposed for 2 days, developed, and stained lightly with Giemsa. 60×. (From Scheller et al., unpublished.)

## 4. PRODUCTS OF TRANSLATION

Biochemical studies of the bag cell precursor, processing, and peptides as well as the atrial peptides have taken place in parallel to the molecular genetic studies. The data regarding the products of translation are in very good agreement with the molecular genetic data.

### 4.1. Precursors

Poly(A+) RNA purified from the bag cells, head ganglia, ovotestis, hepatopancreas, and the atrial gland have ben translated *in vitro* by a lysed rabbit reticulocyte system employing [$^{35}$S]-methionine. The products of translation were then run on SDS–polyacrylamide gels and detected by autoradiography. Translation of total mRNA from each tissue revealed that the bag cells and atrial samples have only a few prominent bands (Fig. 7A); the bag cells synthesize two major species, one 37 kDa in apparent molecular weight and the other 27 kDa. The atrial gland synthesizes one major species of 21 kDa in size. Selection of mRNA by hybridization to the ELH-coding fragment of clone ELH-1 followed by *in vitro* translation revealed four major mRNA species synthesized by the bag cells (Fig. 7B). The most prominent product is a 35–37-kDa species accompanied by others 27, 24, and 19 kDa in size. Presumably, the major species represents the 31-kDa precursor predicted from the nucleotide sequence data although the discrepancy in molecular weights is presently unexplained. Two major species, 21 and 13 kDa in size, were made by the atrial gland sample (Fig. 7B). These findings are in good agreement with the nucleotide sequence data for the atrial gland genes which predict the synthesis of a 19.8-kDa and 13.6-kDa precursor for peptides A and B, respectively.

Further testing of the translation products was accomplished by imunoprecipitating with antibodies directed against ELH or peptide A. These antibodies were raised in rabbits and affinity purified. In enzyme-linked immunosorbant assays (ELISA) the ELH antibodies did not cross-react with peptide A nor the peptide A antibodies with ELH (Rothman et al., unpublished data). Total mRNA translation products of the bag cells precipitated with anti-ELH revealed a prominent 35–37 kDa species and three minor species, 27, 24, and 19 kDa in size (Fig. 7C). The 35–37-kDa species is also precipitated by anti-peptide A. Similarly, the peptide A precursor, 21 kDa in size, is precipitated by both antibodies. In contrast, the B precursor is precipitated only by anti-peptide A. These findings are in perfect agreement with the nucleotide sequence data which predict the presence of peptide-A- or B-like sequences on all three precursors and ELH-like sequences on the bag cell and peptide A precursor. Presumably, the B precursor did not react with anti-ELH because it contains only the first six residues of ELH.

### 4.2. Processing of the Bag Cell Precursor

Like other peptide-secreting systems the bag cells synthesize peptides from a higher molecular weight precursor protein. This protein is cleaved at specific sites to yield the appropriate peptides. Several studies consistently show the

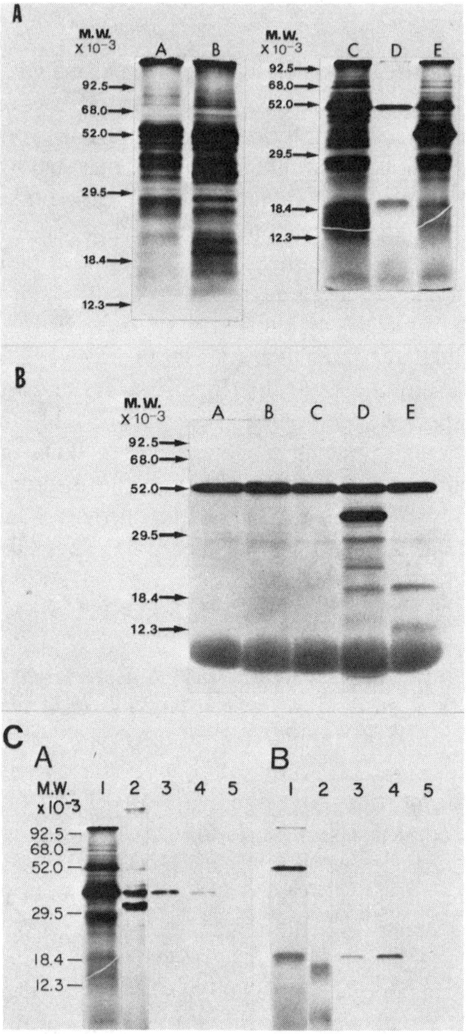


**Figure 7.** Processing and immunoprecipitation of bag cell and atrial gland proteins translated *in vitro*. Poly(A+) RNA was translated *in vitro* by using a rabbit reticulocyte lysate system incorporating [$^{35}$S]-methionine. Proteins were subjected to electrophoresis on SDS–polyacrylamide (12 or 15%) gels, dried, and autoradiographed. A. *In vitro* translations of 1 $\mu$g of total poly(A+) RNA from (A) ovotestis, (B) hepatopancreas, (C) head ganglia, (D) atrial gland, and (E) bag cells. B. Hybridization selection followed by *in vitro* translation of *Aplysia* RNA. ELH-1 DNA was bound to nitrocellulose filters and hybridized to poly(A+) RNA from several tissues. The bound RNA was eluted, precipitated, and used to direct translation *in vitro*. (A) no RNA added, (B) ovotestis RNA, (C) hepatopancreas RNA, (D) bag cell RNA, and (E) atrial gland RNA. C. (A) (1) bag cell poly(A+) RNA, (2) dog pancreas microsomes were included

presence of a 25–29-kDa precursor molecule in bag cell bodies pulse-labeled with radiocactive amino acids (Arch, 1972a, 1972b; Arch et al., 1976a, 1976b; Berry, 1981; Berry et al., 1981; Loh et al., 1975). The results of pulse chase studies in which bag cell proteins were fractionated on SDS–polyacrylamide gels suggest that this precursor is ultimately processed into two major species 13–14.5 kDa in size as well as ELH and an acidic peptide (Arch et al., 1976a; Berry, 1981). ELH and the acidic peptide appear to be cleaved from an intermediate peptide about 6 kDa in size.

These findings are consistent with the nucleotide sequence data for the bag-cell-specific gene. The molecular genetics data predict the cleavage of the precursor protein near the COOH-terminus of the signal sequence yielding (Fig. 3) a 27-kDa precursor. This cleavage almost certainly occurs as the nascent protein is being synthesized, and therefore, the 31-kDa species contianing the signal sequence is never detected in *in vivo* experiments.

Although there is no problem comparing ELH purified by others (Arch et al., 1976a, 1976b; Berry, 1981; Berry et al., 1981; Branton et al., 1978b; Chiu et al., 1979) with the ELH purified by us from bag cell extracts (Mayeri and Rothman, 1982) and found in the nucleotide sequence data (Scheller et al., 1983), identification of the acidic bag cell peptide presents some difficulty. Two acidic peptides similar in size to the one seen by Arch and by Berry are predicted from the nucleotide sequence data (Fig. 3). One (42 residues in length) is located $NH_2$-terminal to the ELH sequence, and the other (27 residues in length) is located COOH-terminal to ELH. Therefore, either could be part of an intermediate peptide containing the ELH sequence. However, only the 42-residue peptide contains a methionine residue as required by Berry's data, which shows that the acidic peptide is labeled with [$^{35}$S]-methionine. It is, therefore, possible that there are two end products that are acidic peptides, the one purified by Arch et al. (1976a) and Berry (1981) and the one purified by us (Scheller et al., 1983a).

Candidates for the 13–14.5-kDa products predicted by Berry (1981) are (1) a mildly basic, 14-kDa peptide that runs from the amino terminus of $\alpha$-BCP to the COOH-terminus of the precursor and (2) an acidic, 11-kDa fragment

---

in the reticulocyte lysate prior to initiation of translation, (3) immunoprecipitation of bag cell poly(A+) RNA translation product with 2 $\mu$g of anti-ELH antibody, (4) immunoprecipitation of bag cell poly(A+) RNA translation product with 2 $\mu$g of anti-peptide A antibody, and (5) immunoprecipitation of bag cell poly(A+) RNA translation product with 2 $\mu$g of rabbit IgG. (B) Same as above with poly(A+) RNA from the atrial gland used to direct the protein synthesis. (From Scheller et al., 1982, 1983a. Copyright 1982 and 1983 M.I.T.)

that runs from the cleavage site after the signal sequence to the start of $\beta$-BCP.

Other parts of the processing scheme proposed by Arch and Berry require additional assumptions to be made. The changes in the pI of ELH after cleavage from a pI of 7 to $> 9$ and for the acidic peptide from 4.1 to 4.9 (Berry, 1981) cannot be explained by cleavages occurring at dibasic residues or removal of these residues but seem to require other chemical modifications such as phosphorylation or dephosphorylation.

A fascinating discovery in the nucleotide sequence data of the ELH precursor is a region that apparently codes for a 21-kDa protein, BC-2, purified from a crude bag cell membrane preparation. Jennings et al. (1982) have shown that BC-2 is bag-cell-specific and have sequenced 17 of its first 23 residues. These residues are in perfect agreement with a region of the ELH precursor that begins 30 residues after the initiation methionine residue and extends 6 residues into the 80-residue insertion coding for $\beta$-BCP, $\gamma$-BCP, and $\delta$-BCP. Cleavage of the precursor at the amino terminus of this sequence and at the cleavage site $NH_2$-terminal to ELH yields a 21-kDa fragment. The amino acid composition of BC-2 is in reasonable agreement with the amino acid composition of this fragment ($r = 0.575$, $P < 0.05$). Jennings et al. (1982) suggest that BC-2 is a membrane protein that is phosphorylated during a bag cell burst discharge or in the presence of cAMP in a crude bag cell membrane preparation. However, the presence of the $NH_2$-terminal sequence of BC-2 on the ELH precursor raises the possibility that BC-2 is a secretory protein or a processing intermediate.

### 4.3. Bag Cell Peptides

Four peptides, ELH (Chiu et al., 1979), acidic bag cell peptide (Rothman et al., unpublished data; Scheller et al., 1983a), $\alpha$-BCP (Rothman et al., 1983), and $\beta$-BCP (Rothman et al., 1981; Scheller et al., 1983b) have been purified from bag cell extracts. Bag cell peptides were purified by surgically removing the bag cell clusters which contain bag cell somata, processes, fibers from other cells and the ganglionic sheath, and extracting them with acetic or formic acid. The soluble material was then fractionated on G50 Sephadex which separates the peptides according to molecular weight (Fig. 8A p. 256). The column eluate was divided into three fractions, corresponding to the void volume ($M_r = > 12$ kDa), the medium molecular weight material ($M_r =$ 2–12 kDa), and the included volume ($M_r < 2$ kDa). The last two fractions were then further purified by reverse-phase high-performance liquid chromatography (HPLC) using a volatile solvent system (Fig. 8B and C) (Rothman et al., 1983a; unpublished data). Peptides were detected by an automated fluorescamine detection system (Stein et al., 1981) which reveals any soluble material containing a primary amino group.

This procedure has permitted us to purify and determine the amino acid sequence of three peptides encoded on the bag cell precursor. The basic strategy has been to lyophilize the material corresponding to each peak and trough of the HPLC profiles and to assay the material for neuronal activity (see below). Material identified as active was then repurified if necessary and its amino acid composition and primary sequence determined by Dr. Jack Shively at the City of Hope Research Center. Two active peptides, ELH (Fig. 8B, peak 12) and α-BCP (Fig. 8C), which is present in three forms (see below) were characterized. A third, the 27-residue acidic peptide (Fig. 8B, peak 11), was characterized because of its abundance. All three peptides were recovered in roughly equimolar amounts from bag cell extracts. The sequence of ELH was already known (Chiu et al., 1979). The primary sequences of the peptides agreed perfectly with the molecular genetic data for the bag cell precursor. A fourth bag cell peptide which was neuronally active was not sequenced, but the amino acid composition of this molecule suggests that it is a pentapeptide. A sequence of five amino acids (β-BCP) with the same amino acid composition is encoded on the bag cell precursor (Fig. 3). These data suggest that the ELH precursor is expressed by the bag cells and that most, if not all, bag cells process the precursor identically to yield one copy of each peptide.

If cleavage of the bag cell precursor occurred at all nine multibasic sites and at the single arginine residues preceeding α-BCP and the $NH_2$-terminus of the B peptide sequence, eleven peptides would be generated ranging in size from 0.6 to 8.9 kDa. The eight other peptides predicted by the molecular genetic data have not yet been isolated, but they are expected to be present in the bag cell extract.

### 4.4. Atrial Peptides

A strategy similar to that used for identifying active bag cell peptides has been used for identifying atrial gland peptides (Rothman et al., 1982, 1984; unpublished data). Atrial extract was fractionated on G50 Sephadex (Fig. 9A p. 258). Medium- ($M_r$ = 6 kDa to 2 kDa, Fig. 9A, bar) and low-molecular-weight material ($M_r$ < 2 kDa, fractions 54–69, Fig. 9A) was lyophilized and further purified by reverse-phase HPLC.

The two-step purification procedure has allowed isolation and identification of the previously described peptides A and B (Heller et al., 1980) (Fig. 9B, peaks 6 and 5, respectively), two novel ELH-like peptides, and a novel acidic peptide homologous to one present in bag cell extract (see below). The existence of large amounts of peptides A and B in atrial extract strongly suggests that, as appears to be the case for α-BCP, the single arginine residue preceding the sequences for these peptides on the precursor molecules is a cleavage site. After elution of peptides B and A, a major peak

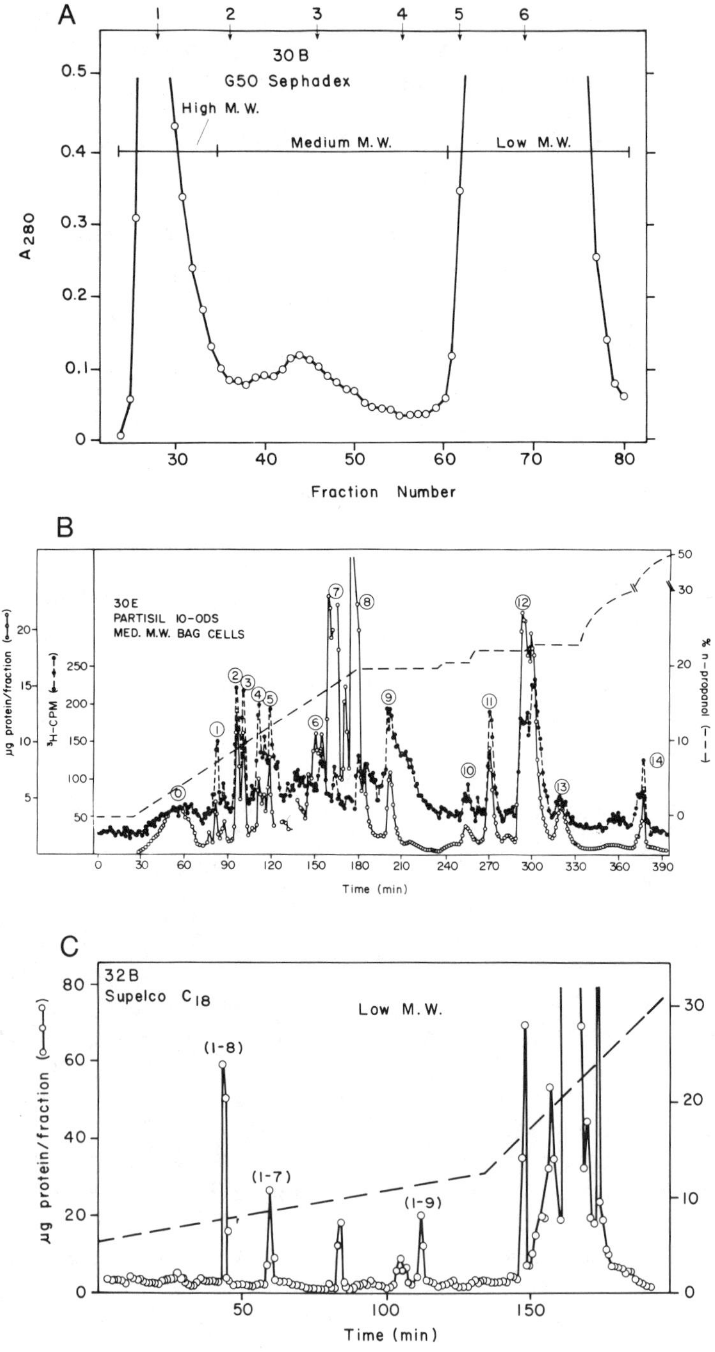
A
30 B
G50 Sephadex
High M. W.
Medium M. W.
Low M. W.
$A_{280}$
Fraction Number
B
30E
PARTISIL 10-ODS
MED. M.W. BAG CELLS
µg protein/fraction
³H-CPM
% n-propanol
Time (min)
C
32B
Supelco $C_{18}$
Low M. W.
(1-8)
(1-7)
(1-9)
µg protein/fraction
% n-propanol
Time (min)

of material (peak 7) was recovered with ELH-like activity on abdominal ganglion neurons (Rothman et al., 1982). This material also cross-reacted with a polyclonal antibody to ELH on an enzyme-linked immunosorbant assay. Its amino acid composition suggests that it is distinct from ELH but its primary sequence has not yet been determined.

The dichrotic peak of material that eluted next (peak 8) has been repurified by reverse-phase HPLC under isocratic conditions (Fig. 9B, inset). Peaks 8A and 8B each contained two peptides that were strongly associated. Sequence analysis of peak 8B revealed two residues at each of the first 17 cycles (Fig. 10 p. 260). One of each pair of residues can be assigned to a sequence homologous to ELH found on the peptide A precursor (A-ELH) and the remaining one of each pair then perfectly matches the first 17 amino acids predicted for a 27-residue acidic peptide (A-AP) on the peptide A precursor (Fig. 3, diagonal bars). The remaining cycles of the sequenator each yielded a unique residue which perfectly matched the rest of the A-ELH sequence. Because reduction with $\beta$-mercaptoethanol and subsequent HPLC fraction-

---

**Figure 8.** Purification of bag cell peptides. A. Fractionation of bag cell extract by gel filtration chromatography. One hunded frozen bag cell clusters plus four pulse chase labeled with [$^3$H]-arginine, -lysine, and -leucine were ground on ice in a glass homogenizer in a total of 20 mL of 1.0 *M* acetic acid containing 0.1% 2-mercaptoethanol, 300 $\mu$g/mL phenylmethylsulfonylfluoride, 25 $\mu$g/mL angiotensin converting enzyme inhibitor, 2 mg/mL phenylalanylalanine, and 250 $\mu$g/mL each of bacitracin, ovomucoid, lima bean trypsin inhibitor, ovoinhibitor, leupetin, and antipain. The homogenate was centrifuged at 12,000*g* for 20 min at 4°C and the supernate applied to a Sephadex G50 (superfine) column (2.5×30 cm) equilibrated at 4°C with 1.0 *M* acetic acid and eluted at a rate of 5 mL/cm$^2$ hr. Fractions of 4 mL each were collected and $A_{280}$ of each fraction was determined (o-o-o). Fifty microliters of each fraction was counted on a Beckman Ls 3133P scintillation counter (o-o-o). The column eluate was separated into high-, medium-, and low-molecular-weight fractions. (From Rothman et al., 1983a.) B. Purification of medium-molecular-weight material by HPLC. Pooled medium-molecular-weight fractions from G50 Sephadex column were applied to $C_{18}$ reverse-phase HPLC column (Whatman Partisil 1O-ODS, 2.0×50 cm). A total of 75 mL was applied to the column equilibrated in pyridine acetate, pH 4.0 (0.09 *M* pyridine, 0.5 *M* acetic acid), and the column washed for 100 min with this buffer. Material was eluted at a rate of 2 mL/min with a gradient of *n*-propanol (- - -) in the same buffer. Protein was determined for 20 $\mu$L of each fraction by an automated fluorescamine detection system (Stein et al., 1981). Radioactivity contained in 50 $\mu$L of each fraction was determined as in A. Acidic bag cell peptide elutes at peak 11 and ELH at peak 12. (From Rothman et al., unpublished.) C. Purification of low-molecular-weight bag cell material by HPLC. Pooled low-molecular-weight fractions (80 mL) from G50 Sephadex column were applied to HPLC column (Supelco, $C_{18}$DB, 4.6×250 mm) at a rate of 2 mL/min. The column was washed with 50 mL of pyridine acetate, pH 4, at a rate of 0.6 mL/min. Material was eluted with two linear gradients (- - -) of *n*-propanol (0–12.5% in 200 min, 12.5–30% in 40 min) in the same buffer and at the same rate as the wash step. Material eluting in the early part of the first gradient is not shown. Peaks corresponding to $\alpha$-BCP (1–7), $\alpha$-BCP (1–8), and $\alpha$-BCP (1–9) are identified. (From Rothman et al., 1983a.)

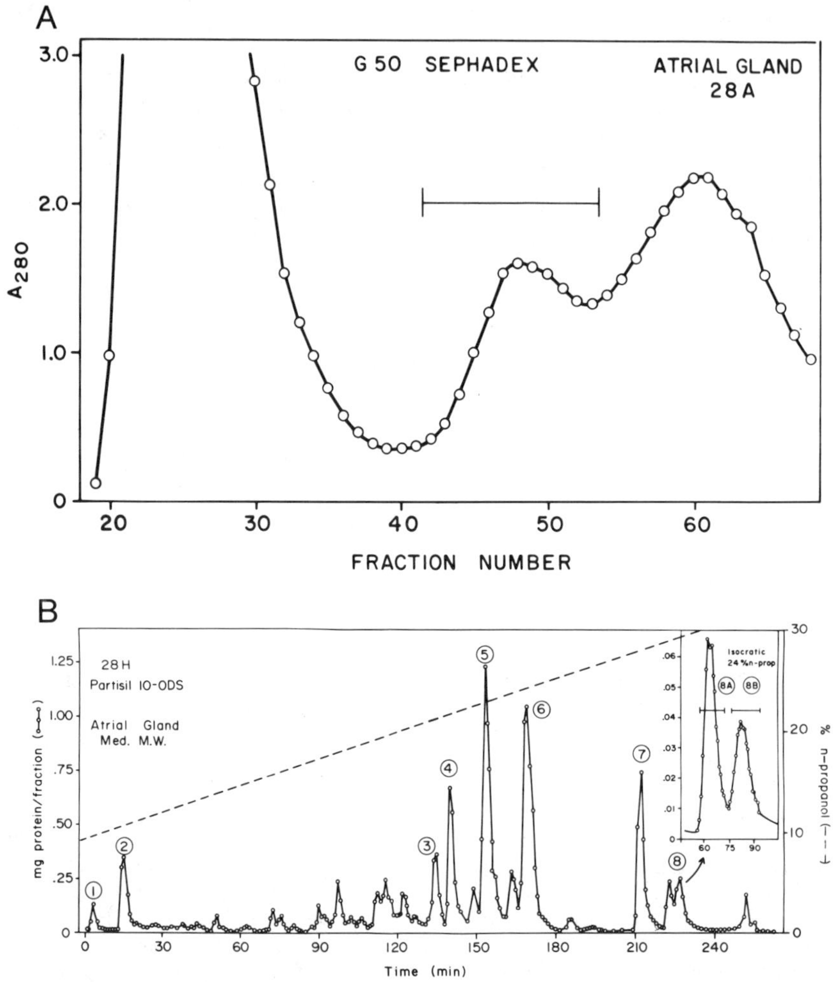


**Figure 9.** Purification of atrial gland peptides. A. Fractionation of atrial gland extract by gel filtration chromatography. Ten frozen (−80°C) atrial glands were ground on ice in a glass homogenizer in a total of 25 mL of 0.5 formic acid containing 300 $\mu$g/mL phenylmethylsulfonylfluoride. The homogenate was treated as in A except that the column was equilibrated with 0.5 *M* formic acid. The medium molecular weight material (bracket, $M_r$=2–6 kDa) was pooled and lyophilized. (From Rothman et al., unpublished.) B. Purification of medium-molecular-weight atrial gland material on HPLC. Twenty-five milligrams of Lyophilized medium-molecular-weight material was dissolved in 2.0 mL of pyridine formate, pH 3.0 (0.17 *M* pyridine, 0.5 *M* formic acid), applied to a Partisil 10-ODS column (2.0×50 cm). The column was washed for 15 min and material eluted with a linear gradient of *n*-propanol (0–30% in 340 min) using the same buffer. Peak 8 was repurified by HPLC on the same resin (4.6×250 mm column) using the same buffer under isocratic conditions of 24% *n*-propanol (see inset). Peaks 5, 6, 8A, and 8B correspond to peptide B, peptide A, A-ELH/A-AP complex (unknown clone), and A-ELH/A-AP complex (clone ELH-18), respectively. (From Rothman et al., unpublished.)

separate peak 8B material into two components and because oxidation with iodoacetamide which preferentially attacks cysteine residues under the condition used converted these peaks into two novel ones, it seems likely that the two peptides are joined by a covalent bond. However, because the molecular genetic data predict that A-AP does not contain a cysteine residue, this hypothesis needs to be further tested by purifying each peptide after reduction and then determining the sequence of each. The material contained in peak 8A yielded almost identical results as peak 8B upon analysis, except that two residues in the A-ELH sequence were different (Fig. 10). This suggests that a gene coding for a second peptide A precursor is expressed in the atrial gland; one possibility is that it is contained in clone ELH-13 (Fig. 2).

A hybrid protein consisting of part of the A or B peptide sequence and part of the ELH sequence has been identified in atrial gland extracts (Schlesinger et al., 1981). This 34-amino-acid peptide is homologous to ELH in 14 of the first 19 positions. Following three divergent amino acids, 11 of the remaining 12 residues are homologous to the A and B peptides (Fig. 10). No gene encoding this peptide has been found in the gene library even after exhaustive screening (Scheller et al., 1983a). It is unlikely that this peptide could be generated from processing of the RNA transcripts derived from the cloned genes. This is because in the hybrid peptide the ELH sequence occupies the $NH_2$-terminal half and the A/B sequence occupies the COOH-terminal half. However, in the cloned precursor genes that have been characterized, the A/B sequences reside $NH_2$-terminal to the sequences encoding ELH.

### 4.5. Sequence Homologies among the Bag Cell and Atrial Peptides

The peptides fall into three classes based on sequence homology and position on their respective precursor proteins: the ELH homology, the A or B peptide homology, and the acidic peptide homology (Figs. 3 and 10). A pattern seems to emerge from the examination of the sequence of the three genes indicating that although significant homology is maintained, the regions of divergence are of profound funcitonal consequences.

The A and B peptides differ at only four amino acid positions (Fig. 10). In the ELH precursor numerous changes have occurred in the A/B peptide region including an 80-amino-acid insertion five residues into the B peptide sequence and the substitutiion of an arginine residue for a methionine that apparently allows processing of $\alpha$-BCP, which has six residues in common with the COOH-terminus of peptides A and B. Furthermore, the tetrapeptide sequence Arg-Leu-Arg-Phe present in peptides A, B, and $\alpha$-BCP is also present in $\beta$- and $\gamma$-BCP (Fig. 10). This suggests that $\beta$- and $\gamma$-BCP arose

## A- and B-like Peptides

```
                                  5                   10                  15                  20                  25                  30
γ-BCP                                                                                                                       Arg Leu Arg Phe Asp
β-BCP                                                                                                                       Arg Leu Arg Phe His
α-BCP[1-9]                                                                                                              Ala Pro Arg Leu Arg Phe Tyr Ser Leu OH
Peptide A    H Ala Val Lys Leu Ser Ser Asp Gly Asn Tyr Pro Phe Asp Leu Ser Lys Glu Asp Gly Ala Gln Pro Tyr Phe Met Thr Pro Arg Leu Arg Phe Tyr Pro Ile NH2
Peptide B    H Ala Val Lys Ser Ser Ser Tyr Glu Lys Tyr Pro Phe Asp Leu Ser Lys Glu Asp Gly Ala Gln Pro Tyr Phe Met Thr Pro Arg Leu Arg Phe Tyr Pro Ile NH2
ERH          H Ile Val Ile Val Ser Leu Phe Lys Ala Ile Thr Asp Met Leu Leu Thr Glu Gln Ile Tyr Ala Asn Tyr Phe Ser Thr Pro Arg Leu Arg Phe Tyr Pro Ile
                   Ser
```

## Acidic Peptides

```
                                          5                   10                  15                  20                  25
Native    Peak 8A         H Asp Ser Asp Val Ser Leu Phe Asn Gly Asp Leu Leu Pro Asn Gly Arg Pro
          Peak 8B         H Asp Ser Asp Val Ser Leu Phe Asn Gly Asp Leu Leu Pro Asn Gly Arg Pro
Expected  Atrial A-AP     H Asp Ser Asp Val Ser Leu Phe Asn Gly Asp Leu Leu Pro Asn Gly Arg Pro Gln Leu Lys Thr Ile Ser Asn Leu Leu Asp OH
          Bag Cell AP     H Ser Ser Gly Val Ser Leu Phe Asn Ser Asn Lys Asp Glu Glu Gln Arg Glu Leu Leu Lys Ala Ile Ser Asn Leu Leu Asp OH
```

## ELH-like Peptides

```
                                          5                   10                  15                  20                  25                  30                  35
Native    Peak 8A         H Ile Ser Ile Asn Gln Asp Leu Lys Ala Ile Thr Asp Met Leu Leu Thr Glu Gln Ile Gln Ala Arg Arg Arg Cys Leu Asp Asp Leu Arg Gln Arg Ser Ser(Lys)
          Peak 8B         H Ile Ser Ile Asn Gln Asp Leu Lys Ala Ile Thr Asp Met Leu Leu Thr Glu Gln Ile Gln Ala Arg Arg Arg Cys Leu Asp Ala Leu Arg Gln Arg Leu Leu
Expected  Atrial A-ELH    H Ile Ser Ile Asn Gln Asp Leu Lys Ala Ile Thr Asp Met Leu Leu Thr Glu Gln Ile Gln Ala Arg Arg Arg Cys Leu Asp Ala Leu Arg Gln Arg Leu Leu Asp Leu OH
          Bag Cell ELH    H Ile Ser Ile Asn Gln Asp Leu Lys Ala Ile Thr Asp Met Leu Leu Thr Glu Gln Ile Arg Glu Arg Gln Arg Tyr Leu Ala Asp Leu Arg Gln Arg Leu Leu Glu Lys OH
```

as duplications of the $\alpha$-BCP region. Thus, in the ELH precursor as many as five peptides contain sequence homologies to peptides A and B. Whether or not these peptides, other than $\alpha$- and $\beta$-BCP, are active remains to be determined.

Based on nucleotide sequence data, the acidic peptide encoded on the peptide A precursor in the atrial gland is expected to have 15 of 27 amino acids in common with the bag cell acidic peptide (Fig. 10). At present only 17 of the residues in the atrial acidic peptide have been sequenced, and they agree perfectly with the genetic data. The peptide B precursor protein does not contain a homologous sequence for the acidic peptide due to the presence of a stop codon in the nucleotide sequence homologous to ELH.

The nucleotide sequence data also predicts the synthesis of an ELH-like peptide from the peptide A precursor (Fig. 10). This ELH-like peptide is expected to share 28 of 36 residues with ELH from the bag cells. The ELH-like peptide has been purified from the atrial gland and a sequence of 34 amino acids determined; it agrees perfectly with the genetic data. A second ELH-like peptide purified from the atrial gland differs from the first by three to four residues and is thought to be the product of a second peptide A precursor gene. The peptide B precursor is expected to express the first six residues of an ELH-like peptide (Fig. 3) at its COOH-terminus. This peptide has not yet been detected in atrial gland extracts.

Comparisons among the groups of homologous peptides permit several interesting conclusions. Four of 34 amino acids in A and B peptides match appropriate positions within the amino acid sequence of ELH. ELH matches the bag cell acidic peptide at 5 of 27 positions while the hybrid peptide matches it at 7 positions. These homologies are above random expectations and suggest that these peptides may have arisen from a triplication of a

---

**Figure 10.** Amino acid sequence homologies among bag cell and atrial gland peptides. Homologous residues are enclosed within boxes. Top: Peptides homologous to peptides A and B. Middle: Peptides homologous to the acidic bag cell peptide. Bottom: Peptides homologous to ELH. Peaks 8A and 8B (Fig. 9B) were separately subjected to amino acid sequence analysis. For peak 8B the first 17 cycles of the sequenator produced two residues per cycle. One of each pair of residues could be assigned to A-ELH and the remaining one to A-AP based on the DNA sequence derived from clone ELH-18. Sequenator cycles 18–34 yielded a single residue per cycle, and these matched the remaining sequence of A-ELH as predicted by the genetic data. Similar results were obtained for peak 8A except that four residues in the A-ELH sequence differed from that of peak 8B suggesting that peptide 8A is encoded by a different gene than peptide 8B. Bag cell ELH and acidic peptide gene sequences are also displayed. Regions of homology among the sequence data are enclosed by lines. (From Rothman et al., unpublished; Scheller et al., 1983a.)

common sequence encoding one ancestral peptide. Smaller duplications are also evident resulting in repeating tetrapeptides which characterize several of the active molecules. Furthermore, at a later time in evolution the entire precursor gene may have duplicated and transposed to generate the small multigene family encoding the egg-laying peptides. These events presumably permitted the evolution of variant peptide sequences without alteration of the original gene. Transposition could allow the introduction of these genes into new positions in the chromosome, perhaps altering the control of their expression. In this manner it may have been possible to create new precursors encoding different peptides expressed in different cells or at different times during development.

## 5. BAG CELL PEPTIDES AS CANDIDATE NEUROTRANSMITTERS

The bag cells are thought to originate in a proliferative zone of the ectodermal lining in the body wall and then migrate to their final position in the rostral margin of the abdominal ganglion (L. B. McAllister, et al., 1983) resulting in an apparently homogeneous population of about 800 neurons (Frazier et al., 1967). The bag cells are electrically coupled, and it is this feature that allows them to fire synchronously (Blankenship and Haskins, 1979; Haskins and Blankenship, 1979; Kupfermann and Kandel, 1970). Their interactions with target neurons in the abdominal ganglion can be studied in the isolated abdominal ganglion (Fig. 11A and B). They communicate with ganglionic neurons by means of granule-filled processes that end blindly in the ganglionic sheath surrounding the target somata (Coggeshall, 1967).

When given a brief electrical stimulus, the bag cells become depolarized and fire a burst discharge that lasts 20–30 min (Fig. 11C). The bag cell discharge produces several types of profound effects on abdominal ganglion neurons (Mayeri et al., 1979a, 1979b). The effects are widespread in the ganglion; perhaps as many as one-half the ganglionic neurons respond to the burst discharge of the bag cells. These responses are specific to individually identified target neurons and are usually prolonged. The responses have been divided into four types (Fig. 11C and D): Burst augmentation, an effect on the bursting pacemaker activity of cell R15; prolonged excitation in which normally silent cells of the left lower quadrant (LLQ) fire tonically; prolonged inhibition in which left upper quadrant (LUQ) cells L2, L3, L4, L6 are hyperpolarized; and transient excitation in which cells L1 and R1 are depolarized, sometimes firing a short train of spikes. The first three responses last as long as several hours while the fourth lasts about 5–10 min.

Candidate transmitters for the mediation of bag-cell-induced effects have been identified by neuroassay of peptides isolated from bag cell clusters. Peptides in sea water solution containing protease inhibitors were arterially perfused into the isolated abdominal ganglion by means of a cannula and the membrane potentials of as many as four target neurons were monitored simultaneously by means of intracellular microelectrodes (Fig. 11A). The electrical activity of the bag cells was monitored by an extracellular electrode.

A summary of the results of these studies is presented in Figure 11D. ELH very likely mediates burst augmentation in cell R15 and prolonged excitation of LLQ cells (Fig. 12). $\alpha$-BCP inhibits LUQ cells and depolarized the bag cells (Fig. 13). A peptide with the amino acid composition of $\beta$-BCP mimics transient excitation of L1 and R1, although less is known about its electrophysiological activities (Fig. 13). The acidic bag cell peptide even at millimolar concentrations has no effect on ganglion neurons, and no function has been reported for this peptide. However, this peptide is released from the bag cells (Rothman and Mayeri, unpublished data; Stuart et al., 1980) and may, therefore, have an important function.

### 5.1. Egg-Laying Hormone (ELH)

Several lines of evidence strongly suggest that ELH is a neurotransmitter (Branton et al., 1978b; Mayeri and Rothman, 1982; Mayeri et al., unpublished data). Arterial perfusion of purified ELH for a 2-min period at a concentration of about $5 \times 10^{-7}$ *M* mimics both the time course and amplitude of prolonged excitation and burst augmentation normally seen following discharge of the bag cells. ELH produces these responses in low-calcium, high-magnesium solutions and on isolated target cell somata indicating that the actions are direct. No other purified fraction of bag cell material mimics these distinctive neurally evoked responses. Combined data indicate that ELH is synthesized (Arch et al., 1976a; Chiu et al., 1979), released (Rothman and Mayeri, unpublished data; Stuart et al., 1980), and localized (Chiu and Strumwasser, 1981; Hopkins et al., unpublished data) to the bag cells. It remains to be shown, however, whether the ionic mechanisms of ELH action are identical to the neurally evoked bag cell actions.

It has been proposed that ELH is a "nonsynaptic" transmitter (alternatively, a locally acting hormone) that is dispersed into the vascular and interstitial spaces of the ganglion to act on those ganglion neurons having receptors for the peptide (Branton et al., 1978a; Mayeri and Rothman, 1982). Correspondingly, many of the physiological properties of ELH resemble those of peripheral hormones rather than those of conventional synaptic transmitters. Compared to the functioning of conventional synaptic trans-

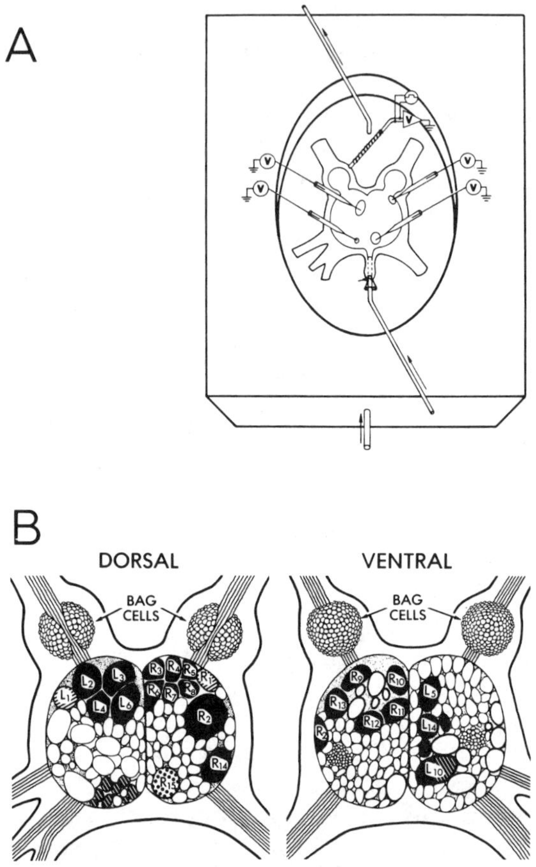


**Figure 11.** Responses of identified abdominal ganglion neurons to bag cell burst discharge. A. Schematic diagram of isolated abdominal ganglion preparation. The ganglion is pinned in a plastic dish (~ 2 mL volume) with a Sylgard floor. The ganglion is constantly perfused (bottom arrow) with sea water. Control and experimental solutions are arterially perfused (middle arrow). Perfusate exits at top arrow. As many as four cells can have their intracellular membrane potential monitored simultaneously. The bag cells are monitored by an extracellular electrode. Electrical signals are amplified (V). B. Schematic diagram of the dorsal and ventral views of the abdominal ganglion of *Aplysia californica* showing the locations of the two bag cell clusters (400 cells each) and identified target cells and cell clusters that respond to bag cell activity. The target cells are textured according to the type of response induced by bag cell activity. (Reproduced with permission from Mayeri et al., 1979b.) C. Intracellular records of a bag cell and four identified neurons showing the types of responses induced by a bag cell burst discharge. Records are taken from more than one experiment and are aligned so that the bag cell burst of each, triggered by a brief electrical stimulus, begins at the arrow. (Reproduced with permission from Mayeri and Rothman, 1982.) D. Table of response types and candidate transmitters. Each response type corresponds to a record in part C. Texture corresponds to cell map in part B. Transmitter candidate is listed at right.

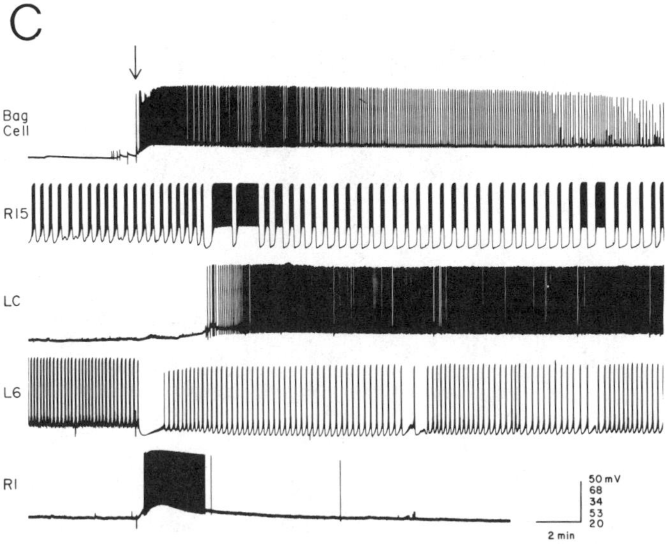


D

| | NEURONAL RESPONSE | TRANSMITTER CANDIDATE |
|---|---|---|
| | Bag Cell Excitation | |
| (dotted) | Burst Augmentation | ELH |
| (cross-hatched) | Prolonged Excitation | ELH |
| (solid) | Slow Inhibition | α-BCP |
| (hatched) | Transient Exciation | |

**Figure 11 (continued)**

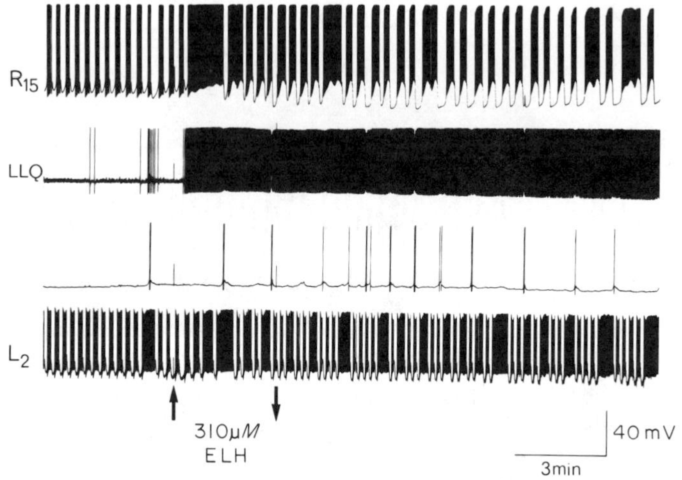


**Figure 12.** Selective actions of ELH. A ganglion was arterially perfused with ELH at a concentration of 310 *µM* (arrows), which is two orders of magnitude higher than the estimated concentration of ELH in the ganglion following a bag cell burst discharge. Intracellular activity of four cells was monitored simultaneously. ELH duplicated two types of bag-cell-induced responses: burst augmentation of cell R15 and prolonged excitation of an LLQ cell. Even at the high ELH concentration used, and unidentified cell (third trace) on the left side of the ganglion was unaffected except for periodic bursts of spike activity due to synaptic excitation by interneuron II. Cell L2 which is normally inhibited by a bag cell burst discharge evidenced mild excitation. (From Mayeri et al., unpublished.)

mitters such as acetylcholine at the vertebrate neuromuscular junction, ELH is released in greater quantities, has more prolonged actions, is effective at lower concentrations, and appears to be inactivated more slowly.

In keeping with a non-synaptic mode of communication mediated by ELH, the peptide has selective actions when perfused arterially into the abdominal ganglion. Even when perfused at high concentration, ELH affects only a subgroup of the cells affected by bag cell burst discharge (Fig. 12) and does not affect cells that are not influenced by bag cell activity (Mayeri et al., unpublished data).

## 5.2. α-Bag Cell Peptides (α-BCP)

Arterial perfusion or somatic application of ELH even at high concentrations has no effect on cells that undergo transient excitation (L1, R1) (Fig. 13) and has little or no effect on cells that undergo prolonged inhibition (LUQ cells)

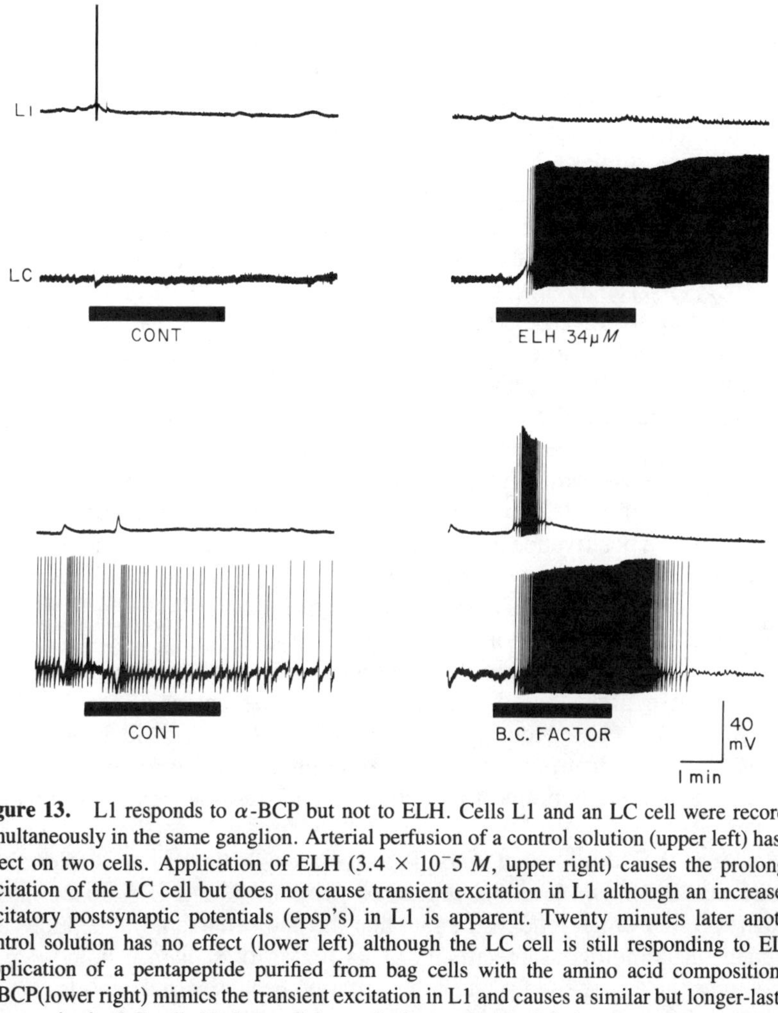


**Figure 13.** L1 responds to $\alpha$-BCP but not to ELH. Cells L1 and an LC cell were recorded simultaneously in the same ganglion. Arterial perfusion of a control solution (upper left) has no effect on two cells. Application of ELH ($3.4 \times 10^{-5}$ *M*, upper right) causes the prolonged excitation of the LC cell but does not cause transient excitation in L1 although an increase in excitatory postsynaptic potentials (epsp's) in L1 is apparent. Twenty minutes later another control solution has no effect (lower left) although the LC cell is still responding to ELH. Application of a pentapeptide purified from bag cells with the amino acid composition of $\beta$-BCP(lower right) mimics the transient excitation in L1 and causes a similar but longer-lasting response in the LC cell. No bag cell burst discharge occurred during the entire experiment. (From Mayeri et al., unpublished.)

(Fig. 12) suggesting that ELH does not directly or indirectly mediate bag cell actions on these neurons. Hence, other bag cell tranmitters are likely to mediate these responses.

$\alpha$-BCP is a second candidate transmitter from the bag cells that appears to have a dual function, to inhibit LUQ neurons and depolarize the bag cells (Rothman et al., 1983a; Sigvardt et al., 1983). Three forms of $\alpha$-BCP have been purified from bag cell extracts: α-BCP (1–9), the nine residue peptide encoded on the ELH precursor and two $NH_2$-terminal fragments, α-BCP (1–8) and α-BCP (1–7). Synthetic α-BCP (1–7) acts directly on LUQ cells and mimics the duration of the bag-cell-induced response in two of the LUQs (L2 and L4). Synthetic and native α-BCP (1–7) are identical in dose–response studies (Fig. 14B) and on HPLC under isocratic conditions (Fig. 15). However, the neurally evoked responses in the other two LUQs (L3 and L6) last longer than the effects of α-BCP (Fig. 14A) suggesting that an additional bag cell transmitter participates in the latter response. Application of α-BCP (1–7) at high concentration produces desensitization of the inhibitory response. Electrical stimulation of a bag cell discharge under these conditions produces little or no inhibitory response in the LUQ neurons suggesting that α-BCP is binding to the same receptors as the endogenous transmitter. The seven residue peptide depolarized the bag cells suggesting that α-BCP serves as an excitatory transmitter among the bag cells themselves (Fig. 16). According to this view, initial depolarization of the bag cells causes release of α-BCP which in turn causes further depolarization and release. Thus, a positive feedback mechanism may contribute to the sustained depolarization underlying the repetitive discharge of the bag cells. This mechanism may explain why the bag cell discharge far outlasts the 1–2-sec electrical stimulus necessary to initiate it.

In assays on LUQ cells the 9-residue peptide is one-thirtieth as potent as the 8-residue peptide (not shown) and one-third as potent as the 7-residue form (Fig. 14B). Our working hypothesis is that the 9-residue peptide is originally cleaved from the precursor, and either before or after release the peptide is activated by proteolysis to the 8- and 7-residue forms (Rothman et al., 1983a).

Residues 2–7 in all three α-BCP forms are identical to a 6-residue sequence in the COOH-terminal region of peptides A and B from the atrial gland (Fig. 10) which reflects the fact that the bag cell and atrial peptides are encoded on homologous precursors. This region of sequence homology is likely to be the active site for bag cell excitation because atrial peptides A and B also excite the bag cells (Heller et al., 1980).

### 5.3. Bag Cells as a Multitransmitter System

The combined electrophysiological data strongly suggest that the bag cells are a multitransmitter system that communicates with target neurons by the release of two or more peptides. Because one copy of each peptide is encoded on a precursor protein and three of the peptides are recovered from bag cell clusters in equimolar amounts, the simplest possibility is that the peptides are

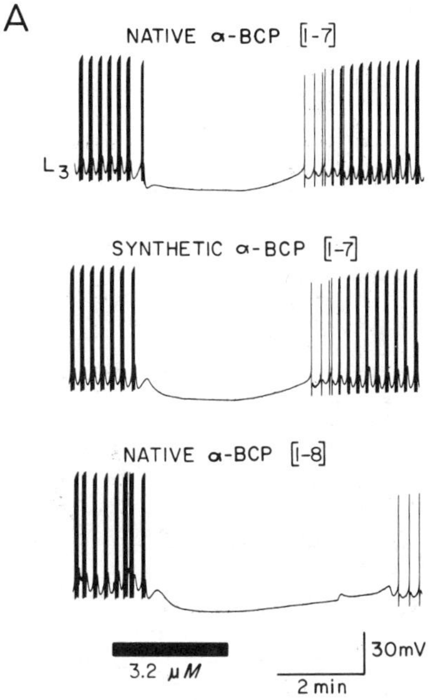


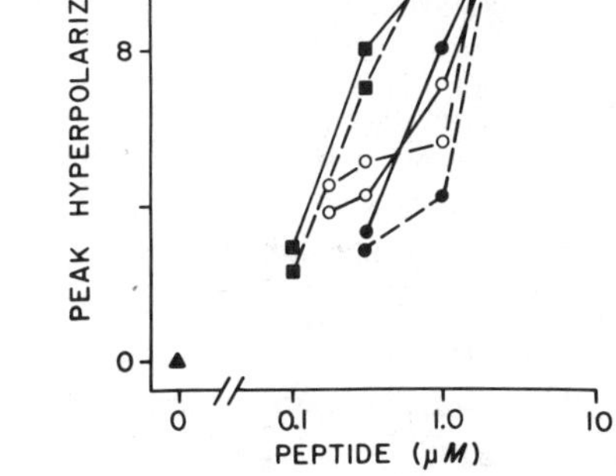


**Figure 14.** Effects of $\alpha$-BCP on left upper quadrant neurons. A. Inhibition of LUQ cell by $\alpha$-BCP. Intracellular recording from cell L3 showing hyperpolarization caused by $\alpha$-BCP. The 3.2 $\mu M$ $\alpha$-BCP solutions were perfused arterially at a rate of 3 $\mu$L/min for 2.67 min (bar). Synthetic and native $\alpha$-BCP (1–7) have identical effects, while native $\alpha$-BCP (1–8) caused a larger, long-lasting hyperpolarization. (From Rothman et al., 1983a.) B. Dose-response relationship for $\alpha$-BCP on cells L3 and L6 from a single abdominal ganglion. Response is measured as peak hyperpolarization after arterial perfusion of peptide (same preparation as in part A). Synthetic and native $\alpha$-BCP (1–7) have identical dose-response curves. Native $\alpha$-BCP (1–8) is about three times as potent as $\alpha$-BCP (1–7). (From Rothman et al., 1983a.)

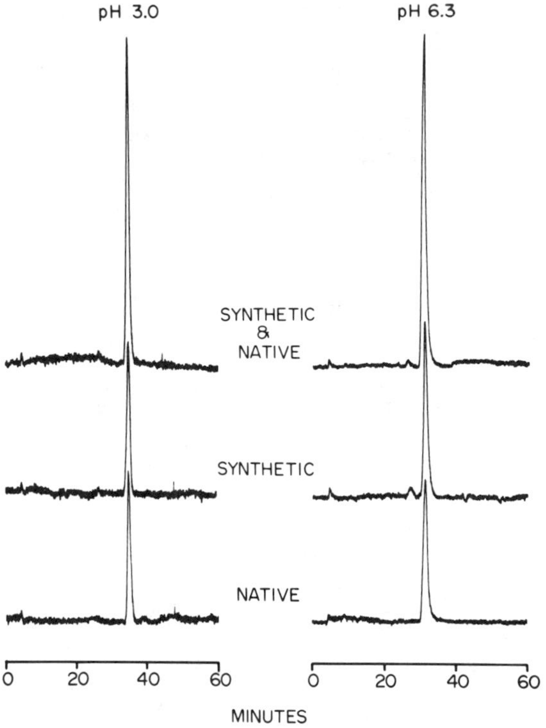


**Figure 15.** Comparison of synthetic and native α-BCP (1–7) by analytical reverse-phase HPLC (Supelco $C_{18}$ DB column, 4.6 × 250 mm, 0.6 mL/min flow rate). Vertical axis represents fluorescence detected as entire column eluate was derivitized with fluorescamine and passed through Gilson Spectraglo flow-through fluorimeter. Left: Peptides run isocratically in pyridine formate, pH 3.0, containing 9% (v/v) *n*-propanol. Synthetic peptide and native peptide (0.9 μg each) were applied together (top trace) or separately (middle and bottom traces). The volume of each sample was 46 μL. Right: Peptides run in pyridine acetate, pH 6.3, containing 7% (v/v) *n*-propanol. Synthetic (1.4 μg) and native (1.1 μg) peptide were applied together (top trace) or separately (middle and bottom traces). The volume of each sample was 64 μL. (From Rothman et al., unpublished.)

released in equimolar amounts. At present it is possible by perfusing a combination of bag cell peptides to mimic several of the bag-cell-induced responses. Arterial perfusion of the ganglion with a sea water solution containing equimolar amounts of ELH, α-BCP (1–7), and the acidic bag cell peptide mimics prolonged excitation of an LLQ cell (LC), the short lasting component of slow inhibition in an LUQ cell (L6), burst augmentation (R15), and depolarization of the bag cells (Sigvardt et al., 1983). However, transient excitation which occurs in cell L1 and R1 following a bag cell burst discharge does not occur. Ultimately, it may be possible to mimic the entire set of bag

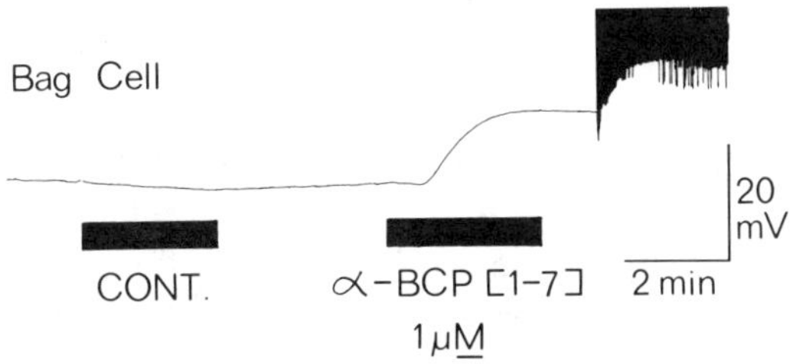


**Figure 16.** Excitation of a bag cell by $\alpha$-BCP (1–7). Intracellular recording of a bag cell soma. A control solution (8 $\mu$L) (sea water plus protease inhibitors) followed by 8 $\mu$L of a solution containing 1 $\mu M$ synthetic $\alpha$-BCP (1–7) were each perfused arterially for 2.67 min. $\alpha$-BCP caused the bag cell to depolarize and later initiate a burst discharge. An extracellular monitor of the contralateral bag cell cluster showed that the burst discharge began in both clusters at approximately the same time. (From Rothman et al., 1983a.)

cell actions by applying a combination of bag cell peptides to the abdominal ganglion.

## 6. ROLE OF BAG CELL PEPTIDES IN EGG-LAYING BEHAVIOR

Presently available data suggests that the bag cells mediate egg-laying behavior through the release of peptides (Mayeri, 1979). ELH is released by the bag cells (Stuart et al., 1980) and is known to modify the activity of neurons located in the abdominal ganglion (Branton et al., 1978b; Mayeri and Rothman, 1982) and the head ganglia (Stuart and Strumwasser, 1980). Target neurons in the abdominal ganglion regulate respiration, cardiac output, and other visceromotor functions while the head ganglia are thought to control locomotion, feeding, and head movements (Kandel, 1976). ELH is also known to release eggs directly from fragments of the ovotestis (Rothman et al., 1983b). These activities suggest that ELH functions both as a neurotransmitter within the abdominal ganglion and as a neurohormone outside of the ganglion. The other peptides, $\alpha$-BCP and $\beta$-BCP, appear to act as neurotransmitters only since their activity does not survive perfusion through the abdominal ganglion without protease inhibitors present in the medium (Mayeri et al., unpublished data). The role of the acidic peptide in egg laying, if any, is unknown; however, it is possible that this highly negatively charged protein may serve as a carrier for one or more of the positively charged (basic) peptides in the precursor. Charge neutralization is critical for dense packing of peptides within a vesicle and maintaining an appropriately neutral intragranule pH.

## 7. CONCLUDING REMARKS

The bag cell system as described seems a reasonable way to regulate egg-laying behavior. However, it is important to consider what advantages are conferred on the system by the use of multiple transmitters that are encoded on a common precursor protein. Although the following points are not a complete answer to this question, they provide a beginning for discussion on this issue.

First, ELH and $\alpha$-BCP have different durations of action and this is likely to be important in coordinating long- and short-lasting events in the egg-laying program. The actions of ELH last for hours while those of $\alpha$-BCP last for about 5–10 min. One possible source of these differences may be the structure of the peptides. ELH contains 36 amino acids and may be amidated at its COOH-terminus and is, therefore, likely to fold into a stable tertiary configuration and be protected from COOH-terminal proteolysis. In contrast $\alpha$-BCP is only seven to nine amino acids in length and is not amidated at the COOH-terminus. It is, therefore, likely to have a shorter half-life in extracellular and vascular spaces which would contribute to its shorter duration of action.

Second, it may be especially adaptive that the multiple transmitters occur on a common precursor. The use of a common precursor ensures that all transmitters are synthesized and released together, yet might still allow a certain degree of plasticity in the system. During development or as a result of a physiological event, an alteration of the processing of the precursor in the post-translational modification of the resultant peptides or in the expression of the member genes might produce changes in the activities, or ratios of peptides released and corresponding changes in electrical signaling.

Third, the use of multiple transmitters encoded on a single precursor may have provided through the process of evolution a means for increasing the complexity of the behavior pattern. Perhaps the present behavioral program evolved from a more simple one that was regulated by a single peptide transmitter. The present complexity of behavior might, therefore, be caused by new peptide sequences being added to the precursor as a result of duplication, mutation, and translocation and new neuronal activities being brought into play by the effects of the new transmitters.

## REFERENCES

Angerer, R. C., E. H. Davidson, and R. J. Britten, *Cell* **6,** 29 (1975).
Arch, S., *J. Gen. Physiol.* **59,** 47 (1972a).
Arch, S., *J. Gen. Physiol.* **60,** 102 (1972b).

Arch, S. and T. Smock, *Behav. Biol.* **19,** 45 (1977).

Arch, S., P. Earley, and T. Smock, *J. Gen. Physiol.* **68,** 197 (1976a).

Arch, S., T. Smock, and P. Earley, *J. Gen. Physiol.* **68,** 211 (1976b).

Arch, S., T. Smock, R. Gurvis, and C. McCarthy, *J. Comp. Physiol.* **128,** 67 (1978).

Arch, S., J. Lupatkin, T. Smock, and M. Beard, *J. Comp. Physiol.* **141,** 131 (1980).

Berry, R. W., *Biochemistry* **20,** 6200 (1981).

Berry, R. W., M.J. Trump, and J. T. Baylen, *Biochemistry* **20,** 6206 (1981).

Blankenship, J. E. and J. T. Haskins, *J. Neurophysiol.* **42,** 347 (1979).

Blankenship, J. E., M. K. Rock, L. C. Robbins, C. A. Livingston, and H. K. Lehman, *Fed. Proc.* **42,** 96 (1983).

Blobel, G. and B. Dobberstein, *J. Cell Biol.* **67,** 835 (1975).

Bradbury, A. F., M. D. A. Finnie, and D. G. Smyth, *Nature* **298,** 686 (1982).

Branton, W. D., E. Mayeri, P. Brownell, and S. B. Simon, *Nature* **274,** 70 (1978a).

Branton, W. D., S. Arch, T. Smock, and E. Mayeri, *Proc. Natl. Acad. Sci. USA* **75,** 5732 (1978b).

Brownell, P. H. and M. E. Schaefer, *Neurosci. Abst.* **8,** 736 (1982).

Chiu, A. Y., Ph.D. thesis, California Institute of Technology, Pasadena (1981).

Chiu, A. Y. and F. Strumwasser, *J. Neurosci.* **8,** 812 (1981).

Chiu, A. Y., M. W. Hunkapiller, E. Heller, D. K. Stuart, L. E. Hood, and F. Strumwasser, *Proc. Natl. Acad. Sci. USA* **76,** 6656 (1979).

Cobbs, J. S. and H. M. Pinsker, *J. Comp. Physiol.* **147,** 523 (1982a).

Cobbs, J. S. and H. M. Pinsker, *J. Comp. Physiol.* **147,** 537 (1982b).

Coggeshall, R. E., *J. Neurophysiol.* **30,** 1263 (1967).

Coggeshall, R. E., B. A. Yaksta, and F. J. Schwartz, *Chromosoma* **32,** 205 (1971).

Dudek, F. E., J. S. Cobbs, and H. M. Pinsker, *J. Neurophysiol.* **42,** 804 (1979).

Frazier, W. T., E. R. Kandel, I. Kupfermann, R. Waziri, and R. E. Coggeshall, *J. Neurophysiol.* **30,** 1288 (1967).

Haskins, J. T. and J. E. Blankenship, *J. Neurophysiol.* **42,** 356 (1979).

Heller, E., L. K. Kaczmarek, M. W. Hunkapiller, L. E. Hood, and F. Strumwasser, *Proc. Natl. Acad. Sci. USA* **77,** 2328 (1980).

Herbert, E., N. Birnberg, J. C. Lissitsky, O. Civelli, and M. Uhler, *Neurosci. Comm.* **1,** 16 (1981).

Jennings, K. R., L. K. Kaczmarek, R. M. Hewick, W. J. Dreyer, and F. Strumwasser, *J. Neurosci.* **2,** 158 (1982).

Kandel, E. R., in E. R. Kandel, Ed., *The Cellular Basis of Behavior,* W. H. Freeman, San Francisco, CA (1976).

Kupfermann, I., *Nature* **216,** 814 (1967).

Kupfermann, I., *J. Neurophysiol.* **33,** 877 (1970).

Kupfermann, I. and E. R. Kandel, *J. Neurophysiol.* **33,** 865 (1970).

Land, H., G. Schultz, H. Schmak, and D. Richter, *Nature* **295,** 299 (1982).

Lasek, R. J. and W. J. Dower, *Science* **172,** 278 (1971).

Loh, Y. P., Y. Sarne, and H. Gainer, *J. Comp. Physiol.* **100,** 283 (1975).

Mackey, S. and T. J. Carew, *J. Neurosci.* **3,** 1469 (1983).

Maniatis, T., R. C. Hardison, E. Lacy, J. Lauer, C. O'Connell, D. Quon, G. K. Sim, and A. Efstratiadis, *Cell* **15,** 687 (1978).

Maxam, A. and W. Gilbert, *Methods Enzymol.* 65, 499 (1980).

Mayeri, E., *Fed. Proc.* **38,** 2103 (1979).

Mayeri, E. and B. S. Rothman, in D. S. Farner and K. Lederis, Eds., *Neurosecretion—Molecules, Cells, and Systems,* Plenum, New York, p. 307 (1982).

Mayeri, E., P. H. Brownell, W. D. Branton, and S. B. Simon, *J. Neurophysiol.* **42,** 1165 (1979a).

Mayeri, E., P. H. Brownell, and W. D. Branton, *J. Neurophysiol.* **42,** 1185 (1979b).

McAllister, L.B., Scheller, R.H., Kandel, E.R., and Axel, R., *Science* **222,** 800 (1983).

Pinsker, H. M. and F. E. Dudek, *Science* **197,** 490 (1977).

Rothman, B. S., L. Padgett, and E. Mayeri, *Neurosci. Abstr.* **7,** 636 (1981).

Rothman, B. S., R. O. Brown, E. Mayeri, and J. E. Shively, *Neurosci. Abstr.* **8,** 14 (1982).

Rothman, B. S., G. Weir, and F. E. Dudek, *Gen. Comp. Endocrinol.* **52,** 134 (1983b).

Rothman, B. S., E. Mayeri, R. O. Brown, P. M. Yuan, and J. E. Shively, *Proc. Natl. Acad. Sci. USA* **80,** 5753 (1983a).

Rothman, B. S., Shively, J., Hawkes, D., Brown, R. O., and Mayeri, E., *Trans. Amer. Soc. Neurochem.* **15,** 119 (1984).

Scheller, R. H., J. F. Jackson, L. B. McAllister, J. H. Schwartz, E. R. Kandel, and R. Axel, *Cell* **28,** 707 (1982).

Scheller, R. H., J. F. Jackson, L. B. McAllister, B. S. Rothman, E. Mayeri, and R. Axel, *Cell* **32,** 7 (1983a).

Scheller, R. H., B. S. Rothman, and E. Mayeri, *Trends in Neurosci.* **6,** 340 (1983b).

Schlesinger, D. H., S. B. Babirak, and J. E. Blankenship, in D. H. Schlesinger, Ed., *Neurohypophyseal Peptide Hormones and Other Biologically Active Peptides,* Elsevier/North Holland, New York, p. 137 (1981).

Sigvardt, K., Rothman, B. S., and Mayeri, E. *Neurosci. Abstr.* **9,** 311 (1983).

Stein, S., in E. Gross and J. Meinhofer, Eds., *The Peptides: Analysis, Synthesis, Biology* Vol 4, Pierce Chem. Co., Rockford, IL, p. 185 (1981).

Steiner, D. F., P. S. Quinn, S. J. Chan, J. Makh, and H. S. Tager, *Ann. NY Acad. Sci.* **343,** 399 (1980).

Strumwasser, F., J. W. Jacklet, and R. B. Alvarez, *Comp. Biochem. Physiol.* **29,** 197 (1969).

Stuart, D. K. and F. Strumwasser, *J. Neurophysiol.* **43,** 499 (1980).

Stuart, D. K., A. Y. Chiu, and F. Strumwasser, *J. Neurophysiol.* **43,** 488 (1980).

Toevs, L. A. and R. W. Brackenbury, *Comp. Biochem. Physiol.* **29,** 207 (1969).

# 10

# *Vasoactive Intestinal Polypeptide: From Gene to Peptide*

***Illana Gozes***



In the mammalian brain a major regulatory polypeptide is vasoactive intestinal polypeptide (VIP). Most of the past work on this peptide dealt with its distribution and abundance, the polypeptide structure of the porcine peptide, and its possible physiological roles. We are now interested in the mechanisms regulating its expression and its gene structure. Using the state-of-art methodology of molecular genetics and monoclonal antibody production, we seek insight into the regulation of the VIP gene in normal and pathophysiological situations. To identify the VIP precursor(s) we are using monoclonal and polyclonal anti-VIP antibodies recognizing discrete sites on the VIP molecule. To identify and characterize the VIP mRNA, we use synthetic oligonucleotides with relatively unambiguous nucleotide sequences predicted from the VIP amino acid sequence and poly (A+) mRNA from brain to generate possible VIP-specific cDNA employing the enzyme reverse transcriptase. Our results suggest several high-molecular-weight precursors to VIP in the brain that may differ in part from those in a VIP-producing tumor. These studies are providing a body of information concerning the genetic regulation of VIP that has not previously been available for most brain regulatory peptides.

## 1. VASOACTIVE INTESTINAL POLYPEPTIDE

The 28-amino-acid polypeptide, VIP (Said and Mutt, 1970) appears to be involved in a wide variety of biological activities including systemic vasodilation, increased cardiac output, and hyperglycemia (Said and Mutt, 1970), smooth muscle relaxation (Piper et al., 1970), some differential effects on secretory processes in the gastrointestinal tract (Barbezat and Grossman, 1971; Makhlouf et al., 1978; Said, 1982), and glycogenolysis in liver slices (Kerins and Said, 1973) and in cerebral cortex slices (Magistretti et al., 1981). VIP may also mediate the watery diarrhea syndrome that occurs in patients with pancreatic, bronchogenic, and other tumors (Bloom and Polak, 1982; Said and Falona, 1975).

The multiple actions of VIP are correlated with its widespread distribution throughout the body. Indeed, VIP can be found in both endocrine and nerve cells, in parasympathetic ganglia, and in the brain. In the brain the peptide has a discrete regional distribution with the highest levels in the cerebral cortex, hypothalamus, amygdala, hippocampus, and corpus striatum (Said, 1980). VIP-positive fibers were found to be present in all of these areas and VIP-positive cell bodies were demonstrated in the cerebral cortex and amygdaloid cortex. VIP nerve terminals also appear in the walls of cerebral pial blood vessels, the intestinal wall, especially at the sphincter zones, the pancreas, and in the parasympathetic ganglia. VIP-containing nerves are also found in the gall bladder, urogential organs, tracheobronchial tree, and in peripheral nerves such as the sciatic and vagus nerves. VIP is also found in copious amounts in the placenta, adrenal medulla, mast cells, and platelets. Organs with little or no VIP include liver and skeletal muscle (for reviews see Potter et al., 1981; Said, 1980).

In addition to its presence in neurons, other evidence supporting a possible neurotransmitter function for VIP include its localization in synaptosomal preparations (Gicchetti et al., 1977) and potassium-induced release from these preparations (Emson et al., 1978; Gicchetti et al., 1977). Furthermore, it has been localized to synaptic vesicles in the rat hypothalamus (Emson et al., 1978), and VIP-like immunoreactivity was found in large dense core vesicles in nerve terminals innervating exocrine glands (Johansson and Lundberg, 1981). Specific receptor sites for radiolabeled VIP have been demonstrated in rat brain (Taylor and Pert, 1979) and guinea pig brain (Robberecht et al., 1978). Moreover, a VIP-stimulated adenylate cyclase has been identified in various areas of the central nervous system (Deschodt-Lanckman et al., 1977; Quik et al., 1978).

VIP may not exist as the exclusive "neurotransmitter" in the neurons where it has been found. Coexistence of VIP and acetylcholine esterase has been observed in lumbrosacral sympathetic ganglia which contain a well-defined cholinergic neuron population and are assumed to be involved in the regulation of sweat secretion (for reviews see Hokfelt et al., 1980; Potter et al., 1981). Recent evidence suggests that VIP in cholinergic neurons may be a general characteristic of secretomotor neurons. It is now postulated that secretion and vasodilation in exocrine glands are induced by acetylcholine and VIP, respectively, released from the same nerve terminal. In this particular system the two putative transmitters would act on receptors located on two separate cell types.

Multiple functions of VIP neurons may also occur in the cerebral cortex, although there is as yet no direct evidence. Morphologically, it has been demonstrated that an individual VIP neuron may span the entire vertical thickness of the cerebral cortex receiving synaptic input in all cortical laminae and impinging upon other cortical elements in a narrowly defined, radial column (Morrison et al., 1981). Given the effect of VIP on cortical carbohydrate metabolism (Magistretti et al., 1981) the VIP neuron may be an important component of the cortical column, modulating the availability of energy substrates as neuronal activity increases locally.

The well-characterized VIP molecule (from porcine gastrointestine) shows structural similarities with other gastrointestinal hormones such as secretin, glucagon, and to some extent the gastric inhibitory polypeptide (GIP) (Said, 1980). However, major differences exist between members of the secretin–VIP family which may argue against simple evolution from common ancestors (Bodanszky, 1975). In addition, VIP exhibits structural similarities with the two recently discovered peptides PHI (peptide having N-terminal histidine and C-terminal isoleucine amide, Tatemoto and Mutt, 1980) and a growth-hormone-releasing hormone (Guillemin et al., 1982). The entire sequence of VIP is not required for significant hormonal activity. Thus, the biological activity for vasodilation is present in the C-terminal 11-amino-acid peptide, VIP [18–28], and this activity increases with increasing N-terminal chain length. The synthetic N-terminal peptides VIP [1–16] and VIP [1–10] also show distinct although weak activity. The presence of VIP-like activity in non-overlapping N- and C-terminal parts of the sequence implies the coexistence of two "command sequences" (Bodanszky, 1977). VIP in brain, in peripheral nerves, and in other extra-intestinal organs is probably present in one form which is similar in molecular size to the 28-amino-acid porcine

VIP (Said, 1980). Molecular forms which varied in charge, however, have been found in endocrine cells, for example, intestinal mucosa (Dimaline and Dockray, 1979).

The synthesis of VIP in some cell types may be hormonally regulated. For example, in ganglio-neuroblastoma cells (Yanaihara et al., 1981) cAMP induces a substantial increase in VIP immunoreactivity. In dog, cat, and human, bombesin induces a decrease in VIP release (Melchiori, 1978). It is also interesting to note that the levels of VIP in the brain may be regulated during growth and development. In the brain of the newborn rat, the average level of VIP was found to be 17 pmol/g wet weight. This level remained constant during the first 2 weeks after birth and then increased progressively to 40 pmol/g wet weight at 20 days, a value comparable to that found in adult animals (Robberecht et al., 1979). It is not yet known at what level of gene expression the changes in cellular VIP content are regulated.

Although some oligopeptides may be synthesized on non-ribosomal polyenzyme templates (Lipmann, 1973), it is likely that VIP, like all other characterized peptide hormones, is synthesized on polyribosomes as a large protein precursor. A portion of the sequence of the precursor protein is likely to be required for the interaction involved in the compartmentalization of proteins and their activation for secretion (for review see Herbert et al., 1981). A precursor protein may give rise to several peptides each having differential biological activities such as the pro-opiomelanocortin, the precursor for corticotropin (ACTH), $\beta$-endorphin, and $\alpha$-, $\beta$- and $\gamma$-melanotropin (MSH) (e.g., Chretien et al., 1979: Nakanishi et al., 1979, 1980; Rubinstein et al., 1978; and for review see Herbert el al., 1981). The nucleotide sequence of a cloned cDNA insert coding for pro-opiomelanocortin indicates that the precursor protein consists of repetitive units. Pairs of lysine and arginine residues separate the component peptides of the precursor molecule and presumably provide signals for polypeptide processing (Nakanishi et al., 1979). Similar active peptide sequences can also be contained in two different precursors; such is the case for Met-enkephalin which is structurally represented in pro-opiomelanocortin in the pituitary and also in different unrelated peptide precursors in the adrenal medulla which may contain multiple enkephalin sequences (Rossier, 1982). However, there is as yet only very preliminary information on the possible precursors for VIP (Said, 1982) and no information concerning VIP-mRNA, its gene structure, and its regulation of expression.

Combining methods of molecular genetics and monoclonal antibody production we have initiated studies toward an understanding of the regulation of the VIP biosynthesis. The significance of research on the regulation of the VIP biosynthesis lies in the presumed functional importance of VIP in processes such as physiological regulation of energy metabolism. Better understanding of the VIP gene structure in relation to the regulation of its expression may thus provide deeper insights into its normal function.

## 2. RESEARCH STRATEGIES

Identification of a protein precursor containing a peptide sequence can be achieved using immunological probes, namely specific antibodies raised against the final peptide product but which may still recognize the precursor. Alternatively, the problem may be attacked at the mRNA level: specific nucleic acid probes may be developed and used in hybridization and rapid sequence analysis to identify and characterize the mRNA for the peptide precursor.

The immunological approach has been used on cell extracts fractionated by molecular sieving to identify higher-molecular-weight materials that cross-react with the anti-peptide antibodies (e.g., pro-opiomelanocortin, Mains et al., 1977). Furthermore, precursor molecules have also been identified by a strategy in which *in vitro* translation of mRNA (usually of high abundance) is followed by immunoprecipitation using specific anti-peptide antisera. A few examples where precursor molecules were identified employing this strategy are luteinizing hormone (Godine et al., 1981), thyroid-stimulating hormone (Chin et al., 1981), glucagon (Lund et al., 1981), follicle-stimulating hormone (Godine et al., 1980), and propressophysin (Schmale and Richter, 1981). In the last example mentioned a common precursor is postulated to give rise to both an active peptide and its carrier protein (Russel et al., 1980).

The hybridoma technique introduced by Milstein and co-workers can provide unique antibody reagents for this latter approach. Immune spleen cells are fused with a plasmacytoma cell line to produce antibody-secreting hybrids which can thereafter be cloned to homogeneity (Galfre et al., 1977; Kohler and Milstein, 1975). In this way specific monoclonal antibodies are produced by each positive clonal cell line for identification of peptide precursors. Monoclonal antibodies against substance-P have been produced (Cuello et al., 1979), and we have produced monoclonal antibodies against VIP (Gozes et al., 1981). We have further introduced the technique in which spleen cells are sensitized with the peptide antigen *in vitro* as this strategy may give higher titer and higher affinity antibodies (Gozes et al., in press-a, Luben et al., 1982). Additional monoclonal antibodies directed to different regions of a peptide molecule may provide critical tools in following diffential processing within branches of the same neuron or among different neurons using both immunohistochemistry and radioimmunoassays.

A second approach for peptide precursor characterization is molecular cloning and nucleotide sequencing of the cDNA corresponding to the m RNA for this precursor. Using this approach, recent advances in restriction endonucleases (Nathans and Smith, 1975) and nucleic acid sequencing (Maxam and Gilbert, 1977; Sanger et al., 1977) make molecular cloning both feasible and efficient. One could clone a cDNA made by reverse transcription from a purified mRNA template. Even if the mRNA is not abundant, it can be enriched from a general population of mRNA's and the muliple resultant clones screened for those of interest. In the case of pro-opiomelanocortin, which is an abundant peptide precursor in the neurointermediate lobe of the pituitary gland, the respective mRNA was purified to homogeneity by physicochemical methods (Kita et al., 1979). The purified mRNA was reverse transcribed into a single-stranded cDNA; the second DNA strand was then synthesized with DNA polymerase and single-stranded DNA was removed by S1 nuclease (from *Aspergillus oryzae*). Finally, homopolymeric poly (dC) were added as "tails" by termninal deoxynucleotide transferase (Villa-Komaroff et al., 1978). For cloning, pBR322 was cleaved with *Pst 1*, (a restriction enzyme from *Providencia stuartii* 164 which cleaves CTGCA ↓ G), terminated by poly (dG) "tails" as above, and annealed to the double-stranded cDNA (Bolivar et al., 1977). The chimeric plasmid was then used to transform *E. coli*. Bacteria containing the cDNA vector were selected on the basis of tetracycline resistance and ampicillin sensitivity and further identified using *in situ* colony hybridization with [$^{32}$P]-labeled purified mRNA as a probe. Final characterization was obtained by DNA sequencing (Nakanishi et al., 1978, 1979).

A similar strategy was followed to clone the precursor for calcitonin (Jacobs et al., 1981). This peptide is primarily synthesized by the mammalian thyroid; tumors of the thyroid were used as an enriched source for the calcitonin mRNA. The clones containing calcitonin cDNA were identified by hybridization to thyroid mRNA and selective translation (Ricciardi et al., 1979) followed by immunoprecipitation of the calcitonin precursor (Jacobs et al., 1981). This study indicated that calcitonin also arises from a large precursor protein that may contain other active peptide components.

In the case of rare polypeptides it is possible to devise an oligodeoxynucleotide probe of defined sequence if some of the amino acid sequence is known. This is the appraoch we have selected for the VIP gene. The use of oligodeoxynucleotide probes for the detection of low-abundance mRNAs has been documented (e.g., Houghton et al., 1980; Hudson et al., 1981a; Noyes et al., 1979; Sood et al., 1981). A span of about four unique or low-degeneracy codons (Jukes, 1980) are chosen and the oligodeoxynucleotide is chemically synthesized using, for example, the diester approach (Agarwal et al., 1972). A general method for detection and characterization of an mRNA using such a probe is outlined by Agarwal et al.,

(1981). The probe is used in reverse transcription of mRNA as described (e.g., Sood et al., 1981) and then cloned as above. The most direct approach for elucidating the mRNA structure is to sequence its cDNA (Maxam and Gilbert, 1977) and then compare it to the gene struture.

One of the major unexpected general developments of molecular genetic research and DNA sequencing has been that RNA is spliced (see review, Sharp, 1981); thus, non-protein coding sequences (introns) that intervene in the DNA coding sequence (exons) are not found in the final mRNA transacript. RNA splicing might have been used to generate new genes rapidly during evolution of eukaryotic cells. Moreover, intervening sequences in the genome may act in the control of gene expression (e.g., Wever et al., 1981).

Another unexpected result has been that somatic cell mutation and gene rearrangement may occur in the case of eukaryotic cell differentiation (Crews et al., 1981). Given the several probably active products of peptide precursors such as the pro-opiomelanocortin precursor, one might have anticipated that each active zone is coded separately in the gene and brought together as the pro-mRNA is transcribed and processed. The isolation and characterization of the pro-opiomelanocortin gene (Nakanishi et al., 1980) was achieved by restriction mapping of calf thymus DNA and identification of the gene fragment by hybridization (Southern, 1975) to cDNA labeled with [$^{32}$P]. Studies of the genomic organization have revealed an intron between the signal peptide and the remainder of the precursor molecule and a second intron in the 5′ non-coding region of the mRNA (Chang et al., 1980: Nakanishi et al., 1980). The different active products of pro-opiomelanocortin, however, are coded.by a continuous nucleotide domain of the gene. This suggests that the evolutionary changes which have occurred have conserved the hormonally active components of the entire precursor molecule as an intact genetic unit. In the case of VIP it would be interesting to find out whether such evolutionary constraints have operated and whether the VIP gene(s) are similarly expressed in the wide variety of cell types where VIP immunoreactivity has been described. Such studies could be extended to monitor possible developmental reorganization and perhaps even pathological mechanism.

The cDNA clones for specific peptide precursors have been useful as probes in several different strategies to analyze the expression of the gene coding for the precursor. These approaches include isolation and characterization of the appropriate genomic DNA, analysis of gene expression by quantification of mRNA, and tissue localization of the mRNA by *in situ* hybridization. We have dealt with gene isolation and characterization above.

As for mRNA quantification such studies were performed using [$^{32}$P]-labeled calcitonin cDNA and hybridization onto RNA, separated on agarose gel, and transferred to diazotized paper (Rosenfeld et al., 1981). Thereafter, the resulting autoradiographs were densitometrically scanned and the RNA characterized and quantified. Both molecular alterations and quantitative changes in the calcitonin mRNA were found when a high-calcitonin-producing tumor was compared to a lower-producing tumor. The gene and the mRNA may also be localized to specific cell populations by hybridization histochemistry as was done for growth hormone and endorphin (Hudson et al., 1981b) and other polypeptides (e.g., Harrison et al., 1973; Pochet et al., 1981; and for review see Watson and Akil, 1981). One advantage of such localization over the conventional immunohistochemical methods for peptide localizations is that the detection of a particular mRNA species in a cell strongly supports its synthesis there. Thus, by employing both immunocytochemistry and mRNA localization we may obtain a more secure understanding of brain functional organization especially with regard to peptides found to affect many cell types. It is of interest to add that genes may also be localized on specific chromosomes by direct hybridization *in situ* in metaphase cells (Gerhard et al., 1981). This question is of particular interest in conjunction with the coexistence of the genes for growth hormone chorionic somato-mamotropin and a growth-hormone-like gene on chromosome 17 in humans; these studies employed somatic cell hybrid methods (Owerbach et al., 1980). Localization to a specific chromosome can also be achieved by chromosomal franctionation (Hughes et al., 1979).

Still dealing with the VIP gene, its known sequence may allow the use of a new strategy for discovery of novel neuropeptides cosynthesized with VIP on the same high-molecular-weight precursor. One can synthesize predicted peptides chemically (Ling et al., 1979). Immune sera raised against these presumptive peptides (Lerner et al., 1981) have led to assays capable of establishing the independent existence of the neuropeptides *in vitro* (Sutcliffe et al., 1980).

Finally, cloned genes have been inserted into mammalian cells in order to study the regulation of their expression. Insertions are achieved by binding to a virus, by intracellular injection, by permeation of the membrane by $Ca^{2+}$ ions (Anderson and Diacumakes, 1981), or by liposomes (e.g., Schaefer-Ridder et al., 1982). Specific DNA may be inserted together with a gene for thymidine kinase (tk) into cells which lack the enzyme, such as tk-L cells (Wigler et al., 1978). Cells containing a foreign DNA insert may contain the tk gene and can be selected for in culture media which would not allow growth of the tk-minus cells (Littlefield, 1964). The surviving cells could be cloned and then assayed for the specific insert expression and regulation (Kurtz, 1981). Structural regulation of gene expression can be studied by directed *in vitro* mutagenesis followed by a change in the gene activity. This would allow more precise mapping of DNA stretches responsible for specific gene activities. However, results must be interpreted with caution since transfection of cells with foreign DNA and induction of expression may raise questions concerning the initial inability of the cells to express the normal endogenous gene.

I have tried to give some examples of the vast array of strategies that are available. In the sections below the particular techniques used and the results obtained are documented, and the application of molecular genetics and hybridoma technology to neuroscience in general and to VIP in particular are elaborated.

## 3. VIP BIOSYNTHESIS

### 3.1. Immunological Approach

VIP gene expression can be localized, quantitated, and characterized using polyclonal or monoclonal antibodies. Development of specific antibodies against VIP may lead to the discovery of possible VIP precursors.

A general strategy for the use of anti-VIP antibodies is in radioimmunoassay. At the moment we have rabbit anti-VIP antibodies (a gift from Dr. Robert Benoit). To obtain maximal binding to VIP these antibodies seem to require the entire molecule.

By way of introduction, one of the major technological breakthroughs of the last decade is the development of the methods for producing monoclonal antibodies. These may be useful as specific probes to study synthesis, structure and distribution of the antigen in question. We have recently started to develop monoclonal anti-VIP antibodies using either the conventional immunization procedure (Kohler and Milstein, 1975) or the *in vitro* sensitization of spleen cells (Luben et al., 1982). The latter technique seems to be more efficient in our hands for VIP (Gozes et al., 1981, in press-a). This method offers a few advantages over the commonly used *in vivo* immunizations. The process is quick as within a few days the cells are secreting antibodies and the process is selective for secreting cells. Non-stimulated cells probably die in the *in vitro* conditions and thus the fusion efficiency for getting positive hybridomas (to myeloma cells) may be higher. In addition, only minute quantities of antigen are required in contrast to repeated *in vivo* immunization. Since the immune response obtained is a primary response, IgM-type antibodies are predominantly elicited. These may be useful in immunocytochemistry to detect quantities of antigen which may not be detected by the much smaller IgG antibody. However, IgM antibodies may have difficulties in permeating the cells to reach their target antigens in immunocytochemistry due to their size. Moreover, due to the multiple binding sites of the IgM antibodies, it requires more antigen to inhibit their specific binding in competition experiments.

Synthetic VIP was purchased from Peninsula Laboratories and coupled to KLH (keyhole limpet hemocyanin) using meta-maleimidobenzoyl-N-hydroxy succinimide ester (Lerner et al., 1981) at 100 μg VIP per 4 mg KLH. It was further used for *in vitro* immunization at a concentration of 50 μg conjugate per $10^7$ spleen cells. Spleen cells were incubated in the presence of thymus conditioned media and antigen for four days and then fused with SP2/0 tumor cells in the presence of polyethylene glycol 1500 (Baker). Antibody-secreting colonies were screened using a solid-phase binding assay with VIP coupled to ovalbumin (as above) as an antigen (Gozes et al., in press-a).

Two monoclonal antibodies thus obtained (IgM type) were characterized for the specific antigenic determinants they recognize on the VIP molecule. We employed short fragments of VIP to inhibit the binding of the antibodies to native VIP which was coupled to ovalbumin, then the VIP-ovalbumin complex was adsorbed onto microtiter plates. For detection a secondary iodinated or enzyme-conjugated antibody was used (Gozes et al., in press-a). The VIP fragments used consisted of residue 6–28, 11–28, 16–28, and 21–28. The results (Fig. 1) were obtained in the presence of 5.3 nmol fragment (Gozes et al., in press-a). Both antibodies 2A2-G9 and 8E8-G3 are maximally inhibited by the VIP fragment 6–28. Antibody 2A2-G9 requires the full sequence 6–28 while antibody 8E8-G3 can recognize 11–28 as well. Thus, monoclonal antibodies that recognize discrete sites on the VIP mole-

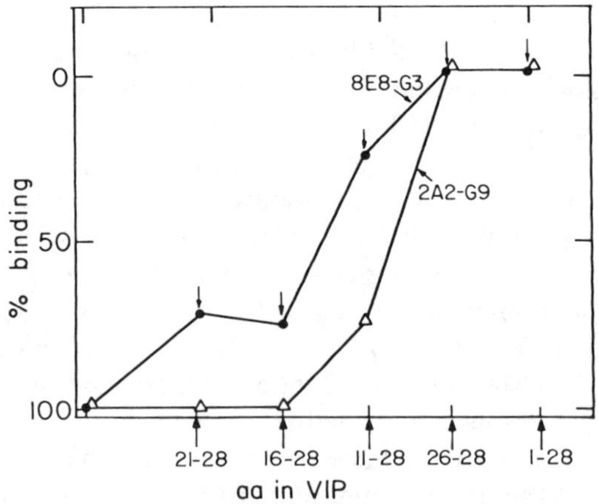


**Figure 1.** Monoclonal antibodies against VIP characterization of binding sites: Inhibition assays. Culture supernatants were mixed with various amounts of VIP fragments (a gift from Dr. N. Ling). The length of the various fragments is indicated on the graph. After at least 3 hr incubation at 4°C these samples were tested for VIP-binding activity by a radioactively labeled second antibody or by the ELISA procedure (enzyme-linked immunosorbant assay) (Gozes et al., in press-a).

cule can be obtained and may be further utilized in the studies of VIP synthesis and localization.

**3.1.1. VIP Biosynthesis as Studied Using Anti-VIP Antibodies.** Anti-VIP antibodies can be used in radioimmunoassay with molecular sieving column chromatography to identify possible high-molecular-weight VIP precursors. We performed several such experiments. In one experiment we used Sephadex G50 column chromatography of a pheochromocytoma tumor chromaffin granules lysate. While part of the VIP immunoreactivity coeluted with the [$^{125}$I]-labeled VIP tracer during gel filtration, a substantial part of the VIP immunoreactivity eluted in the void volume as a high-molecular-weight material (Gozes et al., in press-b).

To characterize this high-molecular-weight immunoreactive material further, we subjected the tumor homogenate to $NaDodSO_4$ polyacrylamide gel electrophoresis followed by electroblotting onto nitrocellulose filters and immunochemical detection. This technique may allow the identification of minor immunoreactive material in the cellular extract by virtue of amplification of the protein stain using a primary antibody and a secondary antibody coupled to horseradish peroxidase (Towbin et al., 1979). Employing this method one may identify specific high-molecular-weight proteins cross-reactive with the anti-VIP antibodies, the binding of which is inhibited by VIP itself. Indeed, we identified such a 70,000 $M_r$ protein in the pheochromocytoma tumor (Gozes et al., in press-b; Fig. 2). We identified a similar protein in brain (Gozes et al., in press -a). However, lower-molecular-weight forms (e.g., a 60,000 $M_r$ protein) were also identified which suggests that the processing into VIP may be more efficient or multiple precursors exist (Fig. 2; Gozes, in press).

The lower-molecular-weight 60,000 $M_r$ protein is discriminated and specifically recognized by monoclonal anti-VIP antibodies (Fig. 2). The additional high-molecular-weight protein bands detected by the polyclonal anti-VIP antibodies may either contain conformations or primary sequences not recognized by the monoclonal antibodies used. It should be noted that the protein transferred to nitrocellulose paper may partially renature allowing recognition of specific conformations by the antibodies (Gozes et al., in press-a).

To verify the results obtained immunologically we subjected the presumed high-molecular-weight precursors and VIP to proteolytic cleavage followed by chromatography and electrophoresis. Prior to the proteolytic cleavage the proteins and the VIP were radiolabeled with [$^{125}$I]. Our preliminary results with the 70,000 $M_r$ protein indicate that it contains cleavage peptides similar to the VIP cleavage peptides.

Further studies on the identity of those proteins are required, and the

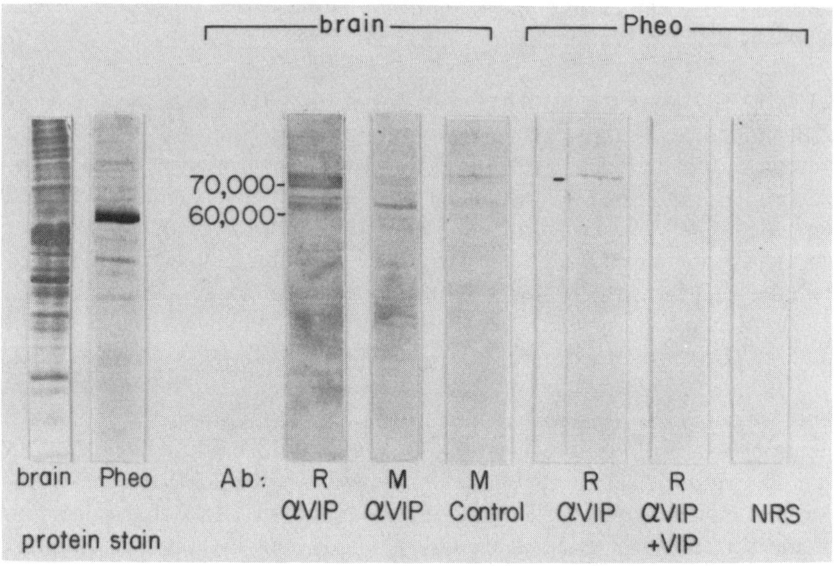


**Figure 2.** High-molecular-weight proteins cross-reactive with anti-VIP. Proteins were subjected to NaDodSO$_4$ 7–20% gradient polyacrylamide gel electrophoresis, electroblotted to nitrocellulose paper, reacted with antibodies and detected using a secondary antibody conjugated to horseradish peroxidase (Laemmli, 1970; Maizel, 1971; Towbin et al., 1979). Tissue homogenates were obtained from rat brain or human pheochromocytoma. Abbreviations used in the figure legend describe the different gel lanes: R$\alpha$VIP: Rabbit anti-VIP antibodies at 1:20 dilution (a gift from Dr. R. Benoit). M$\alpha$VIP: monoclonal anti-VIP antibodies, (IgM type) culture supernatant (1 $\mu$g antibodies per mL). M control: IgM secreting irrelevant hybridoma. Pheo: Human pheochromocytoma. R$\alpha$VIP + VIP: Rabbit anti-VIP antibodies preincubated with an excess of exogenous synthetic VIP (0.1 mg/20 mL non-diluted antisera) for 24 hr at 4°C and then reacted with the nitrocellulose paper. NRS: Normal rabit serum. Note the essential absence of VIP immunoreactive material with control antibodies or in the presence of VIP. (Adapted from Gozes, in press, and Gozes et al., in press-a and in press-b)

monoclonal antibodies should be valuable tools in such studies as well as in studies designed to isolate and characterize the molecular biology of the VIP-gene and its products.

**3.1.2. Immunocytochemical Studies with Monoclonal Anti-VIP Antibodies.** When monoclonal anti-VIP antibodies were used to stain the pancreas (using goat anti-mouse immunoglobulins conjugated to horseradish peroxidase for detection), staining was obtained in islet cells (Fig. 3A) as expected from previous studies conducted with polyclonal antibodies (Said, 1980). A perplexing result was obtained when these monoclonal antibodies were used for immunocytochemistry of brain. In general, we observed no staining comparable to that obtained with anti-VIP polyclonal serum in which

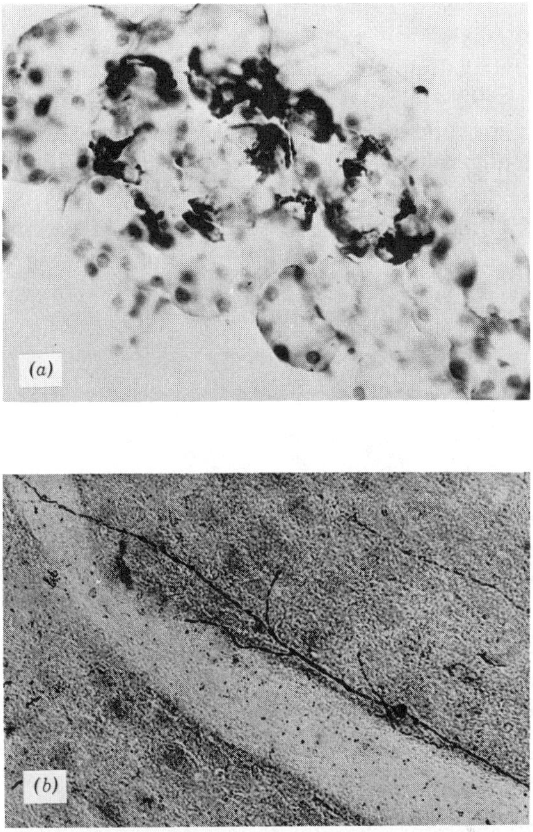

**Figure 3.** (A) Immunocytochemistry of pancreas using monoclonal anti-VIP antibodies. Immunocytochemistry was performed as follows. Rats are anesthetized with chloral hydrate (350 mg/kg interperitoneally). Animals are then perfused transcardially with freshly prepared 4 or 5% paraformaldehyde in phosphate buffer, chilled to 4°C. (Perfusion under pressure should deliver about 200–300 mL/120 sec.) Tissues are removed, cut into slabs, and post-fixed for 3 hr in cold fixative (as above). Tissue slabs are rinsed through a graded series of sucrose solutions in phosphate-buffered saline (12%, 16% 18% and can be stored at 4°C in 18% sucrose until used for histochemistry). Tissue slabs are then frozen on brass chucks in powdered dry ice and equilibrated in a −18°C cryostat for 20 min. Frozen sections (30 $\mu$m) are mounted on cold gelatinized slides, thawed, and air dried for 45 min. Sections are preincubated for 5 min at room temperature in 5% human albumin in phosphate buffer saline (PBS) with 0.3% Triton X-100 (TX100). Sections are then briefly rinsed in PBS and incubated in primary antibody for 18–24 hr at 4°C followed by PBS rinse for 30 min (three rinses 10 min each). Sections are then incubated in goat anti-rabbit IgG conjugated to horseradish peroxidase (Cappel) diluted 1:100 with PBS containing TX100 for 2 hr at room temperature and rinsed as above. Finally, sections are reacted with 0.05% diamino benzidine (DAB) in phosphate buffer containing 0.003% $H_2O_2$ for 10–20 min at room temperature, rinsed, and inspected by light microscopy. For better identification of individual cells, sections can be counterstained with cresyl violet. The dark grains (A) represent the specific stain. The staining showed is of the pancreas using 8E8-G3 antibody. An unrelated IgM-secreting hybridoma did not give the above pattern (data not shown). (B) Immunocytochemistry using polyclonal anti-VIP antibodies. Immunocytochemistry was performed as above on brain sections using a rabbit polyclonal anti-VIP antibody, a gift from Dr. R. Benoit. (Adapted from Gozes et al., in press-a.)

there was pronounced staining of bipolar cortical interneurons (Fig. 3B; Morrison et al., 1981). The absence of pronounced immunocytochemical staining of bipolar neurons in the cerebral cortex using the monoclonal antibodies either could be technical (although under similar conditions rabbit polyclonal antibodies stain nerve cells and nerve fibers) or more likely indicates different conformation and exposure of antigenic determinants of VIP and VIP-related antigens in neurons in the cerebral cortex in comparison to islet cells in the pancreas. It would be interesting to investigate the VIP precursor(s) in the pancreas using either the immunological approach as above or a molecular biology approach.

### 3.2. Molecular Biology Approach

Our approach to identify and characterize the mRNA for VIP depends on the use of a synthetic oligonucleotide with a nucleotide sequence complementary to part of the presumptive sequence of VIP mRNA (Gozes, Milner, Lerner, and Bloom, unpublished results). This approach has been used successfully to clone several mRNAs. The oligodeoxynucleotide we are using was derived by the rules of Agarwal et al. (1981) and is shown below.

5 10 15
HisSerAspAlaValPheThrAspAsnTyrThrArgLeuArgLeuArgLysGlnMetAla

20 25
ValLysLysTyr LeuAsnSerIleLeuAsn-$NH_2$ ⟶

| | 13 20 |
|---|---|
| Peptide: | $H_2N$...Leu.Arg.Lys.Gln.Met.Ala.Val.Lys...COOH |
| mRNA: | 5′............AA$^{A}_{G}$CA$^{A}_{G}$AUGGCN...3′ |
| Probes: | 3′............TTTGTTTACCG...5′ |

N = any nucleotide

This oligodeoxynucleotide probe was custom-synthesized by Collaborative Research (Waltham, MA). This probe can be used directly as a hybridization probe to detect complementary sequences in RNA and DNA and also as a primer for cDNA synthesis using the enzyme reverse transcriptase. Due to the extensive redundancy of the genetic code, we are also using as primers mixtures of oligonucleotide probes to avoid mismatches in hybridization (Suggs et al., 1981).

In collaboration with Dr. M. Fridkin, M. Bodner, and H. Schwartz, we synthesized such oligonucleotide probes using the phosphite-coupled approach (Beaucage and Caruthers, 1981) as outlined below:

| | | |
|---|---|---|
| | | 15 |
| Peptide: | | ...Lys.Gln.Met.Ala.Val.Lys.Lys... |
| Probes: | 1 | 3′.TT$^{T}_{C}$.GT$^{T}_{C}$.TAC.CGT.CAN.TT$^{T}_{C}$.TT...5′ |
| | 2 | -A- |
| | 3 | -C- |
| | 4 | -G- |

These oligodeoxynucleotides are longer than the probe presented above and contain all possible DNA sequences. We prepared four mixtures each containing 32 different sequences, and we analyzed the sequence by the Maxam and Gilbert chemical approach (Maxam and Gilbert, 1977).

Another set of probes which we prepared contain base sequences complementary to regions of the VIP-mRNA not recognized by the above oligonucleotides which should be included in the cDNA transcript if those previous oligonucleotides are used as primers in reverse transcription.

These probes are as follows:

| | | |
|---|---|---|
| | | 6 |
| Peptide: | | ...Phe.Thr.Asp.Asn.Tyr.Thr... |
| Probes: | 5 | 3′...AA$^{A}_{G}$.TGT.CT$^{A}_{G}$.TT$^{A}_{G}$AT$^{A}_{G}$.TG..5′ |
| | 6 | -A- |
| | 7 | -C- |
| | 8 | -G- |

These oligodeoxynucleotides are prepared in four separate mixtures each containing 16 different sequences as confirmed by DNA sequencing (Maxam and Gilbert, 1977). Furthermore, together with our collaborators from San Diego we have obtained additional synthetic oligonucleotides from McNeil Laboratories.

Pilot experiments have been conducted to prepare cDNA using the above oligonucleotide probes as primers in reverse transcription of brain mRNA according to the technique of Sood et al. (1981) which limits the heterogeneity of the cDNA transcripts. The cDNA obtained using radioactively end-labeled probe 8 (Richardson, 1965) as a primer and brain poly(A+) mRNA (Gozes et al., 1975) in reverse transcription gave a single band of approximately 1100 bases on a denaturing polyacrylamide gel (Maxam and Gilbert, 1977). The size of the cDNA transcript obtained is in agreement with a high-molecular-weight precursor for VIP. It would not be the complete size, of course, since the primer initiates the synthesis from some midpoint on the VIP-mRNA. We are now testing the other probe mixtures and plan to synthesize unique sequences and use the most specific one for cloning and screening. Results obtained using brain mRNA hybridization experiments

(Northern blots, Rosenfeld et al., 1981) using [$^{32}$P]-end-labeled probes indicated hybridizing RNA bands of identical electrophoretic migration using probe 8 and probe 1 (Gozes, Bodner, and Fridkin, unpublished results). Molecular weight determinations are in agreement with a 70,000 $M_r$ possible precursor to VIP.

Preliminary experiments were performed with similarly prepared cDNA (using the unique 11 base probe which also gives large cDNA transcripts) to infect *E. coli* with pBR322 as a vector (see Section 2). We obtained 60 colonies that grow on tetracycline but not on ampicillin and thus may contain a presumptive VIP-related insert. A few of these 60 colonies gave positive hybridization with the [$^{32}$P]-labeled oligonucleotide probes using colony hybridization methodology (Dalbadie-McFarland et al., 1982; Grunstein and Hogness, 1975; Suggs et al., 1981). An example of the experiment using probe 8 is outlined in Figure 4. Several of the hybridization positive colonies showed substantial inserts (1100 bases long) after plasmid preparation followed by enzymatic digestion with the restriction enzyme *Pst 1* and gel electrophoresis. Further studies are required to characterize these bacterial clones.

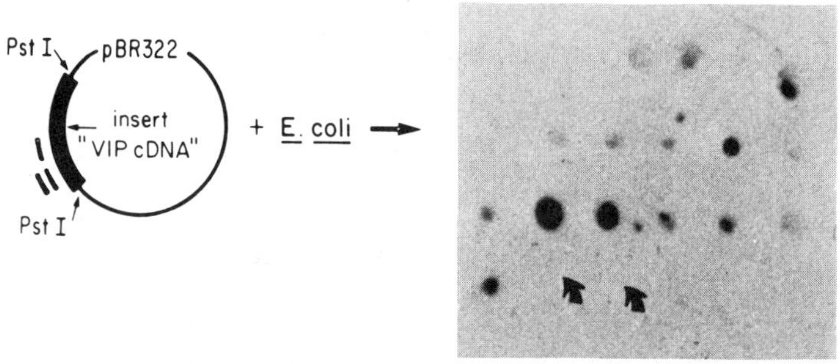


**Figure 4.** Colony hybridization. For colony hybridization the technique developed by Grunstein and Hogness (1975) has been adapted as follows. The colonies to be screened are grown on nitrocellulose filters, lysed, and their DNA is denatured in 0.5 *M* NaOH. The filters are neutralized and baked at 80°C *in vacuo* for 2 hr. Hybridization is performed with [$^{32}$P]-labeled cDNA primed with a VIP-probe using the enzyme reverse transcriptase (prepared as described above) or with an appropriate oligodeoxynucleotide probe labeled with [$^{32}$P] by polynucleotide kinase (Richardson, 1965). Hybridization with the [$^{32}$P]-labeled cDNA is performed at 65°C for 12–18 hr; hybridization with the [$^{32}$P]-labeled probe is performed by incubating the filters (preferably Whatman 541 filters containing colonies lifted from the agar plate) following the empirical rules developed by Suggs et al. (1981) for oligonucleotide hybridization on Southern blots and colony screens. One estimates the apparent melting point (td) to be the sum of 2°C per A:T base pair and 4°C for each G:C base pair when hybridization and washes are done in 0.9 *M* NaCl/6 m*M* EDTA/0.18 *M* Tris-HCl, pH 8.0. Hybridization is carried out at least 5°C below the calculated td followed by a series of washes and autoradiography, first at 5°C and then at higher temperatures (Dalbadie-McFarland et al., 1982; Wallace et al., 1981). Note two hybridization positive clones. (Gozes, Milner, Lerner, Bloom, Bodner, Schwartz, and Fridkin, unpublished results; Gozes, in press)

## 4. VIP BIOSYNTHESIS: PAST AND FUTURE

VIP is widely distributed in neurons in the central nervous system, the peripheral nervous system, and endocrine cells. In view of its wide distribution the question arises whether it is synthesized and processed similarly in various cells or are there mutiple precursors processed differently. Our preliminary data, comparing a neural tumor of pheochromocytoma to brain, suggest the possibility of either additional precursors or more efficient processing in the brain (Fig. 2). Another tissue variability is in the immunocytochemistry results. While our polyclonal anti-VIP antibodies stain bipolar neurons in the cerebral cortex (Fig. 3) and islet cells in the pancreas, our monoclonal antibodies (those assayed thus far) do not stain the brain neurons but stain islet cells suggesting a possible heterogeneity in VIP and/or VIP precursor packaging.

We thus propose to study in depth VIP gene expression not only in brain but also in the pancreas, tumors producing VIP such as the one described above, VIPomas, and VIP-producing ganglion cells such as the sphenopalatine ganglion (Said, 1982) or superior cervical ganglion (Ip et al., 1982). Concerning the super cervical ganglion we have previously cultured those cells which can undergo phenotypic changes upon media variation (Gozes and Sweadner, 1981; Gozes et al., 1983; Patterson, 1978). The phenotypic change involves transition from adrenergic to cholinergic neurons (Patterson, 1978), and it is interesting to determine whether the VIP gene is induced upon transition. Another culture-producing VIP, which is a cell line rather than the primary cell culture outlined above, is the neuroblastoma NB-OK-1 cell line. In this cell line VIP synthesis is induced by cAMP and pulse chase experiments can follow the fate of the precursor and its processing into VIP (Yanaihara et al., 1981).

VIP gene expression can be monitored at the level of the immunoreactive putative precursor as was done for the brain and the neural tumor (Fig. 2). Variations could be detected both by $NaDodSO_4$ gel electrophoresis and by isoelectric focusing as performed with other proteins (e.g., Gozes et al., 1980).

Once having a VIP-cDNA probe in hand we could compare transcription efficiency of VIP-mRNA in the various cell types. We could both quantitate the amount of VIP-mRNA and study possible mRNA heterogeneity as was found for the peptide calcitonin (Amara et al., 1982). Still at the mRNA and protein level, we shall screen to check the evolution of VIP gene expression and try to trace it to lower organisms as well. Such studies may provide clues to an ancestral role of the peptide.

A deeper insight into VIP gene regulation will be obtained from restriction sites analysis of DNA coding for VIP in the various cells and tissues described above; we may find multiple VIP genes containing structural variations which can be sequenced as was found for the tubulin genes (e.g., Cowan et al., 1981). Such studies may reveal species differences in cells and tissue variability. In view of the latest discoveries of gene translocations as playing a role in malignant transformation (Dalla-Favera et al., 1982; Taub et al., 1982; Neel et al., 1982), the question arises as to whether such events play a role in gene activation or inactivation in the course of normal differentiation. Indeed, gene rearrangement was discovered in the case of immunoglobulin (Crews et al., 1981), but is it to be found with other genes? One way to answer this question concerning the VIP gene is to follow its restriction map in the sphenopalatine ganglion neurons over a range of time points during development following quantification by radioimmunoassays. Similar studies could be performed in the neuroblastoma cell line, producing VIP upon cAMP stimulation (see above). In comparison cells such as liver cells that do not produce vast amounts of VIP could be assayed as well, and the VIP gene function may be correlated with structural specificities.

Eventually chromosomal localization of the VIP gene would be necessary to determine whether extensive VIP gene expression is correlated with a specific chromosomal localization; for such studies one may apply somatic hybrids as described (Dalla-Favera et al., 1982). Mental retardation may be associated with chromosomal aberration (Schmickel and Weiss, 1981). Since VIP may be associated with the regulation of energy metabolism in the brain (Magistretti et al., 1981), it may very well be associated with diseases of the nervous system. This idea could be followed by analysis of lymphoid cells, chromosomes and DNA from afflicted people. Such studies may also be performed using somatic cell genetics (Dalla-Favera et al., 1982).

Finally, it would be of great interest to find structural correlates at the level of the genome in the VIP-secretin gene family where structural similarity is observed at the peptide level.

Understanding the regulation of VIP biosynthesis may shed light on the regulation of other peptides in the brain and may give us a better understanding of how the individual cells in a complicated cellular array such as is present in the brain operate at the genomic level.

## ACKNOWLEDGMENTS

I would like to thank Professor F. E. Bloom for constant encouragement and inspiration. I am very grateful to all my collaborators: Professors D. H. Katz, R. A. Lerner, D. T. O'Connor, F. T. Liu, and M. Fridkin, Drs. R. J. Miler and R. Benoit and M. Bodner, H. Schwartz, and D. Warszawski for enthusiastic collaboration. The excellent technical assistance of D. Barry, E. Johnson, E. L. F. Battenberg, M. Lane, Raana Azad, and Yael Shani is gratefully acknowledged. Special thanks are due to Rona Levin for diligent secretarial work. This work is now supported by grants from the United States–Israel Binational Science Foundation (BSF) Jerusalem, a 1982 Bergmann Memorial Research Grant, a grant from The Fund for Basic Research Administered by the Israeli Academy of Sciences and Humanities, the Jewish Agency, and the Leo and Julia Forchheimer Center for Molecular Genetics.

## REFERENCES

Agarwal, K. L., A. Yamazaki, P. J. Cashion, and H. G. Khorana, *Angew Chem. Int. Ed. Eng.* **11,** 451 (1972).

Agarwal, K. L., J. Brunstedt, and B. E. Noyes, *J. Biol. Chem.* **256,** 1023 (1981).

Amara, S. G., V. Jonas, M. G. Rosenfeld, E. S. Ong, and R. M. Evans, *Nature* **298,** 240 (1982).

Anderson, W. F. and E. G. Diacumakes, *Scientific American* **245,** 106 (1981).

Barbezat, G. O. and M. I. Grossman, *Science* **174,** 422 (1971).

Beaucage, S. L. and M. H. Caruthers, *Tetrahedron Lett.* **22,** 1859 (1981).

Bloom, F. E. and J. M. Polak, in S. I. Said, Ed., *Vasoactive Intestinal Peptides,* Raven Press, New York, p. 457 (1982).

Bodanszky, M., in J. C. thompson, Ed., *Gastrointestinal Hormones,* University of Texas Press, Austin, p. 507 (1975).

Bodanszky, M., in S. Bonfils, P. Fromageot, and G. Rosselin, Eds., *The First Internatl. Symp. on Hormonal Receptors in Digestive Tract Physiology, INSERM Symposium No. 3,* Elsevier/North Holland Biomedical Press, New York, p. 13 (1977).

Bolivar, F., R. L. Rodriguez, P. J. Greene, M. C. Betlach, H. L. Heyneker, and H. W. Boger, *Gene* **2,** 95 (1977).

Chang, A. C. Y., M. Cochet, and S. N. Cohen, *Proc. Natl. Acad. Sci. USA* **77,** 4890 (1980).

Chin, W. W., F. Maloof, and J. F. Habener, *J. Biol. Chem.* **256,** 3059 (1981).

Chretien, M., S. Benjaunet, F. Gossard, C. Gianoulakis, P. Crime, M. Lis, and N. G. Seidah, *Canadian J. Biochem.* **57,** 1111 (1979).

Cowan, N. J., C. D. Wilde, L. T. Chow, and F. C. Wefald, *Proc. Natl. Acad. Sci. USA* **78,** 4877 (1981).

Crews, S., J. Griffin, H. Huang, K. Calame, and L. Hood, *Cell* **25,** 59 (1981).

Cuello, A. C., G. Galfre, and C. Milstein, *Proc. Natl. Acad. Sci. USA* **76,** 3532 (1979).

Dalbadie-McFarland, G., L. W. Cohen, A. D. Riggs, C. Morin, K. Itakura, and J. H. Richards, *Proc. Natl. Acad. Sci. USA* **79,** 6409 (1982).

Dalla-Favera, R., M. Bregni, J. Erikson, D. Patterson, R. C. Gallo, and C. M. Croce, *Proc. Natl. Acad. Sci. USA* **79,** 7824 (1982).

Deschodt-Lanckman, M., P. Robberecht, and J. P. Christophe, *FEBS Lett.* **83,** 76 (1977).

Dimaline, R. and G. J. Dockray, *Life Sci.* **25,** 1893 (1979).

Emson, P. C., J. Fahrenkrug, O. B. Schaffalitzky de Muckadell, T. M. Jessel, and L. L. Iversen, *Brain Res.* **143,** 174 (1978).

Galfre, C., S. C. Howe, C. Milstein, G. W. Butcher, and J. C. Howard, *Nature* **266,** 550 (1977).

Gerhard, D. S., E. S. Kawasaki, F. C. Bancroft, and P. Szabo, *Proc. Natl. Acad. Sci. USA* **78,** 3755 (1981).

Gicchetti, A., S. I. Said, R. C. Raynolds, and F. C. Koniges, *Proc. Natl. Acad. Sci. USA* **74,** 3424 (1977).

Godine, J. E., W. W. Chin, and J. F. Habener, *J. Biol. Chem.* **255,** 8780 (1980).

Godine, J. E., W. W. Chin, and J. F. Habener, *J. Biol. Chem.* **256,** 2475 (1981).

Gozes, I., in H. Soreq, Ed., *Molecular Biology Approach to the Nervous System* (in press).

Gozes, I. and K. J. Sweadner, *Nature* **294,** 477 (1981).

Gozes, I., H. Schmitt, and U. Z. Littauer, *Proc. Natl. Acad. Sci. USA* **73,** 701 (1975).

Gozes, I., A. de Baetselier, and U. Z. Littauer, *Eur. J. Biochem.* **103,** 13 (1980).

Gozes, I., D. Barry, R. Benoit, F. T. Liu, D. H. Katz, R. J. Milner, and F. E. Bloom, *Abst. Soc. Neurosci.* **7,** 98 (1981).

Gozes, I., R. J. Milner, F. T. Liu, E. Johnson, E. L. F. Battenberg, D. H. Katz, and F. E. Bloom, *J. Neurochem.* (in press-a).

Gozes, I., D. T. O'Connor, and F. E. Bloom, *Regulatory Peptides,* (in press-b).

Gozes, Y., M. A. Moskowitz, T. B. Strom, and I. Gozes, *Dev. Brain Res.* **6,** 93 (1983).

Grunstein, M. and D. S. Hogness, *Proc. Natl. Acad. Sci. USA* **72,** 3961 (1975).

Guillemin, R., P. Brazeau, P. Böhlen, F. Esch, N. Ling, and W. B. Wehrenberg, *Science* **218,** 585 (1982).

Harrison, P. R., D. Conkie, J. Paul, and K. Jones, *FEBS Lett.* **32,** 109 (1973).

Herbert, E., N. Birnberg, J. C. Lissitsky, O. Civelli, and M. Uhler, *Neurosci. Commentaries* **1,** 16 (1981).

Hokfelt, T., O. Johansson, A. Ljungdahl, J. M. Lundberg, and M. Schultzberg, *Nature* **284,** 515 (1980).

Houghton, M., A. G. Stewart, S. M. Doel, J. S. Emtage, M. A. W. Eaton, J. C. Smith, T. P. Patel, H. M. Lewis, A. G. Porater, J. R. Birch, T. Cartwright, and N. H. Carey, *Nucleic Acids Res.* **8,** 1913 (1980).

Hudson, P., J. Haley, M. Cronk, J. Shine, and H. Niall, *Nature* **291,** 127 (1981a).

Hudson, P., J. Penschow, J. Shine, G. Ryan, H. Niall, and J. Coghlan, *Endocrinology* **108,** 353 (1981b).

Hughes, S. H., E. Stubblefield, F. Payvar, J. D. Engel, J. B. Dodgson, D. Spector, B. Cordell, R. T. Schimke, and H. E. Varmus, *Proc. Natl. Acad. Sci. USA* **76,** 1348 (1979).

Ip, N. Y., C. K. Ho, and R. E. Zigmond, *Proc. Natl. Acad. Sci. USA* **79,** 7566 (1982).

Jaacobs, J. W., R. H. Goodman, W. W. Chin, P. C. Dee, J. F. Habener, N. H. Bell, and J. T. Potts, Jr., *Science* **213,** 457 (1981).

Johansson, O. and J. M. Lundberg, *Neuroscience* **6,** 847 (1981).

Jukes, T. H., *Science* **210,** 973 (1980).

Kerins, C. and S. I. Said, *Proc. Soc. Exp. Biol. Med.* **142,** 1014 (1973).

Kita, T., A. Inoue, S. Nakanishi, and S. Numa, *Eur. J. Biochem.* **93,** 213 (1979).

Kohler, G. and C. Milstein, *Nature* **256,** 495 (1975).

Kurtz, D. T., *Nature* **291,** 629 (1981).

Laemmli, U. K., *Nature* **277,** 680 (1970).

Lerner, R. A., N. Green, H. Alexander, F. T. Liu, J. G. Sutcliffe, and T. M. Shinnick, *Proc. Natl. Acad. Sci. USA* **78,** 3403 (1981).

Ling, N., S. Ying, S. Minick, and R. Guillemin, *Life Sci.* **25,** 1773 (1979).

Lipmann, F., *Acc. Chem. Res.* **6,** 361 (1973).

Littlefield, J. W., *Science* **145,** 709 (1964).

Luben, R. A., P. Brazeau, P. Böhlen, and R. Guillemin, *Science* **218,** 887 (1982).

Lund, P. K., R. H. Goodman, J. W. Jacobs, and J. F. Habener, *Diabetes* **29,** 583 (1980).

Magistretti, P. J., J. H. Morrison, W. J. Shoemaker, V. Sapin, and F. E. Bloom, *Proc. Natl. Acad. Sci. USA* **78,** 6535 (1981).

Mains, R. A., B. A. Eipper, and N. Ling, *Proc. Natl. Acad. Sci. USA* **74,** 3014 (1977).

Maizel, J. V., Jr., in K. Maramorosch and H. Koprowski, Eds., *Methods in Virology,* Vol. V, Academic, New York, p. 179 (1971).

Makhlouf, G. M., A. M. Zfoss, S. I. Said, and M. Schebalin, *Proc. Soc. Exp. Biol. Med.* **175,** 565 (1978).

Maxam, A. and W. Gilbert, *Proc. Natl. Acad. Sci. USA* **74,** 560 (1977).

Melchiori, P., in S. R. Bloom, Ed., *Gut. Hormones,* Churchill Livingston, Edinburgh, p. 534 (1978).

Morrison, J. H., P. J. Magistretti, R. Benoit, and F. E. Bloom, *Abstr. Soc. Neurosci.* **7,** 99 (1981).

Nakanishi, S., A. Inoue, T. Kita, S. Numa, A. C. Y. Chang, S. N. Chen, J. Nunberg, and R. T. Schimke, *Proc. Natl. Acad. Sci.* **USA 75,** 6021 (1978).

Nakanishi, S., A. Inoue, T. Kita, M. Nakamura, A. C. Y. Chang, S. N. Cohen, and S. Numa, *Nature* **278,** 423 (1979).

Nakanishi, S., Y. Teranishi, M. Noda, M. Notake, Y. Watanabe, H. Kakidani, H. Jingami, and S. Numa, *Nature* **287,** 752 (1980).

Nathans, D. and H. O. Smith, *Ann. Rev. Biochem.* **44,** 273 (1975).

Neel, B. G., S. C. Jhanwar, R. S. K. Chaganti, and W. S. Hayward, *Proc. Natl. Acad. Sci. USA* **79,** 7842 (1982).

Noyes, B. E., M. Mevarech, R. Stein, and K. L. Agarwal, *Proc. Natl. Acad. Sci. USA* **76,** 1770 (1979).

Owerbach, D., W. J. Rutter, J. A. Martial, J. D. Baxter, and T. B. Shows, *Science* **209,** 290 (1980).

Patterson, P. H., *Ann. Rev. Neurosci.* **1,** 1 (1978).

Piper, P. J., S. I. Said, and J. R. Vane, *Nature* **225,** 1144 (1970).

Pochet, R., H. Brocas, G. Vassart, G. Toubeau, H. Seo, S. Refetoff, J. E. Dumont, and J. L. Pasteels, *Brain Res.* **211,** 433 (1981).

Potter, P. D., E. J. Furshpan, and S. C. Landis, *Neurosci. Comm.* **1,** 1 (1981).

Quik, M., L. L. Iversen, and S. R. Bloom, *Biochem. Pharmacol.* **27,** 2209 (1978).

Ricciardi, R. P., J. S. Miller, and B. E. Roberts, *Proc. Natl. Acad. Sci. USA* **76,** 4927 (1979).

Richardson, C. C., *Proc. Natl. Acad. Sci. USA* **54,** 158 (1965).

Robberecht, P., P. De Neef, M. Lammens, M. Deschodt-Lanckman, and J. P. Christophe, *Eur. J. Biochem.* **90,**147 (1978).

Robberecht, P., M. Deschodt-Lanckman, P. De Neef, and J. P. Christophe, *Life Sci.* **25,** 1001 (1979).

Rosenfeld, M. G., S. G. Amara, B. A. Ross, E. S. Ong, and R. M. Evans, *Nature* **290,** 63 (1981).

Rossier, J., *Nature* **298,** 221 (1982).

Rubinstein, M., S. Stein, and S. Udenfriend, *Proc. Natl. Acad. Sci. USA* **75,** 669 (1978).

Russel, J. T., M. J. Brownstein, and H. Gainer, *Endocrinology* **107,** 1880 (1980).

Said, S. I., in G. B. Jerzy Glass, Ed., *Gastrointestinal Hormones,* Raven Press, New York, p. 245 (1980).

Said, S. I. Ed., *Vasoactive Intestinal Polypeptide: Advances in Peptide Hormone Research,* Raven Press, New York (1982).

Said, S. I. and G. R. Faloona, *N. Engl. J. Med.* **293,** 155 (1975).

Said, S. I. and V. Mutt, *Science* **169,** 1217 (1970).

Sanger, F., S. Nicklen, and A. R. Coulson, *Proc. Natl. Acad. Sci. USA* **74,** 5463 (1977).

Schaefer-Ridder, M., Y. Wang, and P. H. Hofschneider, *Science* **215,** 166 (1982).

Schmale, H. and D. Richter, *Proc. Natl. Acad. Sci. USA* **78,** 766 (1981).

Schmickel, R. D. and L. Weiss, *Am. J. Mental Defic.* **85,** 335 (1981).

Sharp, P. A., *Cell* **23,** 643 (1981).

Sood, A. K., D. Pereira, and S. M. Weissman, *Proc. Natl. Acad. Sci. USA* **78,** 616 (1981).

Southern, E. M., *J. Mol. Biol.* **98,** 503 (1975).

Suggs, S. V., T. Hirose, T. Miyake, E. H., Kawashima, M. J. Johnson, K. Itakura, and R. B. Wallace, in D. D. Brown and C. F. Fox, Eds., *Developmental Biology Using Purified Genes, Academic, New York, p. 683 (1981).*

Sutcliffe, J. G., T. M. Shinick, N. Green, F. T. Liu, H. L. Niman, and R. A. Lerner, Nature **287,** 801 (1980).

Tatemoto, K. and V. Mutt, *Nature* **285,** 417 (1980).

Taub, R., I. Kirsch, C. Morton, G. Lenoir, D. Swan, S. Tronich, S. Aaronson, and P. Leder, *Proc. Natl. Acad. Sci. USA* **79,** 7837 (1982).

Taylor, D. and C. B. Pert, *Proc. Natl. Acad. Sci. USA* **76,** 660 (1979).

Towbin, H., T. Staehelin, and J. Gordon, *Proc. Natl. Acad. Sci. USA* **76,** 4350 (1979).

Villa-Komaroff, L., A. Efstratiadis, S. Broome, P. Lomedico, R. Tizard, S. P. Naber, W. L. Chick, and W. Gilbert, *Proc. Natl. Acad. Sci. USA* **75,** 3727 (1978).

Wallace, R. B., M. Schold, M. J. Johnson, P. Dembek, and K. Itakura, *Nucleic Acids Res.* **9,** 3647 (1981).

Watson, S. J. and H. Akil, *Neurosci. Comm.* **1,** 10 (1981).

Wever, S., M. B. Comer, C. L. Jahn, C. A. Hutchinson, and H. W. Marshall, *Cell* **24,** 403 (1981).

Wigler, M., A. Pellicer, S. Silverstein, and R. Axel, *Cell* **14,** 725 (1978).

Yanaihara, N., S. Toshimutsu, H. Sato, O. Koshino, and C. Yanaihara, *Biomed. Res.* **2,** 728 (1981).

# Index